EXTREME TRIBOLOGY
Fundamentals and Challenges

Ahmed Abdelbary

Ph.D., M.Sc., B.Sc.(Eng.), EGTRIB Mem
Chief Engineer and Tribology Consultant
Egyptian Government
Alexandria, Egypt

CRC Press
Taylor & Francis Group
Boca Raton London New York

CRC Press is an imprint of the
Taylor & Francis Group, an **informa** business

A SCIENCE PUBLISHERS BOOK

CRC Press
Taylor & Francis Group
6000 Broken Sound Parkway NW, Suite 300
Boca Raton, FL 33487-2742

First issued in paperback 2021

© 2020 by Taylor & Francis Group, LLC
CRC Press is an imprint of Taylor & Francis Group, an Informa business

No claim to original U.S. Government works

ISBN-13: 978-1-138-32815-0 (hbk)
ISBN-13: 978-1-03-217599-7 (pbk)
DOI: 10.1201/9780429448867

This book is dedicated to the memory of my father

Foreword

Tribology is the science and engineering of interacting surfaces in relative motion. It includes the study and application of the principles of friction, lubrication and wear. This field is highly interdisciplinary in nature and draws upon several academic areas, including physics, chemistry, materials science and engineering (Wikipedia, 2019). The term "tribology" became widely used following 'The Jost Report', published in 1966 (Jost, 1966). The report highlighted the huge cost of friction, wear and corrosion to the UK economy. Despite considerable research since then, the global impact of friction and wear on energy consumption and carbon dioxide emissions is still considerable. In 2017, Kenneth Holmberg and Ali Erdemir (Holmberg, 2017) attempted to quantify their worldwide impact on economic aspects. They considered the four main energy consuming sectors: Transportation, manufacturing, power generation and residential.

The following facts were concluded:

- In total, ~ 23% of the world's total energy consumption originates from tribological contacts. Of that, 20% is used to overcome friction and 3% is used to remanufacture worn parts and spare equipment due to wear and wear-related failures.

- By taking advantage of the new surface, materials, and lubrication technologies for friction reduction and wear protection in vehicles, machinery and other equipment worldwide, energy losses due to friction and wear could potentially be reduced by 40% in the long term (15 years) and by 18% in the short term (8 years). On a global scale, these savings would amount to 1.4% of GDP annually and 8.7% of the total energy consumption in the long term.

- The largest short term energy savings are envisioned in transportation (25%) and in the power generation (20%) while the potential savings in the manufacturing and residential sectors are estimated to be ~ 10%. In the longer term, the savings would be 55%, 40%, 25% and 20%, respectively.

- Implementing advanced tribological technologies can also reduce global carbon dioxide emissions by as much as 1460 metric tons of carbon dioxide equivalent (Mt CO_2) and result in 450,000 million Euros cost savings in the short term. In the long term, the reduction could be as large as 3140 Mt CO_2 and the cost savings 970,000 million Euros (Friedrich, 2018).

The new book written by Professor Ahmed Abdelbary on "Extreme Tribology" focuses especially on tribological situations under extreme operating conditions. The latter, as defined in the book, can be related to high loads and/or temperatures, or severe environments, such as in space. Also, they may be related to high transitory contact conditions (e.g., wheel/rail interface), or to situations with near-impossible monitoring and maintenance opportunities (e.g., mechanical sub-sea oil pipe repair connectors). In general, extreme conditions can typically be categorized as involving abnormally high or excessive exposure to cold, heat, pressure, vacuum, voltage, corrosive chemicals, vibration, or dust.

Operation in such extreme conditions is a great challenge for tribologists to develop tribosystems that could meet these extreme requirements. Often, only multifunctional materials fulfill such requirements (Friedrich, 2015). However, more work needs to be done in order to reveal the physical and chemical nature of these extreme tribological characteristics and to generate reliable data for design. So far, most of the publications in this field have been written for engineers and scientists who are already active in the field of tribology and want to broaden and update their knowledge towards more complicated loading situations. On the other hand, there are not too many papers or books that are especially dedicated to the education of senior and graduate students interested in these important issues.

The new textbook of the author is, therefore, an ideal solution to this problem. After a comprehensive introduction to the field of tribology in general, in which the effects of the volume properties of the tribo-materials, their surface properties and the use of possible lubricants on friction and wear of tribo-systems are also described, the author focusses the attention of the reader on the theory of friction, factors affecting the latter, measurement techniques, and friction related heat and temperature effects. Two other important subjects follow: (a) A description of possible wear mechanisms, including corresponding wear measurements, and (b) the lubrication of tribo-systems by fluidic, gaseous, grease or solid media, including various types of additives incorporated in them. In the following, Dr. Abdelbary dedicates two other chapters to the tribology of polymers and their composites, and to the tribology of automotive components. Especially in this field, polymer composites have been very successfully used in recent times. It would be hard to find a polymer for automotive tribo-applications which is not in a composite or a polymer/metal hybrid form. His work is then concentrated on friction and wear under other extreme conditions, where he especially emphasizes the mechanisms and effects that occur at high or low cryogenic environments, in a vacuum, and at high speeds. Also, abrasion aspects in mining, mineral processing, and treatment of solid concrete are touched upon. This is followed by a chapter on lubrication and coating challenges in extreme conditions, which also considers vacuum effects in space tribology and lubrication in artificial joints, to mention only a few. The two concluding chapters deal with the simulation and modelling of tribo-systems, and with failures in particular types of wear components, like gears, bearings, wheels and artificial implants.

The book is not only intended to teach postgraduate students in the field of tribology in general, but also under extreme conditions in particular. Nevertheless, it will be also of interest as a balanced textbook for different levels of readers: (1) those who are active or intend to become active in research on materials' tribology in general (material scientists, physical chemists, mechanical engineers); (2) those who have encountered a practical friction or wear problem and wish to learn more methods of solving such problems (designers, engineers and technologists in industries dealing with selection, processing and application of engineering materials); (3) professors at universities, who want to set-up new courses in this field. This is very important, since by now, the average mechanical engineer receives only a few hours of instruction on wear during his/her university studies. Therefore, the book is highly recommendable for a broad engineering community.

Kaiserslautern, 28th June 2019

Prof. Dr.-Ing. Dr. hc Klaus Friedrich
Institute for Composite Materials (IVW GmbH)
67663 Kaiserslautern, Germany

References

Friedrich, K. 2018. Polymer composites for tribological applications (A review). Advanced Industrial and Engineering Polymer Research 1: 3–39.

Friedrich, K. and Breuer, U. (eds.). 2015. Multifunctionality of Polymer Composites, Elsevier, Oxford, UK.

Holmberg, K. and Erdemir, A. 2017. Influence of tribology on global energy consumption. Costs and Emissions, Friction 5: 263–284.

https://en.wikipedia.org/wiki/Tribology.

Jost, H.P. 1966. Lubrication (Tribology)—A Report on the Present Position and Industry's Needs. Department of Education and Science, H.M. Stationary Office, London, UK.

Acknowledgment

First and foremost, I would like to express my sincere gratitude to *Prof. Dr. MN Abouelwafa*, who has been instrumental in completing my postgraduate degrees and has been my role model since the moment I started my research career until his death in 2013.

I would like to record my gratitude and thanks to *Prof. Klaus Friedrich.* His contribution to this work is immense and cannot be acknowledged in words.

I was fortunate to have assistance from *Dr. Maged E Elnady* for proof reading this book. His invaluable mentorship, stimulating discussions and support were instrumental in completing this work.

I would also like to extend my gratitude and thanks to *Dr. Yasser S Mohamed* (Assistant Professor, Faculty of Engineering, Alexandria University) for going through the book and suggesting helpful changes and ideas.

I owe my family the deepest debt. My mother, thank you for your love and constant support. My wife, *Gigi*, and my children, thank you for bearing added stress during the elaboration of this book.

A Abdelbary

Contents

Chapter 1

Introduction to Tribology

1.1 Introduction

Tribology is a relatively new word, although it represents phenomena and problems that have been around since the Stone Age and extend to our present lives. The word "tribology" was introduced in 1964, when a working group chaired by Dr. H. Peter Jost CBE was invited by the UK Department of Education and Science to investigate the state of lubrication education and research, and to give an opinion on the needs of the industry. When the group introduced the report in February 1966, they proposed, for the first time, the word "tribology" to describe this field (Jost, 1966). The word comes from the ancient Greek "tribo" meaning "rubbing" and the suffix "logy" for "the knowledge of", so the literal translation would be "the science of rubbing". Currently, dictionaries define tribology as "the study of friction, wear and lubrication, and the design of bearings" or "the science of interacting surfaces in relative motion" (Oxford, Dictionary).

By definition, tribology is an interfacial phenomenon that is affected by physical and mechanical properties of the two interacting surfaces as well as operational conditions. Since surface interactions are highly complex aspects, their understanding requires studying a variety of fields, including mechanical engineering, material science, surface topography and fluid mechanics. Moreover, there is a further complexity due to the fact that tribology is strongly associated with a wide range of practical applications, with small possibilities of theoretical formulation for a limited range of them.

If we try to explore the principal constituents of tribology of interacting solid surfaces, we realize that friction and wear are very ancient ones. In fact, solid surfaces are in contact with each other in relative motion, that is, surfaces sliding, rolling and rubbing on each other. For effective solution of tribology problems, all three constituents should be considered carefully and equally.

Since time immemorial, efforts have been made in order to minimize wear and control friction. This has triggered the need for lubrication, in which fluids (or solids) are introduced as a lubricant film to reduce the contact between moving surfaces in order to save power and material, Fig. 1.1 represents an illustration for tribology triangle.

The word "friction" came into usage in the middle of the 16th century, it comes from the Latin *frictiō*, meaning "a chafing or rubbing", and from Middle French *friction* as a noun of action from the past participle stem of *fricare*, meaning "to rub". The first record of using friction as "resistance to motion" was in the middle of the 17th century (Oxford, Dictionary). Even though many scientific definitions were introduced in order to describe this phenomenon, most of them have the common essence that there are two regimes of friction: Static friction between non-moving surfaces and kinetic (or dynamic) friction between moving surfaces.

Friction is expressed in quantitative terms as the force generated between the two surfaces in the direction opposite to the direction of motion (for kinetic friction) or potential motion (for static friction). To easily differentiate between the two cases, one can observe that, in static friction, the force is insufficient to cause motion, whereas in dynamic friction, it is sufficient.

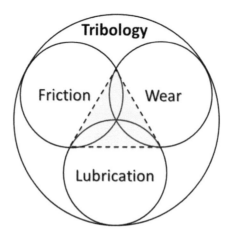

Figure 1.1: Tribology triangle.

In general, the friction force is described in terms of a coefficient of friction (μ) which is the ratio of the friction force (F) to the normal load (W) acting between both surfaces:

$$\mu = \frac{F}{W} \tag{1.1}$$

The coefficient is called either static or dynamic coefficient of friction, depending on the friction case. The main factors contributing to the friction coefficient are: Topography of surfaces in contact, surface area, normal pressure, sliding velocity and the time during which the surfaces remain in contact. Other factors, such as temperature, humidity, etc., may have a considerable influence on the nature of friction (Dowson, 1998).

In the late of the 13th, century the word "wear" was used in reference to clothing in order to express the gradual damage caused by continued use. In material science, wear is related to the removal and deformation of material on a solid surface as a result of mechanical action of the opposite surface (Materials, 2001). Among several definitions of this process, wear can be simply expressed as "progressive loss of substance from the operating surface of a body occurring as a result of relative motion at the surface" (Materials, 1969; Rabinowicz, 1995; Halling, 1979).

Although wear is generally thought of as a harmful phenomenon, this is not totally true. Even though this can be found in the majority of practical cases, there are a considerable number of useful applications of the wear process. For instance, surface production and surface finishing of manufactured objects using abrasion processes and the use of a pencil or a chalk are examples of desirable wear. Thus, practically speaking, wear is not always to be avoided.

Wear can be described by three basic terms: (i) wear mechanism, (ii) wear process, and (iii) wear type. Although our consideration of the various mechanisms of wear is now improving, no unique law (compared to that for friction) has been generalized. The main reason for this is that the wear process involves diverse phenomena interacting in a largely unpredictable manner.

The first record of using the word "lubrication" was in the early 19th century. The Latin *lūbricātus*, past participle of *lūbricāre*, means "to make slippery" (Oxford, Dictionary). Since lubrication can effectively change the performance of tribo-systems (any system containing tribological components), one of the most general definitions is that "a lubricant is a substance capable of altering the nature of the surface interaction between contacting solids". We should emphasize here that the term "substance" in the previous definition is not simply exclusive to oil or grease as it may be thought of, even though they are the most common lubricants in use. In fact, there are other conditions of lubrication, such as fluid lubrication, grease, solid lubrication, and air and gas lubrication.

In Fluid lubrication, any fluid, including water, can be used as a lubricant in the proper application. In fluid lubrication, the lubrication regime is considered according to the condition of the lubricant introduced into a sliding system. The three distinct situations are fluid film lubrication, boundary lubrication, and mixed lubrication. The most important property of the lubricant in fluid lubrication is the viscosity of the lubricant.

Grease is a stabilized mixture of a liquid lubricant (mineral or synthetic fluid) and a thickening agent. Some additives may be introduced to add particular properties.

Solid lubrication occurs when a soft solid film (e.g., Molybdenum disulfide MoS_2) is introduced between sliding surfaces. Also, it may be found as a result of a chemical reaction between the sliding surface and its environment.

Air or gas is now a rather common lubricant in high-speed bearing applications.

For all previous kinds of lubrication, the main task is to reduce friction, wear and surface adhesion. In other cases, lubrication is introduced in tribo-systems in order to control the interfacial temperature. When lubrication breaks down, components can rub destructively against each other, causing heat, local welding, destructive damage and, finally, failure.

1.2 History of Tribology

With the beginning of humankind, the first man used tribology concepts to facilitate his daily activities. The first practical application of tribology was the use of the friction phenomenon in lighting fires, rapidly grinding pieces of solid burnable materials, such as wood, against each other or a hard surface in order to create heat. In this Paleolithic period, archeologists found evidence for using the friction and grinding wear concepts in the development of weapons, tools and construction. As far back as the middle of Mesolithic period in Egypt, records show the use of studded wheels and bearings for stone potter's wheels, reducing friction in transnational motion, Fig. 1.2.

Furthermore, there is some indication that these 5000-year-old wheel bearings were lubricated with bitumen.

Ancient Egyptians were the earliest pioneers in using lubricants in the transportation of large stone building blocks and monuments. Figure 1.3 is an artwork from the tomb of Djehutihotep in El Bersheh (1800 B.C.). It was discovered in the Victorian era and depicts the use of a sledge to transport a heavy statue of the Middle Kingdom. In this transportation, 172 slaves dragged a large statue weighing about 600 kN along a wooden track. Closer examination shows a man, often referred to as the first tribologist, standing on the sledge supporting the statue and pouring a liquid (most likely water) onto the

(a) **(b)**

Figure 1.2: (a) Drawing of a car with two studded wheels, circa 1338 A.D., and (b) Wooden bearing for stone wheels, ancient Egypt.

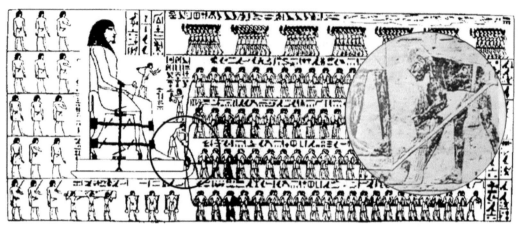

Figure 1.3: Transporting an Egyptian statue, El Bersheh, circa 1900 B.C. (Artificial arrangement of the persons probably arises from the Egyptian artist's inability to draw perspective).

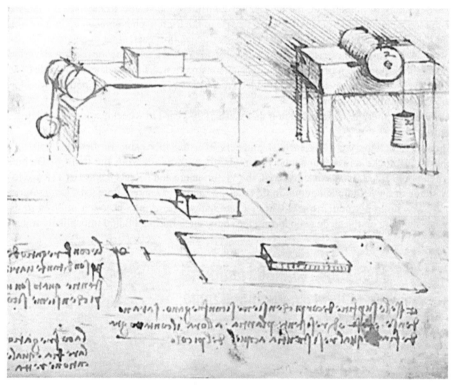

Figure 1.4: Drawing of sled friction test geometry, Leonardo da Vinci (1452–1519).

path of motion (Bhushan, 2013). Dowson assumed that each slave can pull about 800 newtons (Dowson, 1998). Consequently, the total work required should be at least equal to the friction force. This would suggest a coefficient of friction of about 0.23 which is comparable to lubricated sliding of wood. Thus, we can infer that Fig. 1.3 is a true record of what actually occurred (Davison, 1961).

Egyptians also used organic lubricants around the same period. A chariot in an Egyptian tomb still contains some of the original animal-fat lubricant in its wheel bearings. It is interesting to know that this lubricant was contaminated with road dirt in the form of quartz, compounds of aluminum, iron and lime (Halling, 1979).

From a tribological point of view, the wheel has been considered as one of the most important innovations, with the transition from sledges to wheeled vehicles first discovered in about 3500 B.C. The remains of many wheels, dating back to the third millennium, have been found in Europe, Asia and Africa. In 1928, fragments of what looks like a 2000-year-old thrust ball-bearing were found in Italy, near Rome. This ball-bearing was probably used in supporting a statue in a sculptor's workshop.

The scientific study of the friction phenomenon is much more recent than what the abovementioned applications might suggest. During the Renaissance period (1450–1600 A.D.), Leonardo da Vinci (1452–1519) postulated the first scientific approach to study friction in his manuscripts, Fig. 1.3. The most important mathematical result of his studies demonstrated that the friction force is proportional to load and independent of the apparent (or nominal) area of contact. Another important study was a description of a low-friction bearing alloy, early form of ball-bearing cage, and ball-and-roller pivot bearings. Although da Vinci was a pioneer in his findings, his manuscripts had no historical influence because they remained unpublished for hundreds of years.

Another development for wheel bearings was achieved in 1684 by Robert Hooke (1635–1703). He suggested that the combination of steel shafts and metal bushes would be preferable to the direct contact between wooden wheels and iron shafts. Consequently, he presented a series of concepts on bearing design, seals, materials and lubrication.

The German mathematician Leonard Euler (1707–1783) published two important papers concerning friction. He introduced a mathematical definition of the force required to move a weight up a slope inclined to the horizontal plane and defined the coefficient of friction as a function of that slope. The most important outcome of his studies was his differentiation between kinetic and static friction.

However, in 1699, Guillaume Amontons, independently of Leonardo da Vinci, discovered the rules of friction after he studied dry sliding between two flat surfaces.

Another important finding in tribology studies, introduced in 1785 by Charles-Augustin Coulomb, states that once motion starts, the friction force becomes independent of velocity (Amontons, 1699; Coulomb, 1785). It is important to mention here that the above findings of friction are still considered as reasonably true, even today.

In 1920, a resurgent interest in the adhesion hypothesis began and it became readily possible to examine the friction properties of a surface with different degrees of contamination (Hardly and Hardly, 1919; Tomlinson, 1929). Further investigations pointed out that there was a crucial difference between the apparent and the real area of contact, and that it was the real area alone which determined the magnitude of the friction force (Holm, 1938; Bowden and Tabor, 1950). To date, there has been an increasing interest in studying the friction process, especially the way the friction force is produced.

Developments in lubrication were coupled with the petroleum industry. The production of bitumen and oils from petroleum started in Scotland, Canada and the United States in the 1850s. In fact, the scientific understanding of the principles of hydrodynamic lubrication started with the experimental investigations of Beauchamp Tower (Tower, 1884) and the perceptive theoretical interpretation by Reynolds (Osborne, 1886) and related work by Petroff (Petroff, 1883). Since that time, there has been a steady increase of interest in hydrodynamic bearing theory and practice to meet the demand for reliable bearings in new machinery.

Conversely, when we turn to investigate the history of wear and adhesion, we find it a much younger subject than friction and bearing development, although wear phenomena must have been first recognized many years earlier. This unexpected situation is attributed to the historically very late study of wear process and the very recent explanation of its governing laws. The Industrial Revolution period (1750–1850) was the golden age of rapid production machinery development. Consequently, the need for reliable machine components emerged. Since the beginning of the 20th century, enormous industrial growth has lead to high demand for better tribology. Accordingly, knowledge in wear process has expanded tremendously. The earliest substantial contributions to the scientific study of wear were performed in the mid-twentieth century by Holm (Holm, 1946) and followed by Bowden and Tabor (Bowden and Tabor, 1950). Over the past three decades, the understanding of wear mechanisms, especially in extraordinary operation conditions, has developed rapidly with the use of advanced measuring devices and methods. There are now many examples of advanced engineering products whose development and successful use are possible only through the understanding and successful control of wear processes.

1.3 Development and Future Challenges

Since the publication of the Jost Report (Jost, 1966) in 1966, the understanding and ability to model many tribology subjects has increased enormously. The report's promise of saving 5 billion pounds per year in the UK motivated tribologists to work on applying the existing knowledge of friction, wear and lubrication to solve engineering problems. During the past 50 years, tribology proved to represent an important area of technical engineering.

The development of tribology was not an easy task. Many undesired problems began with the definition of the tribological process. There is also a great discrepancy between studying triosystems as a black box system and the traditional way of studying friction, wear and lubrication independently.

Another important challenge is that there is no general model or mathematical tool to be used in tribology. In fact, it is very difficult for tribologists to use the results obtained under one condition in other condition. For example, in studying the tribology process on macro- and micro-scales, many studies failed to relate the macro-mechanisms of friction and wear to what happens on the micro-scale.

It was realized that most of the techniques can be applied only to a specific branch of field for a specific target. However, prediction of friction and wear is still possible in some rather restricted cases.

Over the last three decades, tribology has extended beyond the field of machinery. New areas of tribology have emerged, namely, space tribology, biotribology, green tribology, micro/nano-tribology and others. These multidisciplinary areas involve studying friction, wear and lubrication at different scales. It also covers biological systems and ecological aspects.

Currently, the world trends of tribology development significantly change the subjects of tribological studies. This was aimed at creating tribological materials capable of serving under the most adverse external impact conditions (Sviridenok et al., 2015). Table 1.1 is a summary the world's most popular tendencies in tribology development and areas of application for the next two to three decades.

There has been a great deal of interest in improving lubricants to achieve low wear rates with thinner films. Therefore, during the 1950s to 1970s, the improvment of film lubrication led to the development of the elastro-hydrodynamic lubrication. This was followed by the development of the first biodegradable lubricants which were produced from harvestable fatty materials. Presently, there is also active research in lubricants and lubricant additives which are effective for non-ferrous metals, ceramics and engineering surfaces.

The future challenges in the field of tribology are related to many critical features and performance requirements for modern engineering. Therefore, tribological developments are often focused on improving engineering systems, engineering products and manufacturing processes. Examples for such challenges involve high-speed sliding or rolling, small dimensions, and hostile environments. These also include gas turbine engines, artificial human joints, automotive engines and transmissions, tires and brakes, hard disk drives for data storage and an increasing number of electromechanical devices for domestic and industrial use. Tribological developments of space mechanisms to work in a vacuum and at extreme temperatures also play a major role in the overall success of a space mission.

Finally, It is obvious that the engineers of the future need to be aware of and informed about tribology. The management at all levels also need to appreciate the importance of the application of tribological knowledge.

Table 1.1: World's tribology developments and tendencies in the 21st century. After Sviridenok et al. (2015), with permission from Springer.

General lines of tribology research	Problems and goals
General requirements of tribology in 21st century	– miniaturization of tribo-technical devices; – greening (reduction of noise and environmental pollution); – significant decrease of energy consumption when translating motion and power; – significant prolonging of the operating life at high rates of loading, speed, and temperature during operation in abrasive environments; – extending life of human joints and other organs functioning in friction and lubrication conditions; – use of new physical, chemical, and biological principles and effects when solving the problems of tribology; – estimation of the role of friction in snow and stone avalanches; – development of theory and methods of study of tribo-couplings in deep space, at high doses of radiation exposure, hypersonic speeds, and high pressure at extreme temperatures above and below zero.
Triboanalysis (tribophysics, tribochemistry, tribobiology, tribomechanics)	– switching from micro- to nano-level of analysis of nature and frictional activity mechanisms; – development of discrete frictional contact mechanics with regard to the three level model (macro, micro, nano); – development of thermal calculations of frictional contact in order to measure the mutual influence between basic force, speed, and structural characteristics of friction materials and surfaces; – development of theory of frictional adhesion in discrete zones of micro- and nano-contacts; – expanding the field of research of friction behavior features in moving biological objects in order to apply biological control methods of friction properties created by nature and implemented in human articulation joints; – studying the mechanisms and modelling of resistance against the motion of fish and sea creatures.
Tribological Material Science	– creation of multipurpose tribo-technical materials adapting to external impact; – development of new micro- and nano-filled lubrication agents of high temperature resistance and load bearing capacity; – creation of lubricants of extraordinarily low resistance (friction factor) in ultrathin layers to ensure functioning of new micro- and nano-mechanisms; – creation of a new class of low degrading friction (braking) materials; – creation of a new class of extra hard friction materials for cutting instruments, including drilling tools.
Tribotechnology	– development of additive technologies of creation of multilayer and multipurpose hardening coatings for tribosurfaces; – development of theory and methods of frictional properties control during pressure treatment of metals, recycling into polymer and composite items, during jet transportation of bulk products, etc.; – development of application methods for thin ultra-lubricant coatings; improving durability of cutting tools during processing of materials at a high speed.
Tribotechnics	– analysis and creation of friction units for micro- and nano-electromechanical systems, nano-computers, nano-manipulators, etc.; – development of the 3D prototyping method in terms of modeling complex friction units; – development of new principles of consumption (use) of vehicle braking friction energy in order to increase their effectiveness and operation life and to reduce noise and environmental pollution; – creation of new tribo-devices of increased operation life having been exposed to ionizing radiation and high temperatures, operating in the void and in sea water; – development of CAD systems for the majority of unified tribojoints; – development of new structures of tribo-joints with a provision for computer-based control of friction parameters.

Table 1.1 contd. ...

...Table 1.1 contd.

General lines of tribology research	Problems and goals
Tribomonitoring and Tribodiagnostics	– development of methods and instruments of multi parameter tribo-diagnostics and active use of controlling signals in order to improve; – frictional properties of critical tribo-joints; – development of new methods and instruments to study the nature and mechanisms of friction activities at the micro- and nano-levels; – development of direct measuring methods of friction and wear of tribo-couplings in biological objects; – promotion of participation in the development of international research standardization for friction and wear.
Triboinformatics	– creation of tribo-technical data banks and sharing computer based access to them with broad researcher and student audiences; – initiating the creation of an international association of tribology related journals in order to rapidly communicate the latest information; – issuing high-quality reading tutorials and textbooks dedicated to the basics of tribology; – introduction of the main tribology-related editions issued in the Russian language in the list of peer-reviewed international journals.

1.4 Knowledge of Tribology: Education, Publications and Researches

Since the publication of the Lubrication Report, there has been an increasing awareness of the knowledge of tribology. The details of tribological knowledge relating to the basic sciences, to lubricants, and to wear or machine elements, have been recognized in numerous aspects.

In 1987, the ASME/STLE Tribology Conference was concerned that some of engineering students may not be learning enough about the fundamentals of tribology (Jahanmir and Kennedy, 1991). For instance, in the USA, there were approximately 250 universities and colleges offering engineering degrees and only 30% offered tribology courses. Furthermore, there is no single widely-accepted textbook for a comprehensive tribology course. Since knowledge of tribology is required for the design of reliable and efficient mechanical systems, it was recommended that tribology fundamentals should be included in the education of future engineers, especially mechanical engineers. Now, tribology is universally recognized as a generic technology underlying many industrial sections. Tribology courses are now included in the curricula of most engineering colleges and universities, in all countries, with textbooks covering all aspects of tribology. Subsequently, many research institutes and societies were established all over the world in order to study the fields of friction, wear, lubrication and material science. A valuable contribution to the development of tribology is made by research groups of such institutes. Table 1.2, introduces samples of these facilities and their frameworks and Table 1.3 introduces samples of advocacy and professional societies (Jahanmir and Kennedy, 1991).

Many international conferences and symposia in tribology are held to provide opportunities to academic scientists, researchers and research scholars and those working in related fields to share, discuss and assimilate their experiences on all aspects of tribology. Table 1.4 provides data of the most famous tribology conferences.

Scientific publications in the field of tribology are numerous; several reports and books are aimed at providing the academic community and industry with many subjects related to tribology. Besides, a number of international journals are dedicated to publishing papers to provide an archival resource for scientists from all backgrounds. The journals facilitate reporting experimental and theoretical studies and address the fundamentals of friction, lubrication, wear and adhesion. Tables 1.5 and 1.6 introduce a number of classical and modern publications, books and journals, in the field of tribology.

Table 1.2: Tribology higher education and facilities (Sviridenok, 2015; Jahanmir, 1991) (listed in Alphabetical order).

Country	University/Institute	Facility	Website
Australia	Curtin University	Tribology Laboratory	http://www.curtin.edu.au
	University of New South Wales	Tribology and Machine Condition Monitoring	http://www.unsw.edu.au
Belgium	Ghent University	Soete Laboratory	http://www.ugent.be
Belarus	National Academy of Sciences of Belarus	V.A. Belyi Metal Polymer Research Institute	http://www.nasb.gov.by/rus
	Belarusian National Technical University	Mechanical Engineering Technology	http://www.en.bntu.by
	Russian Academy of Sciences	Interdepartmental research council on tribology	http://www.ras.ru/en/index.aspx
Brazil	University of Sao Paulo	Surface Phenomena Laboratory	http://www.usp.br
Canada	Dalhousie University	Advanced Tribology Lab	http://dal.ca
	University of Waterloo	Tribology Research Group	http://www.uwaterloo.ca
	University of Windsor	Tribology of Materials Research Group	http://www.uwindsor.ca
China	Chinese Academy of Sciences	State Key Laboratory of Solid Lubrication	http://www. cas.ac.cn
	Hefei University of Technology	Institute of Tribology	http://www.hfut.edu.cn
	Tsinghua University	State Key Laboratory of Tribology	http://www.tsinghua.edu.cn
France	Ecole Centrale de Lyon	Laboratory of Tribology and Dynamics of Systems	http://www.ec-lyon.fr/en
	Institut Supérieur de Mécanique de Paris	Laboratoire d'Ingénierie des Systèmes Mécaniques et des Matériaux	http://www.supmeca.fr
	INSA Lyon/University of Lyon	Contact and Structural Mechanics Laboratory	http://www.insa-lyon.fr
	University of Poitiers	Laboratory of Solids Mechanics	http://www.univ-poitiers.fr
Germany	Karlsruhe Institute of Technology	Microtribology Centre	http://www.kit.edu
	RWTH Aachen University	Institute for Machine Elements and Machine Design	http://www.rwth-aachen.de
	Technical University of Berlin	Fachgebiet Systemdynamik und Reibungsphysik	http://www.tu-berlin.de
	University of Technology Munich	Forschungsstelle für Zahnräder und Getriebebau	http://www.tum.de

Table 1.2 contd. ...

...Table 1.2 contd.

Country	University/Institute	Facility	Website
Japan	Kanazawa University	Tribology Laboratory	http://www.kanazawa-u.ac.jp/
	Nagoya University	Tribology Laboratory	http://www.nagoya-u.ac.jp
	Niigata University	Tribology Laboratory	http://www.niigata-u.ac.jp
S. Korea	KAIST	Precision Machine Elements and Tribology Laboratory	http://www.kaist.edu
	Korea University	Functional Materials Research Laboratory	http://www.korea.edu
	Yonsei University	Center for Nano-Wear	http://www.yonsei.ac.kr/en_sc/
South Africa	The University of Pretoria	Tribology Laboratory	http://www.up.ac.za
	South African Institute of Tribology		http://www.sait.org.za
Malaysia	International Islamic University Malaysia	Dr. Maleque	http://www.iium.edu.my
	Malaysia-Japan International Institute of Technology	Tribology and Precision Machining i-Kohza	http://www.mjiit.utm.my/
	Universiti Malaya	Tribology Researcher	http://www.um.edu.my
	Universiti Kebangsaan Malaysia	Advanced Materials Processing & Integrity Group	http://www.ukm.my
	Universiti Teknologi MARA	Tribology Researcher	http://www.uitm.edu.my
	Universiti Teknikal Malaysia Melaka	Green Tribology & Engine Performance Research Group	http://www.utem.edu.my
	Universiti Teknologi Malaysia	Tribology Researcher	http://www.utm.my
	Universiti Sains Malaysia	Nanofabrication & Functional Materials Research Group	http://www.usm.my
Netherlands	Delft University of Technology	Tribology Research Group	http://www.tudelft.nl/en
	University of Twente	Laboratory for Surface Technology and Tribology	http://www.utwente.nl/en
Portugal	Aveiro University	Machining & Tribology Research Group	http://www.ua.pt
Sweden	Luleå University of Technology	Machine Elements Research Group	http://www.ltu.se
	Uppsala University	Tribomaterials Research Group	http://www.uu.se

Country	Institution	Research Group / Centre	URL
United Kingdom	Bournemouth University	Sustainable Design Research Centre	http://www.bournemouth.ac.uk
	Cardiff University	Tribology and Contact Mechanics Research Group	http://www.cardiff.ac.uk
	Imperial College London	Tribology Group	http://www.imperial.ac.uk
	University of Cambridge	Tribology Research Group	http://www.cam.ac.uk
	University of Central Lancashire	Jost Institute for Tribotechnology	http://www.uclan.ac.uk/
	University of Leeds	Institute of Functional Surfaces	http://www.leeds.ac.uk
	University of Loughborough	Dynamics Research Group	http://www.lboro.ac.uk
	University of Sheffield	The Leonardo Tribology Centre	http://www.sheffield.ac.uk
	University of Southampton	National Centre for Advanced Tribology	http://www.southampton.ac.uk
	University of Strathclyde	Tribology Group	https://www.strath.ac.uk/
United States of America	Auburn University	Multiscale Tribology Laboratory	http://www.auburn.edu
	Georgia Institute of Technology	Tribology Research Group	http://www.gatech.edu
	Lehigh University	Surface Interfaces and Materials Tribology Laboratory	http://www.lehigh.edu
	Louisiana State University	Center for Rotating Machinery	http://www.lsu.edu
	Northwestern University	Center for Surface Engineering and Tribology	http://www.northwestern.edu
	Purdue University	Materials Processing and Tribology Research Group	http://www.purdue.edu
	Texas A&M University	Rotor Dynamics Laboratory Tribology Group	http://www.tamu.edu
	University of Akron	Timken Engineered Surfaces Laboratories	http://www.uakron.edu
	University of Dayton	Tribology Group	http://www.udayton.edu
	University of Delaware	Materials Tribology Laboratory	http://www.udel.edu
	University of Illinois	Tribology and Microtribodynamics Laboratory	http://www.illinois.edu
	University of Texas at Arlington	Turbomachinery and Energy Systems Laboratory	http://www.uta.edu
	University of Utah	Nanotribology and Precision Engineering Laboratory	http://www.utah.edu

Table 1.3: Advocacy and professional societies.

Title	Region served	Website
American Bearing Manufacturers Association (ABMA)	USA	http://www.americanbearings.org
American Gear Manufacturers Association (AGMA)	Worldwide	http://www.agma.org
American Society of Mechanical Engineers (ASME) - Tribology Group	USA	https://www.asme.org
Asociación Argentina de Tribología	Argentina	http://www.aatribologia.org.ar
Austrian Tribology Society	Australia	http://www.oetg.at/startseite
Egyptian Society of Tribology	Egypt	http://www.egtrib.org
Finnish Society for Tribology	Worldwide	http://www.tribologysociety.fi
French Association for Mechanics - Tribology Group	France	http://www.afm.asso.fr
Gesellschaft für Tribologie e. V.	Germany	https://www.gft-ev.de/en/the-gft
International Federation for the Promotion of Mechanism and Machine Science (IFToMM)	Worldwide	http://www.iftomm.net
Institution of Engineering and Technology - Tribology Network (UK)	UK–Worldwide	http://www.heiet.org
Institution of Mechanical Engineers (IMechE) - Tribology Group (UK)	Worldwide	http://www.imeche.org
Institute of Physics - Tribology Group (UK)	UK	http://www.iop.org
International Tribology Council (ITC)	Worldwide	http://www.itctribology.net
Italian Tribology Association (ATI)	Italy	http://www.aitrib.it
Japanese Society of Tribologists (JAST)	Japan	http://www.tribology.jp
Korean Society of Tribologists and Lubrication Engineers (KSTLE)	S. Korea	http://www.kstle.or.kr
Malaysian Tribology Society (MYTRIBOS)	Malaysia	http://www.mytribos.org
Society of Tribologists and Lubrication Engineers (STLE) (USA)	USA	http://www.stle.org
Serbian Tribology Society (STS)	Serbia	http://www.sts.fink.rs
Society of Bulgarian Tribologists	Bulgaria	http://www.bultrib.com
South African Institute of Tribology	South Africa	http://www.sait.org.za
Tribology Institute of Chinese Mechanical Engineering Society	China	http://www.cntribo.org/en.asp
Tribology Society of India	India	http://www.tribologyindia.org
UK Tribology Network	UK	http://www.uktribology.net

Table 1.4: Data of the most famous tribology conferences.

Title	Location	Organizer	Website
Leeds-Lyon Symposium in Tribology	UK-France	University of Leeds-INSA de Lyon	
World Tribology Congress WTC	Worldwide	Worldwide	
International Conference on Wear of Materials WOM	USA	Elsevier	http://www.wearofmaterialsconference.com
International Conference on Industrial Tribology	India	Tribology Society of Tribology	http://www.tribologyindia.org
International Colloquium Tribology	Germany	Technische Akademie Esslingen	https://www.tae.de
International Conference on Tribology	Turkey	J. of the Balkan Trib. Ass.	http://www.scibulcom.net
International Conference on Tribology	Serbia	Faculty of Eng. Univ. of Kragujevac & Serbian Tribology Society	http://www.serbiatrib.mfkg.rs
Young Tribological Researcher Symposium	Germany	Gesellschaft für Schmiertechnik GFT	https://www.gft-ev.de
National Tribology Conference	India	Tribology Society of India (TSI)	http://www.tribologyindia.org
STLE/ASME International Joint Tribology Conference	USA	American Society of Mechanical Engineers	http://proceedings.asmedigitalcollection.asme.org
STLE Annual meeting	USA	Society of Tribologists and Lubrication Engineers	http://www.stle.org
International Conference on Tribology ROTRIB	Romania	Romanian Tribology Association	
International Conference on Biotribology	Worldwide	Elsevier	https://www.elsevier.com
Tribology Conference of Győr	Hungary	Széchenyi István Egyetem	http://www.gytt.hu
Malaysian International Tribology Conference	Malaysia	Malaysian Tribology Society (MYTRIBOS)	https://www.mytribos.org
European Space Mechanisms and Tribology Symposium	EU	Airbus Defence and Space	http://www.esmats.eu
International Tribology Conference	Japan	Japansese Society of Tribologists	http://www.tribology.jp
TriboUK	UK	Students from a different UK university each year	http://www.uktribology.net
International conference on polymer tribology, PolyTrib	Slovenia	Slovenian Society for Tribology	https://www.tint-polytrib.com
Int. Conference on Tribology in Manufacturing Processes & Joining by Plastic Deformation (ICTMP)	Worldwide	Technical University of Denmark	http://www.conferencemanager.dk

Table 1.4 contd. ...

...Table 1.4 contd.

Title	Location	Organizer	Website
Tribocorrosion Symposium	UK	University of Strathclyde, UK	https://www.strath.ac.uk
Annual Cambridge Tribology Course	UK	University of Cambridge, Dept. of Engineering	https://www.ifm.eng.cam.ac.uk
International Conference on Advanced Tribology	Singapore	Dept. of Mechanical Engineering, National University of Singapore	http://www.nus.edu.sg
China International Symposium on Tribology	China	Chinese Academy of Sciences (CAS), Lanzhou Inst. of Chemical Physics (LICP)	http://english.licp.cas.cn
International Conference on Tribochemistry	Poland	University of Lodz, Dept. of Chemical Tech. and Environmental Protection	http://chemia.p.lodz.pl
International Conference on Advances in Tribology and Engineering Systems	UK	Int. Aerospace and Mech. Eng. Committee	http://www. waset.org

Table 1.5: Suggested Tribology publications, Books (listed in chronological order).

Author(s)/Editor(s)	Title	Year	Publisher	ISBN
Ian Hutchings, Philip Shipway	Tribology	2017	Elsevier	978-008-1009-109
Ming Qiu, Chen Y., Yingchun Li, Jiafei Yan	Bearing Tribology: Principles and Applications	2016	Springer	978-366-253-097-9
Abdelbary, A.	Wear of Polymers and Composites	2015	Woodhead Publishing	978-1-78242-177-1
Straffelini, G.	Friction and Wear: Methodologies for Design and Control	2015	Springer	978-331-905-894-8
Lieng-Huang Lee	Advances in Polymer Friction and Wear	2013	Springer	978-146-139-942-1
Ioan D., Marinescu W., Brian Rowe, Boris Dimitrov, Hitoshi Ohmori	Tribology of Abrasive Machining Processes	2012	William Andrew	978-143-773-467-6
Friedrich, K.	Friction and Wear of Polymer Composites	2012	Elsevier	978-044-459-711-3
Chang-Hung Kuo	Tribology - Lubricants and Lubrication	2011	InTech	978-953-307-371-2
Davim, J. Paulo	Tribology for Engineers	2011	Woodhead Publishing	978-085-709-144-4
Sinha, K. and Briscoe, J.	Polymer Tribology	2009	Imperial College Press	1-84816-202-2
Chand, N. and Fahim, M.	Tribology of Natural Fiber Polymer Composites	2008	Woodhead Publishing	978-184-569-393-0
Mellor, B.G.	Surface Coatings for Protection Against Wear	2006	Woodhead Publishing	978-185-573-767-9
Stachowiak, G.W.	Wear – materials, mechanisms and practice	2005	John Wiley and Sons	0-470-01628-0
Bayer, Raymond G.	Mechanical Wear Fundamentals and Testing	2004	Marcel Dekker	0-8247-4620-1
Bhushan, B.	Introduction to Tribology	2002	John Wiley and Sons	978-1-119-94453-9
Bhushan, B.	Modern Tribology Handbook	2000	CRC Press	978-084-937-787-7
Buckley, D.	Surface effects in adhesion, friction, wear, and lubrication	2000	Elsevier	978-008-087-569-9
Grainger, S.	Engineering Coatings	1998	Woodhead Publishing	978-185-573-369-5
Halling, J.	Principles of Tribology	1997	Macmillan Press	978-1-118-06289-0
Booser, E. Richard	Tribology Data Handbook	1997	CRC Press	978-142-005-047-9
Ludema, K.C.	Friction, Wear, Lubrication	1996	CRC Press	0-8493-2685-0

Table 1.5 contd. ...

...Table 1.5 contd.

Author(s)/Editor(s)	Title	Year	Publisher	ISBN
Rabinowicz, E.	Friction and Wear of Materials	1995	John Wiley and Sons	978-0-471-83084-9
Dalmaz, G., Childs, T.H.C. Dowson, D., Taylor C.M.	Lubricants and Lubrication	1995	Elsevier	978-008-087-594-1
Holmberg, K. and Matthews, A.	Coatings Tribology	1994	Elsevier	978-0-444-52750-9
Williams, J.	Engineering Tribology	1994	Cambridge Univ. Press	978-052-160-988-3
Stachowiak, G.W. and Batchelor, A.W.	Engineering Tribology	1993	Butterworth-Heinemann	978-008-087-588-0
Dalmaz, G., Childs, T.H.C., Dowson, D., Godet, M., Taylor, M.	Thin Films in Tribology	1993	Elsevier	978-008-087-589-7
Glaeser, W.	Materials for Tribology	1992	Elsevier	978-008-087-584-2
Hutchings, I.	Tribology: Friction and Wear of Engineering Materials	1992	Butterworth-Heinemann	978-034-056-184-3
Stolarski, T.A.	Tribology in Machine Design	1990	Butterworth-Heinemann	0-7506-3623-8
Kajdas, C., Wilusz, E., Harvey, S.	Encyclopedia of Tribology	1990	North Holland	978-008-087-579-8
Yamaguchi, Y.	Tribology of Plastic Materials	1990	Elsevier	978-008-087-580-4
Jones, M.H. and Scott, D.	Industrial Tribology	1983	North Holland	978-008-087-572-9
Kragelsky, V., Dobychin, M., N. Kombalov, V.S.	Friction and Wear	1982	Pergamon	978-148-314-550-1
Lansdown, A.R.	Lubrication	1982	Pergamon	978-148-313-751-3
Sarkar, A.D.	Friction and Wear	1980	Academic Press	978-012-619-260-5
Moore, Desmond F.	Principles and Applications of Tribology	1975	Pergamon	978-148-315-728-3
Neale, M.G.	The Tribology Handbook	1973	Butterworth-Heinemann	0-7506-11-987
Clauss, Francis J.	Solid Lubricants and Self-Lubricating Solids	1972	Academic Press	978-032-315-822-0
Braithwaite, E.R.	Solid Lubricants and Surfaces	1964	Pergamon	978-148-315-681-1

Table 1.6: Tribology publications, International Journals (listed in Alphabetical order).

Title	Editor/Publisher	ISSN (Print)	Website
Advances in Tribology	Hindawi	1687-5915	https://www.hindawi.com
ASME Journal of Tribology	ASME	0742-4787	http://tribology.asmedigitalcollection.asme.org
Biotribology	Elsevier	2352-5738	http://www.journals.elsevier.com/biotribology/
Biosurface and Biotribology	Elsevier	2405-4518	http://www.journals.elsevier.com/biosurface-and-biotribology/
Friction	Springer	2223-7690	http://link.springer.com/journal/40544
Industrial Lubrication and Tribology	Emerald	0036-8792	http://www.emeraldinsight.com/loi/ilt
International Journal of Surface Science and Engineering	Inderscience	1749-7868	http://www.inderscience.com
Japanese Journal of Tribology	Allerton Press	1045-7828	http://www.allertonpress.com/journals/jst.htm
Journal of Bio- and Tribo-Corrosion	Springer	2198-4220	http://www.springer.com/materials/surfaces+interfaces/journal/40735
Journal of Friction and Wear	Allerton Press	1068-3666	https://link.springer.com/journal/11959
Journal of the Egyptian Society of Tribology	Egyptian Soc. of Tribology	2090-5882	http://www.egtribjournal.com
Journal of the Japanese Society of Tribologists	Japanese Soc. of Tribologists	0915-1168	http://www.tribology.jp/journal_e/
Journal of the Balkan Tribological Association	Balkan Tribological Assoc.	1310-4772	http://www.scibulcom.net/
Journal of Tribology - ASME	ASME	0742-4787	http://www.tribology.asmedigitalcollection.asme.org/journal.aspx
Jurnal Tribologi - MYTRIBOS	Malaysian Tribology Soc.	2289-7232	http://www.jurnaltribologi.mytribos.org/
Journal of Lubrication Technology	American Soc. of Mech. Eng.	0022-2305	http://www.tribology.asmedigitalcollection.asme.org/
Lubrication Science	Wiley	0954-0075	http://www.onlinelibrary.wiley.com/journal/
Lubrication and Lubricants	ACS Publications	2075-4442	http://www.pubs.acs.org/doi/abs/10.1021/ie50016a001
Lubricants	MDPI	2075-4442	http://www.mdpi.com/journal/lubricants
Proceedings of the Institution of Mechanical Engineers, Part J: Journal of Engineering Tribology	Sage	1350-6501	http://www.pij.sagepub.com/
Surface and Coatings Technology	Elsevier	0257-8972	http://www.journals.elsevier.com/surface-and-coatings-technology/

Table 1.6 contd. ....

...Table 1.6 contd.

Title	Editor/Publisher	ISSN (Print)	Website
Surface Engineering	Taylor & Francis	0267-0844	http://www.tandfonline.com/loi/ysue20
Tribologia	Finnish Soc. for Tribology	1797-2531	https://www.journal.fi/tribologia
Tribology and Interface Engineering Series	Elsevier		http://www.sciencedirect.com/science/bookseries/15723364
Tribology and Lubrication Technology	STLE	1545-858X	http://www.stle.org/
Tribology in Industry	Univ. of Kragujevac, Serbia	0354-8996	http://www.tribology.fink.rs/
Tribology International	Elsevier		http://www.journals.elsevier.com/tribology-international/
Tribology Letters - STLE	Springer	1023-8883	http://link.springer.com/journal/11249
Tribology - Materials, Surfaces and Interfaces, Online	Maney Society of Tribologists		http://www.maneyonline.com/loi/trb
Tribology Online - JAST	Japanese Soc. of Tribologists		http://www.tribology.jp/trol/
Tribology Transactions - STLE	Taylor and Francis	1040-2004	http://www.tandfonline.com/loi/utrb20#.Va9V3kY1Nl0
Wear	Elsevier	0043-1648	http://www.journals.elsevier.com/wear/

References

Amontons, G. 1699. De la résistance causée dans les machines (About resistance and force in machines). Mem l'Acedemie R A: 257–282.

Bhushan, B. 2013. Introduction to Tribology. Second edition ed. New York: John Wiley & Sons Inc.

Bowden, F.P. and Tabor, D. 1950. The Friction and Lubrication of Solids. U.K.: Clarendon Press Oxford.

Coulomb, C.A. 1785. Théorie des Machines Simples, en ayant regard au Frottement de leurs Parties, et a la Roideur dews Cordages. Mem. Math. Phys., X, Paris: 161–342.

Davison, C.St.C. 1961. Transporting sixty-ton statues in early Assyria and Egypt. Technology and Culture 2(1): 11–16. Doi: 10.2307/3101295.

Dictionary. http://www.dictionary.com.

Dowson, D. 1998. History of Tribology. Second edition ed. London, U.K: Institution of Mechanical Engineering.

Halling, J. 1979. Principles of Tribology. London: Macmillan Press Ltd.

Hardly, W.B. and Hardly, J.K. 1919. Note on static friction and on lubricating properties of certain chemical substances. Phil. Mag. 6th Series (38): 32–48.

Holm, R. 1938. The friction force over the real area of contact (in German). Wiss. Veröff. Siemens-Werk 17(4): 38–42.

Holm, R. 1946. Electric Contacts. Gerbers, H., ed. Stockholm, Sweden.

Jahanmir, S. and Kennedy, F.E. 1991. Tribology education: Present status and future challenges. Journal of Tribology 113(2): 229–231. Doi: doi:10.1115/1.2920610.

Jost, P. 1966. Lubrication (Tribology) Education and Research, Technical report. H.M.S.O.

Materials, American Society for Testing and Materials. 2001. Standard terminology relating to wear and erosion. In standard G-40-01: American Society for Testing and Materials.

Materials, Organisation for Economic Co-operation and Development. Research Group on Wear of Engineering. 1969. Glossary of Terms and Definitions in the Field of Friction, Wear and Lubrication; Tribology. Edited by M.B. Peterson and W.O. Winer: Organisation for Economic Co-operation and Development.

Osborne, R. 1886. On the theory of lubrication and its application to Mr. Beauchamp Tower's experiments. Phil. Trans. R. Soc.

Oxford. https://en.oxforddictionaries.com.

Petroff, N.P. 1883. Friction in machines and the effects of the lubricant (in Russian). J. St. Petersberg, pp. 1–4.

Rabinowicz, E. 1995. Friction and Wear of Materials. Second edition ed. New York: John Wiley and Sons.

Sviridenok, A.I., Myshkin, N.K. and Kovaleva, I.N. 2015. Latest developments in tribology. Journal of Friction and Wear 36(6): 449–453. Doi: doi.org/10.3103/S106836661506015X.

Tomlinson, G.A. 1929. A molecular theory of friction. Phil. Mag. 7th Series (7): 905–930.

Tower, B. 1884. Report on friction experiments. Proc. Instn. Mech. Engrs.

Chapter 2

Tribo-material Properties

2.1 Introduction

When selecting materials for a specified application, it is important to assign and understand the material properties which influence this implementation. In general, there are more than ten categories of qualitative and quantitative properties to classify, identify, and finalize the material selection. These sets include acoustical, chemical, electrical, environmental, mechanical, optical, and other properties. With respect to tribological applications, engineering materials are subjected to force, heat, vibration, chemical reaction, and various other effects. Since tribology is focused on wear, friction and lubrication of two mating surfaces in relative motion, it is important to identify the key properties that influence interactions of tribo-surfaces.

As might have been anticipated, parameters which affect a tribo-interaction of a surface can be distinguished in two main categories (Rabinowicz, 1995). First, the volume properties that relate to contacting bodies as a whole. These properties include plastic deformation, elastic deformation, hardness, creep, fatigue and fracture. Nevertheless, thermal properties could be also an important parameter in high speed applications. Second, the surface properties which determine the contacting interface of the bodies in the tribo-contact. Among these properties, adhesion, surface reactivity and surface roughness should be mentioned. However, in a lubricated tribo-contact, we should cite the parameters which govern the lubricant action, these include viscosity of the fluid lubricant and surface adsorption of the solid film lubrication.

We should mention that a close relationship exists between many of the volume and surface parameters themselves. This inflicts severe constraints on the material selection of mechanical parts, especially those operating in extreme working conditions. Thus, the following chapter is aimed at outlining and discussing the fundamental tribo-material properties in order to familiarize the reader with the tribo-concepts.

2.2 Volume Properties

2.2.1 Elastic Deformation

After application of load to a specimen in the simple tensile test, the force divided by the cross-sectional area at any time gives the tensile stress (σ), while the lengthening of the specimen divided by the original length represents the strain (ε). Elastic deformation is the reversibility of this action, that is, removal of the load results in return of the specimen to its original dimensions. In a typical stress-strain curve, Fig. 2.1, the initial ascending straight sloped section is known as the elastic region, whereas the subsequent less sloping region represents the plastic region. The relation between stress and strain in the first section of the curve is essentially linear up to the proportional limit (P).

At any point beyond this limit, the total strain is made up of two parts, ε_e, the elastic component, and ε_p, the plastic component. However, the elastic deformation will prevail until the elastic limit or yield point (Y), is reached. In most materials, there is very little difference between the two limits. The

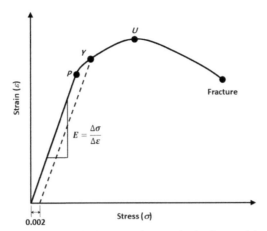

Figure 2.1: A typical stress-strain curve for ductile materials.

slope of the elastic section is called the modulus of elasticity, or Young's modulus (E), and is represented mathematically by Hook's Law,

$$E = \frac{\sigma}{\varepsilon} \tag{2.1}$$

On the other hand, if material is loaded in transverse direction (shear), the resulting shear stress (τ) generates elastic shear strain (γ) which is proportional to the shear stress according to the following relation:

$$G = \frac{\tau}{\gamma} \tag{2.2}$$

where (G) is the shear modulus or modulus of rigidity.

It is important to state that Young's and shear moduli are anisotropic within materials, that is, they vary with the crystallographic direction. Such anisotropy becomes significant in polycrystalline materials (Bhushan, 2013), furthermore, they decrease with temperature.

In practical tribo-applications with a special emphasis on the interaction between solid surfaces, there are two important considerations to be pointed out in regard to the elastic concepts.

First, the microscopic interaction between asperities on the mating surfaces. The first statistical elastic contact model of rough surfaces was presented by Greenwood et al. (Greenwood and Williamson, 1966). In their model, the rough surfaces were characterized by a population of isolated spherical asperities of identical radii. Each individual contacting asperity was assumed to deform elastically according to Hertz theory. This model will be discussed in detail in Chapter 3.

In general, due to the method of formation or machining, surfaces of solid bodies contain microscopic irregularities or deviations from a prescribed geometrical form. The high points on the surfaces are usually referred to as asperities, while the lower ones are referred to as valleys. When two solid surfaces come in contact, they do not make contact everywhere in the apparent contact area, but only at multiple asperity contact spots (junctions). The squeeze at the asperity contacts results in a deformation which can be either elastic and/or elastic–plastic (Bhushan, 1998). The size and number of contacting asperities are dependent on the interface conditions. Both friction and wear are strongly influenced by the deformation that occurs at the contacting asperities. The volume and the degree of deformation are functions of the amount of work or energy that was put into the deformation process and the nature of the material as well. Some materials are much more prone to deformation and work hardening than others.

Second, the macroscopic tribo-practical situation when two elastic solids are squeezed together. For instance, in rolling contact devices, such as ball bearings, loading-unloading cycles may result in a slight departure from elastic linearity and cause a detrimental change of shape of the bearing. Likewise, effects such as strain-rate sensitivity become important with soft materials, and with many non-metals, such as polymers and its composites (Rabinowicz, 1995).

2.2.2 Plastic Deformation

In a stress-strain curve, Fig. 2.1, when the material specimen is subjected to increasing stress beyond the yield point (Y), the material no longer returns to its original shape after releasing the stress. The deformation beyond the yield point is called plastic deformation. The change from elastic to plastic states is very gradual, the point of intersection is rather ambiguous and can be characterized by many parameters: The proportionality limit, the elastic limit, and the 0.2% yield stress. In plastic deformation, the bonds between initially neighboring atoms can be considered broken, and a new set of bonds which are as stable as the original ones is established. On a release of the load, the new configuration will be maintained and the deformation is permanent. However, if new bonds are not established, fracture ensues at a maximum nominal stress, called the ultimate tensile strength (σ_u) or the ultimate strength. The corresponding load is the greatest load that the material will bear, point (U) on stress-strain curve. Furthermore, a plastic deformation commonly causes residual stresses which can either decrease or increase the subsequent resistance of a component to fatigue or environmental cracking, depending on whether the residual stress is tensile or compressive, respectively (Dowling, 2012).

The plastic yield strength of metals can be determined by either a tensile test, or by pressing a hard indenter into a flat surface of material to be tested, and noting the area of indentation produced by a unit load. The property measured by a hardness test is the plastic strength of the material which is the amount of plastic deformation produced by a known force. It is important to mention that the yield criterion, which governs the plastic flow, was assumed to be independent of the rate of strain. However, the plastic flow of some materials is sensitive to the strain rate, which is known as material strain rate sensitivity, or viscoplasticity, which is quantitatively certified by the value of strain rate sensitivity index (m) (Ambrosio, 2001). Material strain rate sensitivity is a material effect and becomes important with soft metals such as lead, as well as many non-metals, such as superplastic materials (e.g., Polymers).

Under sliding conditions, the geometry of a typical surface interaction, in which a surface with rough asperities is squeezed against another, is very similar to the condition during an indentation hardness test. Thus, the plastic strength parameter of tribo-materials can be expressed by its indentation hardness.

With more focusing on tribology, as the normal load between two contacting bodies is applied, asperities on the tribo surfaces are deformed elastically according to their Young's moduli of elasticity. As the load is increased, a local plastic deformation may take place at the tips of the asperities of the lower hardness body. As the normal load is further increased, the plastic zone grows until the entire material surrounding the contact has gone through a plastic deformation.

In this case, for the plastic deformation of asperities under an applied normal load (W), the size of the real contact area (A_r) can be expressed in terms of the hardness (H) of the material as (Sarker, 1976):

$$A_r = \frac{W}{H} \tag{2.3}$$

It should be noted that the hardness term in Equation (2.3) can by replaced by the flow pressure (yield strength) of the material (σ_y) with an incorporated constant.

In practical tribological applications, during the initial traversals of two squeezed surfaces, the predominant mode of deformation is plastic but the component gains hardness to a finite depth below the surface, giving rise to a largely elastic situation. Metallurgical examinations suggested that if a tribo-surface is hard, further plastic flow should not be expected (Ambrosio, 2001). Such phenomena have led to the concept of plasticity index (ψ), which is essentially the ratio of the hardness of the material before sliding under load to that obtained after sliding and is considered as an important quantity predicting the nature of a tribo-contact (Sarker, 1976). The plasticity index was found to be greater than unity in plastic conditions, while the situation at the interface is elastic when ψ is lower than 0.6. The plasticity index is defined as:

$$\psi = \frac{E}{H}\left(\frac{\sigma}{R}\right)^{\frac{1}{2}} \tag{2.4}$$

where (σ) is the standard deviation of asperity heights, (R) is the mean asperity radius curvature, and (E) is the modulus of elasticity of the material.

2.2.3 Creep Deformation

In addition to elastic and plastic deformations as previously discussed, materials may deform by a mechanism called *creep* which is defined as the "Progressive deformation of a material at a constant stress" or simply we can say that creep is the deformation that accumulates with time. Creep deformation may proceed to the point which is called creep rupture, where separation into two pieces occurs. This is similar to a ductile fracture, except that the process is time dependent. Another time dependent situation that should be considered in the engineering design is the deformation of a material under repeated stresses at a high temperature. In this case, a creep-fatigue interaction that harmfully accelerates the fatigue process can occur. The physical mechanisms causing creep of solid materials are diffusion of atoms, vacancies, or dislocations in a time-dependent manner.

A typical strain-time behavior during creep under constant load is illustrated in Fig. 2.2. The curve shows a transient stage of an elastic and perhaps also a plastic strain, followed by a steady-state stage of gradual accumulation of creep strain. At the end of the secondary stage, the deformation increases in an unstable manner as rupture failure approaches.

Creep deformation is probably expected in low strength and stiffness materials due to relatively weak bonds between its chain molecules. In metals and ceramics (crystalline materials), creep is strongly temperature dependent and becomes an important engineering consideration around 0.3 to 0.6 of the absolute melting temperature (Dowling, 1993).

Polymers and other superplastics (noncrystalline materials) are creeping in a more complex behavior in the form of a viscous flow (very thick liquid) which occurs at temperatures substantially above the glass transition temperature and approaching the melting temperature. Particularly in this case, much of the creep deformation may recover and disappear slowly with time after removal of an applied stress.

For soft metals and superplastics, creep has a dominant role in understanding many of the tribology considerations, especially friction, junction growth, real area of contact, stick-slip, and others which will be discussed in detail in this book.

In a dry sliding, it is demonstrated that the interfacial shear strength inherently depends on the competition between the processes of detachment and re-attachment of the asperity junctions (micro-contacts) on the interface. The detachment and re-attachment of the asperities during sliding friction usually occurs in a very short time and it is highly influenced by the sliding velocity (Tian et al., 2016). At a low sliding velocity, the dynamics of the micro-contacts are creep-dominated if the stick time is long enough for the aging of asperities to play a leading role. On the other hand, when the sliding velocity is high enough, the asperities have no time to creep before slip occurs; the dynamics of stick-slip turns to be inertia-dominated. This may also give an explanation to the fact that the static coefficient of friction is higher than the dynamic coefficient. This is because the two surfaces in contact under a load tend to creep and comply with each other and increase the true contact area between them. The coefficient of friction is proportional to contact area, so more time in contact gives higher values.

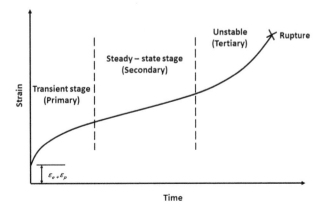

Figure 2.2: A typical strain-time behavior during creep under constant load.

2.2.4 Hardness

Hardness is a material property, which gives it the ability to resist being permanently deformed when a load is applied. Hardness is most commonly defined as "the resistance of a material to plastic deformation", usually by indentation. The greater the hardness of the metal, the greater resistance it has to deformation.

Indentation hardness value for metals is obtained by measuring the depth or the area of the indentation using one of the three most common hardness test methods; they are the Brinell hardness (BH) test (International, 2017d), the Rockwell hardness (HR) test (ASTM, 2017e), and the Vickers hardness (HV) test (ASTM, 2017). The indenter may be a diamond pyramid, a hardened steel ball, or a tungsten carbide ball according to the type of test, Fig. 2.3. However, there are many examples of conversion tables to convert hardness values between the three test methods. The accuracy of these tables depends on the accuracy of the provided data and the resulting curve-fits.

Further to the former implementations, the hardness of ceramic substrates can be determined by the Rockwell hardness test, according to the specifications of ASTM E-18. This test measures the difference in depth caused by two different forces, using a dial gauge. Also, the hardness testing of plastics is most commonly measured by the Rockwell hardness test or Shore hardness (SH) test. Both methods measure the indentation resistance of the plastic. Both scales provide an empirical hardness value that doesn't correlate to other properties or fundamental characteristics. Rockwell hardness is generally chosen for 'harder' plastics such as nylon, polycarbonate, polystyrene, and acetal, where the resiliency or creep of the polymer is less likely to affect the results (Mutton and Watson, 1978).

The indentation hardness test can be the best representative in evaluating materials for tribo-applications. This arises from the fact that the geometry of a typical surface interaction, in which one surface with rough asperities is pressed against another, is quite similar to the geometry which prevails during an indentation hardness test. Thus, the plastic strength parameter, which best characterizes the strength of the material, shows its indentation hardness (Rabinowicz, 1995).

Regarding abrasive wear situations, as will be discussed in the upcoming chapter, a universal quantitative relationship between abrasion resistance and hardness has not yet been established. The reason for this is that hardness is an intrinsic property of the material while abrasion resistance is not an intrinsic property since it may depend on many variables, such as the testing technique, the properties of the abrasive and the environmental conditions. In a case of surface hardened materials or advanced low-alloyed steels where multiple phases are intentionally present, most of the plastic behavior associated with asperity contact should occur within the surface layer. Thus, surface hardness Hs (or hardness of the abrasive particles, H_a) should be considered along with the bulk hardness H (Halling, 1979; Mutton and Watson, 1978; Luyckx and Love, 2004).

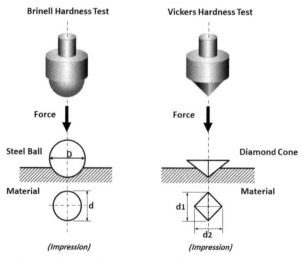

Figure 2.3: Illustration of the indenters in common hardness tests.

On the other hand, based on experimental data of various dry sliding metallic pairs, Holm (Holm, 1946) derived the expression "adhesive wear": The amount of wear by adhesion is generally proportional to the applied load and sliding distance and generally inversely proportional to the hardness of the rubbed surface, this issue will be re-discussed deeply in Chapter 3.

2.2.5 *Fatigue and Fracture Mechanics*

Fatigue is a problem that can affect any operating part or component. An estimation states that more than 80% of mechanical service failures are related to this phenomenon (Dowling, 2012). It was initially recognized as a problem in the early 1800s, when investigators in Europe observed that bridge and railroad components were cracking when subjected to repeated loading. This phenomenon was described for the first time as "fatigue" in 1839 by J.V. Poncelet in his book "Industrial Mechanics" to express the weakening of a structure due to alternating tension and compression.

In a component subjected to a cyclic (or fluctuating) loading (or stress), a fatigue crack can be initiated on a microscopically small scale, followed by crack growth to a macroscopic size, and finally to fatigue failure. Nevertheless, these cracks may be initially present in a component from manufacture, or they may start early in the service life. Failure due to fatigue probably occurs at stress levels that are much lower than the stress required to cause failure during a single application of stress. Nonetheless, fatigue may result from a repeated heating and cooling, causing a cyclic stress due to a differential thermal expansion and contraction, called thermal fatigue.

In fatigue situations, a special approach called fracture mechanics is used to specifically analyze initiation and propagation of fatigue cracks. Prediction and prevention of fatigue failure is a pivotal aspect of machine and structure design that are subjected to a cyclic loading or vibration. This would be achieved by understanding the fatigue mechanism and considering various technical and environmental conditions affecting fatigue life and fatigue crack.

The fatigue life is usually split into two phases: (1) crack initiation and (2) crack propagation (growth). The crack initiation is supposed to include some microcrack growth, but the fatigue cracks are still too small to be visible. In the second phase, the crack is growing rapidly until complete failure. It is important to mention that several practical conditions have a large influence on the crack initiation period, but a limited influence or no influence at all on the crack propagation period (Dowling, 2012).

A fatigue test is usually done by subjecting a test specimen for a given material to a fluctuating (axial, torsion, or bending) stress until a fatigue crack or other damage leading to complete failure is developed. The results of such tests (the magnitude of an alternating stress versus the number of cycles to failure) from a number of different stress levels may be plotted on a logarithmic scale to obtain a stress–life (or S-N) curve which is considered as the classical strength theory in fatigue analysis. Based on the theory, some other researchers brought in plastic strain amplitude as the calculating parameter and deduced the famous Manson-Coffin formula to calculate the fatigue life. Later, Paris (Paris and Erdogan, 1963) applied the theory of fracture mechanics to investigate the fatigue crack propagation. Figure 2.5 shows a typical S-N curve containing three different areas: Plastic, elastic and infinite life regions.

On the figure, there are three important values that distinguished these regions, namely: Ultimate Strength (σ_u), Yield Strength (σ_y), and Endurance Limit (S_e). Another important point to be mentioned here is that the term fatigue strength (S_{Nf}) is used to specify the value of stress at which failure occurs after (N_f) cycles, and fatigue limit (S_f) as the limiting value of stress at which failure occurs.

In many tribological applications, the term surface fatigue wear has long been recognized as an important process that takes place during a repeated rolling or sliding motion over a track. The cyclic stresses (loading and unloading cycles) may induce the initiation of surface and subsurface cracks upon the contact surface which, on further deformation, leads to crack propagation parallel to the surface, and wear debris can be spalled off the surface by continued motion. The new surface of the material also experiences the same cyclic stressing which leads to a progressive process and flaking off of fragments becomes rapid. However, such type of wear can also be recognized by a formation of pits on the contact surfaces (Rabinowicz, 1995).

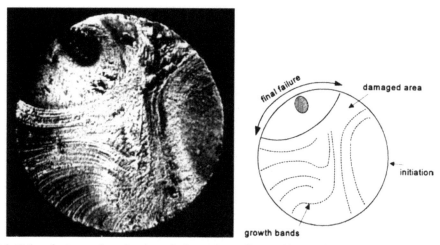

Figure 2.4: Fatigue fracture surface of a rear axel of a motor car, diameter 20 mm. The fatigue failure started as a single dominant fatigue crack. After Schijve (2009) with permission (2009).

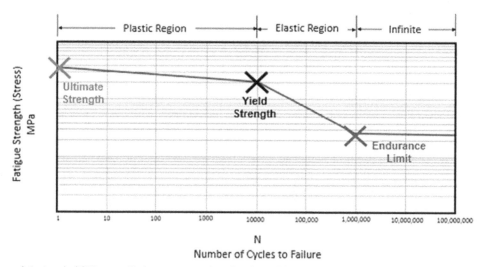

Figure 2.5: A typical S-N curve. Cycle stress vs number of cycles to failure. After Schijve (2009) with permission (2009).

Delamination is a particular form of surface fatigue, which is related to the accumulation of dislocations in a narrow band below the surface. Ratcheting is another particular form of repeated-cycle deformation wear that is based on the incremental plastic flow, the accumulation of plastic strain, and the mechanical shakedown (Bayer, 2004).

Another tribological consideration related to the fatigue phenomenon is the fretting fatigue wear, which is a combination of normal fatigue, which results from cyclic stress, and fretting wear. It occurs when low-amplitude (usually less than a millimeter) oscillatory motion in the tangential direction takes place between contacting surfaces. The oscillatory motion is usually a result of external vibration or cyclic stress which results in early initiation of fatigue cracks and, consequently, more damage.

A practical example of fretting fatigue is the contact between a clamp and a flexing beam. The flexing motion of the beam could result in a slip between the clamp surface and the beam surface. The wear caused by the fretting motion will be accelerated due to the formation of fatigue cracks, which then propagate through the material, leading to fracture of the component (Bhushan, 2013; Bayer, 2004). The next chapter will give an advanced discussion on the different types of wear.

2.3 Thermal Properties

Thermal properties are the properties that describe the behavior of a material in a wide range of parameters characterizing its internal state and structure. Understanding such properties is an essential and demanding task in selecting materials which operate at elevated temperatures and/or wide temperature ranges (Bobilev, 1987). Thermal properties of a material include its heat capacity, thermal conductivity, thermal resistance, thermal expansion, latent heat, thermal stability, and melting and freezing points.

Heat capacity (J/mol-K) is the ability of a material to absorb heat energy under heating and to emit it under cooling. It is defined as "the energy required to change a material's temperature by one degree", whilst the specific heat capacity is the heat capacity per unit mass.

Thermal conductivity (W/m.K.) is a microstructure sensitive property defined as the ability of a material to conduct heat energy from a high temperature region to a low temperature region.

Thermal resistance is the ability of a material to keep its internal structure and strength under sharp changes of temperature.

Thermal expansion is a phenomenon in which the absorption of heat energy results in an increase in the distance between the atoms of the material, which leads to an increase in its dimensions. It is quantified in terms of the thermal expansion coefficient. Linear coefficient of thermal expansion is defined as the change in the dimensions of the material per unit length.

Latent heat (J/kg) is the amount of energy necessary for the melting of a unit mass of crystal material at the freezing point.

Thermal stability is the stability of the material's molecules at high temperatures, i.e., a molecule with more stability has more resistance to decomposition at high temperatures. It also describes the ability of some fluids (e.g., fuels and lubricants) to resist oxidation under high temperature operating conditions.

Melting and freezing points: Crystal materials have characteristic melting points, above which the solid melts to become a liquid. By contrast, the softening of amorphous materials occurs gradually with increasing temperature, evolving into viscous fluids with decreasing viscosity under increasing temperature. On the other hand, the freezing point of liquids is the temperature at which they turn into solids. Theoretically, the melting point of a solid should be the same as the freezing point of the liquid. But, in fact, small differences between these quantities can be observed.

Thermal properties of tribo-components are major contributors in many situations, as well as having an important impact on the tribological behavior and failure, especially of sliding components. During sliding, nearly all of the frictional heating (energy dissipated in frictional contacts) is transformed into heat, resulting in increasing the temperatures of the sliding bodies within the contact region on their sliding surfaces where the temperatures are highest (Uetz and Föhl, 1978). Depending on the overall geometry of contacting bodies, surface temperatures can become high enough to cause changes in the structure and properties of the sliding materials, oxidation of the surface, and possibly even melting of the contacting solids (Bhushan, 2001). The degree to which this phenomenon, which is referred to as thermal effect, occurs can often be responsible for changes in the wear mechanism of the material, which is, in this situation, called thermal wear mechanism. In relatively low melting point materials (e.g., thermoplastics), there is a thermal limit, defined as the maximum allowable temperature for the operating tribo-component, which is determined based on pressure and speed (Halling, 1979).

In lubrication, the thermal stability of a lubricant, which is a fundamental characteristic of the pure lubricant type, influences the maximum useful temperature at which the lubricant can be used. Moreover, the thermal conductivity of a lubricant can govern the desired dissipation of frictional heating.

2.4 Surface Properties

2.4.1 Adhesion

Adhesion is a phenomenon that occurs between solid surfaces whenever one solid surface is sliding over another surface or when two surfaces are pressed together. In these cases, the interactive forces (covalent, ionic, metallic, hydrogen, and van der Waals) between contact surfaces cause the formation of adhesive bonds (or atomic interactions) (Ludema, 1996). Consequently, in order to pull apart or separate the two surfaces either normally or tangentially, a finite adhesive force should be applied to the desired direction, which may result in the production of undesired wear fragments. Figure 2.6 illustrates the mechanism of adhesion.

The ratio of the normal tensile (or shear) force (F_a) required for separation to the normal compressive force initially applied (F_c) is often referred to as the coefficient of adhesion (f_a), and is generally directly proportional to the duration of the static contact and the separation rate (Bhushan, 2013). The coefficient of adhesion can be represented mathematically as:

$$f_a = \frac{F_a}{F_c} \tag{2.5}$$

The phenomenon of adhesion is intimately connected to surface energy (γ), which is defined as the work done in creating a new unit area (J/m^2).

The theory of adhesion was introduced by Bradley (Bradley, 1932) and Bailey (Bailey, 1961) based on the concept of free surface energy and known as "*Free Surface Energy Theory of Adhesion*". Based on their early studies, work of adhesion (W_{ad}) between two materials in contact, per unit area is represented in Dupre's formula:

$$W_{ad} = \Delta\gamma = \gamma_1 + \gamma_2 - \gamma_{12} \tag{2.6}$$

where

γ_1, γ_2 the free surface energies for the two materials in contact, per unit area.

γ_{12} the surface energy of the interface, per unit area.

According to the Theory of Adhesion:

- $\Delta\gamma$ represents the energy required to separate a unit area of the interface or to create new surfaces, thus, it would be expected to have a negative value.

- The higher the surface energy of a solid surface, the stronger the bonds it will form with a mating material.

- In tribological applications, it should be considered to select materials in contact that have a low surface energy and a low $\Delta\gamma$.

A special case for Dupre's equation occurs at the line where liquid, vapor (or gas), and solid phase meet. Such cases occur in many lubrication situations when a liquid droplet is deposited onto a solid surface exposed to air. Dupre's equation, expressing the solid-liquid work of adhesion WSL, can be expressed as:

$$W_{SL} = \gamma_{SV} + \gamma_{LV} - \gamma_{SL} = \gamma_{LV} (1 + Cos\theta) \tag{2.7}$$

where V, L and S denote vapor, liquid and solid surface energies, respectively, and θ is the angle the tangent to the drop plane makes with the surface plane, in stable equilibrium conditions, as shown in Fig. 2.7.

According to Equation 2.7, the contact angle is directly related to the strength of the adhesion between the liquid and the solid and suggests that contact angle measurements can be used to estimate adhesion energies (Rios et al., 2007).

As already mentioned in the previous chapter, there are many practical applications where a strong adhesion is desired in order to bond two surfaces together (e.g., for writing with a pencil, there should be good adhesion between the lead and the paper). It should be emphasized that adhesion is independent of the roughness of the contacting surfaces, even smooth surfaces often give rise to adhesion (Rabinowicz, 1995).

The compatibility for adhesion of two materials in contact is an important factor governing their adhesion properties. The lake of compatibility will result in phase separation of certain components in the adhesion formulation and thereby lower production performance (de María, 2016). Compatibility among two or more materials is governed by their molecular weights, molecular weight distributions, and solubility. For different metal couples, no correlation was found between solubility and adhesion, while a positive correlation with friction and a great correlation with wear were observed (Halling, 1979).

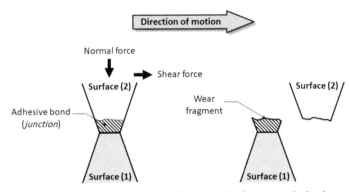

Figure 2.6: An illustration of the adhesion mechanism at asperity level.

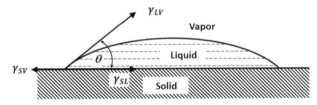

Figure 2.7: Equilibrium of surface energy of a liquid drop on a solid surface.

In contrast, Compatibility in Dupre's formula correlates with the solubility parameter (δ) that is the square root of the cohesive energy density. This parameter provides a numerical estimate of the degree of interaction between materials and could be considered as a good indicator for compatibility, particularly for polymers (de María, 2016).

Rabinowicz (Rabinowicz, 1995) rewrote Equation 2.6 and introduced the compatibility parameter, C_m, in order to describe the extent of the affinity between two materials as:

$$W_{ad} = C_m(\gamma_1 + \gamma_2) \tag{2.8}$$

In the case of metals, C_m lies in the range of 1 to 0, and a greater compatibility is associated with increased solubility, Table 2.1 shows a semi-quantitative classification of materials in contact. Compatibility parameter is often found to be directly proportional to the ratio of the adhesion energy over the hardness of the softer material *Wad/H*.

Moreover, there are other factors having impacts on the interface adhesion, such as the surface energy, where lower surface energies result in lower adhesion. Likewise, the presence of oxides, lubricants and contaminates on contact surfaces reduce surface energy and result in a weak adhesion. Also, adhesion is markedly affected by temperature; high temperatures result in stronger adhesion due to softening, ductility and solubility of solid surfaces. Adhesion force generally increases linearly with an increase in the normal load. Further increase may result if a shear force is applied in addition to the normal load. Plastic flow under the effect of stress combination will result in a growth in the real area of contact, hence, high adhesion. Adhesion is affected by the real area of contact, which is a function of surface roughness and mechanical properties. Material crystal structure, crystallographic orientation, chemical activity and separation of charges, duration of contact, and separation rate are also contributors to adhesion behavior (Bhushan, 2013).

Polymers are usually characterized by a relatively lower adhesion compared to metals. The adhesive bonds formed between polymers and solid surfaces are mainly governed by van der Waals attraction forces (Sinha and Briscoe, 2009). These bonds are followed by junctions appearing on real contact spots. The formation and rupture of the junctions govern the adhesion component of friction (see Sec. 3.1.1). There are other concepts involved in adhesion of polymers. First, polymers are deformed easily in comparison to other hard solid surfaces. Second, inter-diffusion of polymeric chains across the interface may occur, which will increase the adhesive strength.

Table 2.1: Semi-quantitative classification of materials in contact.

Contact	C_m
Identical Metals	1.00
Identical Nonmetals	1.00
Compatible Metals	0.50
Compatible Nonmetals	0.60
Partially Compatible Metals	0.32
Partially Incompatible Metals	0.20
Incompatible Metals	0.12
Incompatible Nonmetals	0.36
Other Solids	0.22

2.4.2 Surface Reactivity

A surface defines the boundary between a material and its surrounding media and affects interactions with that media. At the molecular level, a massive atomic and electronic reconstruction on a clean surface can lead, in many cases, to a surface structure totally different from that projected from the bulk structure. Surface reactivity, or surface chemistry, is an extrinsic property which makes the material surface and interfaces favored media for chemical processes (Gellman and Spencer, 2002). This property takes place at the molecular level where the atoms of the surfaces have a different chemical environment then that in the bulk material. Accordingly, these surface atoms with changed structures display a high propensity to chemical reactions in which at least one of the steps of the reaction mechanism is the adsorption of one or more reactants (Somorjai and Li, 2010).

In chemistry, reactivity is a rather unclear concept that involves both thermodynamic and kinetic factors determining the ability and speed of chemical reactions. Both factors are actually distinct, and commonly depend on temperature.

In tribology, the term surface reactivity (or *Tribochemistry*) is commonly used in boundary lubrication situations (a very thin lubricant layer) in order to define the chemical reactions between the tribo-surface and the lubricant molecules inside a sliding contact.

According to the monolayer theory of boundary lubrication, introduced by Hardy (Hardy and Bircumshaw, 1925), the sliding surfaces were held apart by adsorbed and oriented monolayers of lubricant molecules (See Sec. 2.4.2), which form a plane of low shear strength, thus lowering friction and affording protection of the surfaces, as illustrated in Fig. 2.8. In this situation, the chemical wear mechanism is greatly affected by the chemical reactivity of the lubricant towards the surfaces, thus, not all lubricant formulations have the same tribological behavior on all surfaces. Besides, the ability of the molecules to react with the surface is also an indicator of a potential corrosion problem that is at very high surface reactivity, the high chemical reaction rate would produce corrosive wear (Stachowiak, 2005).

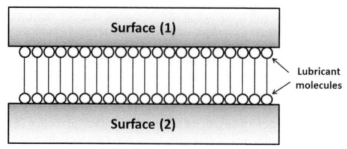

Figure 2.8: An illustration of the boundary lubrication showing monolayers of lubricant molecules adsorbed and oriented upon the mating surfaces.

Further to the former case, free surfaces of almost all metals and alloys may form an oxide film in air as well as nitride, sulfide, and chloride films in other environments (Rabinowicz, 1995). The thickness of such films varies from metal to metal, depending on the chemical reactivity of its surface to the environment and the mechanical properties of the surface. These films have a noticeable impact on the interaction between surfaces, depending on their nature. Favorable tribological effects can be realized in the formation of soft adhered layers which act as a lubricant film and, consequently, separate the contacting surfaces, while hard brittle films are less effective, since they tend to get broken during the load application.

2.4.3 Surface Roughness

Surface texture, including surface roughness, waviness, lay and flaws, is the repetitive or random deviation from the nominal surface that forms the three-dimensional topography of the surface. According to the surface texture illustration in Fig. 2.9, the following terminologies can be realized:

- Surface roughness, represents the surface irregularities, in nano- or micro-scale, which result from different machining, or surface finishing, processes. It is often characterized by the height of the irregularities with respect to a reference line; the width between successive peaks, and cut-off length, which is the greatest spacing of respective surface irregularities to be included in the measurement.
- Waviness refers to the irregularities in macro scale which are outside the cut-off values. The main reason for waviness is the deflection of the tool or work-piece during machining processes.
- Lay represents the principal direction of the predominant surface pattern produced and it is determined by the machining method used to produce it.
- Flaws are unexpected and unwanted unintentional interruptions in the surface texture.
- Mean line, is the reference line about which the profile deviations are measured.
- Peak (or Summit) is the point of maximum height on a portion of a profile that lies above the mean line.
- Valley, is the point of maximum depth on a portion of a profile that lies below the mean line.

As a component of surface texture, Surface roughness is defined as the deviation of the surface shape from some ideal or prescribed form. It can be also expressed as the height of the small irregularities or asperities (peaks or summits) on the surfaces (Rabinowicz, 1995; Bhushan, 2013; Ludema, 1996). In general, surface roughness is often necessary in order to ensure that a surface is fit for a purpose; also, it plays an important role in determining how a real object will interact with its environment. In particular, regarding tribology, roughness is essential to surface interaction because a surface profile has great impact on the real area of contact, friction, wear and lubrication.

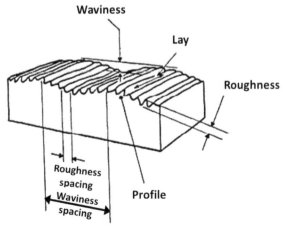

Figure 2.9: Illustration of surface texture.

According to the ISO 4287, 4288, 25178 standards, there are many different parameters to evaluate roughness. Some parameters, namely amplitude parameters, are related to the height of the asperities (*Ra, Rq, Rp, Rt, Rz*), others (*Rsk, Rku, Rmr*) characterize the texture morphology (or height distribution), and others are related to spatial information (*RSm, Rdq*). Hybrid, Functions, Segmentation parameters may also be used, depending on the desired application. It is important to mention that these parameters reduce all of the information in a surface topography to a single number, so special care must be considered in applying and interpreting them (Bhushan, 2013; Bhushan, 2001).

Surface Roughness Average (*Ra*), Arithmetic Average (AA) or Center Line Average (CLA) is the most widely used one-dimensional roughness parameter, which can be determined from the profile chart of the surface as the area between the roughness profile and its mean line, or the integral of the absolute value of the roughness profile height over the evaluation length. A close approximation of the average may also be automatically performed by electronic instruments using appropriate circuitry through a meter or a chart recorder. If *y* is the measured value from the profile meter, then the (*Ra*) value can be calculated from Equation 2.9. Root Mean Square (*Rq*) is a modified parameter which can be calculated using Equation 2.10, and its numerical value is about 11% higher than that of (*Ra*) (Degarmo et al., 2003). Table 2.2 is a common conversion table with roughness grade numbers.

$$R_a = \frac{1}{n} \sum_{i=1}^{n} | y_i |$$ (2.9)

$$R_q = \sqrt{\frac{1}{n} \sum_{i=1}^{n} y_i^2}$$ (2.10)

The desired surface finish is usually specified and appropriate processes are required in order to maintain the quality. Hence, the measurement of surface roughness is very important when assessing the quality of a component. Currently there are various methods and instruments to evaluate features of surface roughness in micro-, nano-, and atomic-scales. The measurement methods can be divided into three types:

(1) Contact (or Direct) Measurements: During the measurement, a component of the measurement instrument contacts the surface to be measured. Profile Meters (Stylus) is the most famous instrument of this category. However, Scanning Probe Microscopes (SPM) could be introduced here, since the distance between the probe (sharp tip) and the specimen surface is 10 nm or less.

• **Profile Meter**

The device consists of a stylus measurement head with a stylus of a rounded tip and a scanning mechanism. In a measurement, the stylus arm is loaded against the sample as the stylus or sample moves. The stylus rides over the sample surface detecting surface deviations by inductive, pizo-electric, phase grating interferometric (PGI), or any other transducer, Fig. 2.10. It produces an analog signal corresponding to the vertical stylus movement. This signal is then amplified electrically or mechanically and usually recorded on a chart. Although this method is considered as the most broadly used technique available to measure surface roughness, it is quite poor at detecting sharp crevasses, and gives a highly distorted reading of sharp ridges. It was suggested that a profile containing many peaks and valleys of radius of curvature of about 1 μm or less or many slopes steeper than 45° would probably be distorted by a stylus (Bhushan, 2001). Stylus load is also considered as a source of errors since the stylus local pressure may exceed the hardness of the tested surface, resulting in a plastic deformation (micro-scratches) of the surface. Thus, it is important to select stylus loads low enough to minimize this source of error.

• **Scanning Probe Microscopes (SPM)**

The development of this family of scanning probe microscopes started with the original invention of Scanning Tunneling Microscopy (STM), in which an electric current is allowed to flow, due to potential difference, between the probe and the metal surface separated by a thin insulating film (Behm and Hösler, 1986). This allows one to image surfaces with exquisite resolution, laterally less than

Table 2.2: Surface roughness conversion table.

Roughness ISO grade number, N	Roughness value				Cut-off Length	
	Ra (or) CLA		*Rq*, RMS (µin.)	*Rt* (µm)		
	(µm)	(µin.)			mm	inch
1	0.025	1	1.1	0.3	0.8	0.003
2	0.05	2	2.2	0.5	0.25	0.01
3	0.1	4	4.4	0.8	0.25	0.01
4	0.2	8	8.8	1.2	0.25	0.01
5	0.4	16	17.6	2	0.25	0.01
6	0.8	32	32.5	4	0.8	0.03
7	1.6	63	64.3	8	0.8	0.03
8	3.2	125	137.5	13	2.5	0.1
9	6.3	250	275	25	2.5	0.1
10	12.5	500	550	50	2.5	0.1
11	25.0	1000	1100	100	8.0	0.3
12	50.0	2000	2200	200	8.0	0.3

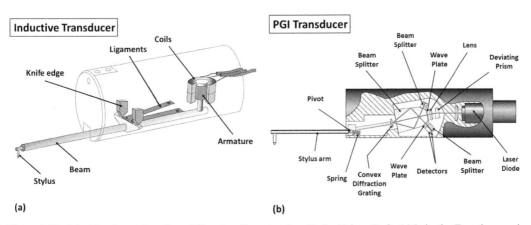

(a)   (b)

Figure 2.10: Principle of operation of two different profile meters from Taylor Hobson Tech.; (a) Inductive Transducer, and (b) PGI transducer.

1 nm and vertically less than 0.1 nm, which is adequate to define the position of single atoms. The relatively high vertical resolution of the STM is obtained because the current varies exponentially with the distance between the probe metal tip and the scanned surface. However, a basic drawback in the application of STM is that is the surface to be measured should be electrically conductive. This limitation was the key factor for Gerd Binnig and his colleagues to develop an instrument capable of investigating surfaces of both conductors and insulators on an atomic scale, called the Atomic Force Microscopy (AFM) (Binnig et al., 1986). The AFM is a combination of the principles of the STM and the stylus, in which the probe (a sharp tip with a nano-scale radius of curvature) is attached to a very flexible and light cantilever made from silicon or silicon nitride, Fig. 2.11.

The tip is brought into contact with a sample surface which is allowed to move in x, y and z directions using piezoelectric scanners. According to the configuration described, the interaction between the tip and a sample is transduced into changes of the motion of a cantilever, consequently, AFM can measure the nano displacement between the cantilever surface and a reference surface.

Figure 2.11: AFM flexible cantilever.

(2) Noncontact (or Indirect) Measurement: As mentioned previously, the measurement of surface roughness using the stylus techniques has several disadvantages. A noncontact measurement utilizing optical techniques becomes demanding for measuring the surface roughness with improved accuracy. The main concept of such techniques is to shine a light beam (electromagnetic radiation) at the engineering surface, and then inspect the reflection of the light wave. According to Lambert's law and Snell's law, this reflection will be either totally specular, totally diffused, or combined specular and diffused, which is also related to the degree of irregularities in the scanned surface. Optical methods may be divided into geometrical (light profile taper and sectioning) and physical (specular and diffuse reflections, speckle pattern, and optical interference) methods (Thomas, 1999). Of these numerous methods, the following examples are introduced:

- **Light Profile Method**

In this method, an optical system is utilized in order to project an incident slit beam and the reflected image is casted and used to study the nature of the surface irregularity. Smooth surfaces usually appear as a straight line, while rough surfaces appear as undulating lines, Fig. 2.12. This method could be a good choice for measuring roughness consisting of parallel ridges and grooves (Rabinowicz, 1995). In some developed automated configurations, the tested surface is fixed to a table which is driven by a stepper motor, while the 2D measurement of surface roughness is performed using a TV camera projected through a microscope (Uchida et al., 1979).

- **Specular Reflection Method**

In this method, an oblique light beam is spotted on the measured surface at an arbitrary incident angle. The angle depends on the level of roughness, and the fraction of light that is not specularly reflected is measured, Fig. 2.13. Hence, a specular or gloss reflectance is a surface property of the material, namely, the refractive index and surface roughness. The main advantages of this method is its ease, simplicity, and speed of analysis, however it has some disadvantages: Relatively low accuracy, a limited reproducibility and dependence on the refractive index of the tested surface (Bhushan, 2001).

(3) Comparison Methods: In these methods, the roughness is roughly evaluated by observation of the surface and comparing it with standard specimens. In fact, these methods are not reliable as they can be misleading if the comparison is not made with surfaces produced by same techniques. There are various methods available as comparison methods such as Touch Inspection, Visual Inspection, Scratch Inspection, Microscopic Inspection, Surface Photographs, and Micro-Interferometer. Figure 2.14 shows examples of Surface Finish Comparison Plates used in this method of surface roughness evaluation.

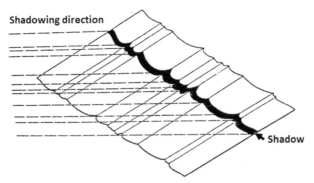

Figure 2.12: Light profile method.

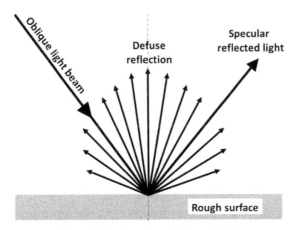

Figure 2.13: Specular reflection method.

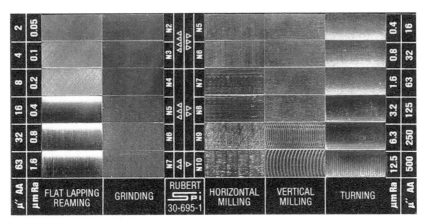

Figure 2.14: Surface finish comparison plates.

2.5 Lubricant Properties

As mentioned previously, in Chapter 1, the principal objective of lubrication is to reduce friction between surfaces in a mutual contact, which ultimately reduces the wear and heat generated. This action is typically accomplished by inserting a third substance called "lubricant" between the two moving tribo-surfaces.

Although oil and grease are the most commonly-used lubricants, the term lubricant is not exclusive to them. In fact, any fluid, including water, and any gas, including air, could be used as lubricants in the right application. Furthermore, soft solids are effective lubricants used in high temperature and extreme contact pressure applications wherever fluids or gases are not suitable. The performance of a lubricant, which is governed by a number of physical, mechanical, and chemical properties, is known as its lubricity. Most of these governing properties will be discussed here.

2.5.1 Density and Specific Gravity

Both density and specific gravity are often used to characterize lubricants. Density (ρ) is a basic physical property of lubricants and it represents the mass per unit volume mathematically. It is commonly expressed in units of kilograms per cubic meters (kg/m^3), grams per cubic centimeters (gm/cc^3) or pounds per cubic feet (lbs/ft^3).

Density is affected by temperature because the kinetic energy of the particles increases with the temperature increase. In liquid lubricants and grease, operating temperature has an impact on their density, but it is much lower than in gases. In practical applications, correlations for lubricating oil density and temperature are usually calculated and represented in tables or in colored charts.

Specific gravity (S), on the other hand, is a measure of the relative density of a substance as compared to the density of water at a standard temperature. The importance of this property arises when using grease as a lubricant. The grease requirement is often specified in grams, while grease dispensers are usually calibrated in cubic centimeters. Thus, we use the specific gravity to convert the units of weight to the units of volume. In the petroleum industry, an API (American Petroleum Institute) unit, which is a derivative of the conventional specific gravity, is used. The API scale is expressed in degrees which, in some cases, are more convenient to use than the specific gravity reading (Stachowiak and Batchelor, 1993). The API degree can be calculated according to the following empirical formula (O'Connor et al., 1968):

$$API = (141.5/S) - 131.5 \qquad (2.11)$$

where (S) is measured at 15.6°C (60°F).

2.5.2 Viscosity

Viscosity can be considered as the most important single property of liquid lubricants. The term 'Viscosity' is the physical property that characterizes the flow resistance of simple fluids. The modern viscosity theory was introduced by Newton in the 17th century based on a two-plate model of fluid flow, Fig. 2.15. Newton's law of viscosity postulated that the shear stress (τ) between adjacent fluid layers is proportional to the negative value of the velocity gradient (du/dy) between the two layers. This law can be expressed mathematically as:

$$\tau = \eta \frac{du}{dy} \qquad (2.12)$$

The constant of proportionality (η) is referred to as dynamic viscosity. The most commonly-used unit of viscosity is the dyne second per square centimeter [dyne s/cm$_2$], which is given the name poise (P). However, the SI unit of viscosity, which is rarely used today, is the pascal second ($Pa\ s$), equivalent to 0.1 pois.

There is another quantity to express this property, called kinematic viscosity (v). It equals the ratio of the dynamic viscosity of a fluid to its density. Kinematic viscosity is a measure of the resistive flow of a fluid under the influence of gravity and can be expressed as in Equation 2.13.

$$v = \frac{\eta}{\rho} \qquad (2.13)$$

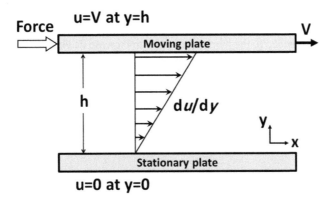

Figure 2.15: Two-plate model of fluid flow.

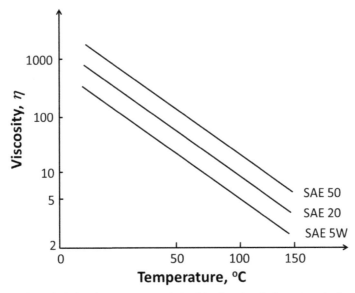

Figure 2.16: Typical viscosity-temperature relationships for mineral oils at atmospheric pressure.

Viscosity-temperature relationship for a lubricant is governed by its type. For greases, the viscosity at elevated temperature is not generally the critical property. In fact, the upper operating temperature limit is generally defined either by stability or volatility.

Liquid lubricants generally have a more pronounced response to a temperature change than grease. The main reason for this is that, at elevated temperatures, the liquid expands; the intermolecular forces decrease and, consequently, the viscosity also decreases. Typical viscosity-temperature curves for a number of liquid lubricants at atmospheric pressure is shown in Fig. 2.16.

Viscosity Index

In many lubricated tribo-systems, lubricant fluids may be exposed to extreme operating conditions, including wide ranges of temperature. At a high temperature, the lubricating film may break due to a drop in the lubricant viscosity. At the other extreme, the lubricant fluid may become too viscous for proper lubrication. Thus, the variation of kinematic viscosity with temperature is a very important property of a lubricant. The less the oil viscosity changes with temperature, the better its lubricating performance at a wide range of temperatures becomes. The term "Viscosity Index" (VI) is often used to express the effect of temperature on kinematic viscosity where, in general, higher VI is preferred for lubricants (Dean and Davis, 1929). Additives are sometimes added to the lubricant in order to increase its viscosity index.

According to the American Society for Testing and Materials, ASTM D2270 is used to calculate the viscosity index (ASTM, 2016). This method covers the procedures for calculating the viscosity index of petroleum products, such as lubricating oils, and related materials from their kinematic viscosities at 40°C and 100°C. Practically, most commercial lubricating oils have VI value around 100, while automobile engine oils have values of the order of 160 because of the wide range and extreme temperatures over which they operate (Halling, 1979).

2.5.3 Thermal Properties

In fact, the function of a lubricant fluid in many mechanical systems is not limited or restricted to mitigating friction and consequently wear. In some tribo-systems, a liquid lubricant importantly serves as a heat-transfer fluid raising the importance of its thermal properties. Specific heat, thermal conductivity, and thermal diffusivity are the most important thermal properties of liquid lubricants and have a significant impact in evaluating their cooling properties.

Specific Heat

Mostly, the specific heat (Cp) (or Heat Capacity per unit mass) of mineral and synthetic hydrocarbon-based lubricants has values in the range of 0.44 to 0.48, i.e., half that of water. Lubricant oils with a larger specific heat value will transfer heat energy more efficiently. The following empirical formula can be used for a rough estimation of the specific heat [J/kgK] at a given temperature (Cameron, 1981):

$$Cp = (1.63 + 0.0034\theta)/S^{0.5} \tag{2.14}$$

where (S) is the specific gravity and (θ) is the temperature of interest [°C].

Thermal Conductivity

Thermal Conductivity k [W/mK] of lubricant oils is very sensitive to temperature through a linear relation. Simply put, the larger the thermal conductivity, the more efficiently the oil will transfer heat.

Equation (2.15) may be used for a rough calculation (Cameron, 1981):

$$k = (0.12/S) \times (1 - 1.667 \times 10^{-4}\theta) \tag{2.15}$$

where (S) is the specific gravity and (θ) is the temperature of interest [°C].

However, it was also reported that the thermal conductivity was doubled as the pressure increased (Larsson and Andersson, 2000).

Thermal Diffusivity

Thermal diffusivity (α) of a lubricant is the parameter that measures its ability to conduct thermal energy relative to the ability to store thermal energy. Put simply, it describes how quickly a lubricant reacts to a change in temperature. The value of (α) [m²/s], at a constant pressure, is calculated by dividing the thermal conductivity by density and specific heat:

$$\alpha = \frac{k}{\rho C_p} \tag{2.16}$$

where (k) is the thermal conductivity [W/mK], (ρ) is the density [kg/m₃], and (Cp) is the specific heat [J/kgK].

Values of thermal properties for some typical materials are tabulated in Table 2.3 (Stachowiak and Batchelor, 1993).

Table 2.3: The values of thermal properties for some typical materials.

Material	Specific heat, C_p at 20°C [J/kgK]	Thermal conductivity, k at 100°C [W/mK]	Thermal diffusivity, α at 100°C [m²/s]
Aluminum	870	230	101.7×10^{-6}
Brass	380	80–105	$23–31 \times 10^{-6}$
Bronze	380	50–65	$15–19 \times 10^{-6}$
Steel	460	46.7	13.02×10^{-6}
Mineral Oil	1670	0.14	$0.059–0.102 \times 10^{-6}$
Water	4184	0.58	0.16×10^{-6}

2.5.4 Other Properties

Flash Point

By definition, flash point is the temperature at which a combustible liquid lubricant must be heated to give off a flammable vapor. The mixture of the emitted vapor and air should burn momentarily in the presence of a flame and under specific conditions. This property indicates the fire hazard potential because it is a measure of the temperature at which a flammable vapor is released.

Pour Point

It is the temperature below which the lubricant loses its viscosity. This property is used to define the lower working temperature limit of petroleum lubricants. It is also used to predict the proper circulation of the lubricating system at low temperatures. ASTM D79 is the standard test method often used to measure pour point (ASTM, 2017a).

Alkalinity

The alkalinity of a lubricant is an important parameter of liquid lubricants, especially petroleum oils; it varies according to the desired application. This property is expressed as the Total Base Number TBN. It is the amount of milligrams of potassium hydroxide per one gram of lubricant and is expressed in [mg KOH/g]. It is often measured by titration through an acid ASTM D2896 (ASTM, 2015). In modern lubricants, TBN generally ranges from 6 to 80 mg KOH/g.

Acidity

The acidity of liquid lubricants is a measure of an acid concentration present in a lubricant. It expressed as the neutralization number or the Total Acid Number TAN of the lubricant. This value is determined by the amount of milligrams of potassium hydroxide required to neutralize the acids in one gram of the lubricant. TAN is also represented in [mg KOH/g]. The acidity of lubricants can result from the presence of additives, degradation, and oxidation. Sharp change in the lubricant TAN value is an important sign of a harsh oxidation and, consequently, the potential for corrosion problems. The slandered test method ASTM D664 is used to measure this property (ASTM, 2017c).

Foaming

Lubricant foaming is an undesired property in lubricating systems. It can cause severe operational problems, especially in high speed gearing, crank splash lubrication, and pumping systems. It probably leads to inadequate lubrication, cavitation and other problems. ASTM D892 test method covers the determination of the foaming characteristics of lubricating oils at specified temperature limits (ASTM, 2013).

Fire Resistance

This property is extremely important when the lubricant fluid comes in contact with an ignition source, sparks, or very hot engineering surfaces. In such cases, the fire resistance of the selected liquid lubricant is an important consideration. Therefore, for example, non-flammable aerospace hydraulic fluids are currently used strictly to reduce the fire hazards in aircraft hydraulic systems. The International Standards Organization (ISO) further classifies these fluids as follows (ISO, 1999):

Table 2.4: Fire-resistant categories and applicable operating temperatures.

ISO class	Composition	Operating temperature range
HFAE	Oil-in-water (emulsions)	+5 to +50°C
HFAS	synthetic aqueous fluids	+5 to +50°C
HFB	Water-in-oil (invert emulsions)	+5 to +50°C
HFC	Water polymer solutions	−20 to +50°C
HFDR	synthetic anhydrous fluids (composed of phosphate esters)	−20 to +70°C
HFDU	synthetic anhydrous fluids (other than phosphate esters)	−20 to +70°C

Compatibility

This property defines the acceptance of a lubricant fluid for a specified application. In particular, it must be compatible with all of the materials with which it will come into contact. Significant effects could be expected if the lubricant is not compatible with individual parts of the system, e.g., O-rings, gaskets, and other seals (O'Connor et al., 1968). The same is true in the case of lubricant mixtures. Generally, lubricant fluids that are either designed to be mixed or that may be mixed with no harmful effects, are referred to as compatible. In some cases, the mixing of two lubricating oils can produce a substance markedly inferior to either of its constituent materials. The compatibility of mixtures of turbine lubricating oils of the same ISO grade and type can be evaluated by applying ASTM D7155 (ASTM, 2011).

Consistency

It is a grease property that defines its resistance to deformation under an applied force or, in other words, its relative hardness. This property is an important parameter characterizing its suitability for a desired application. Grease that is too soft may not stay in place, while grease that is too hard will not flow properly. The two cases result in either a poor lubrication or difficulties in dispensing tribo-systems. The National Lubricating Grease Institute (NLGI) has established a consistency property (sometimes called "NLGI grade") from 000 to 6 that represents very low to very high viscosity. ASTM D217 test method is applied for measuring the consistency of lubricating greases (ASTM, 2017b). The procedures of this test method include the penetration of a cone of specified dimensions, mass, and finish. The penetration is measured in tenths of a millimeter.

Surface Adsorption

It is the property by which atoms or molecules of a liquid lubricant become attached to a solid surface. There are two classes of adsorption: Physical and chemical (Ludema, 1996). Physical adsorption involves Van Der Waals forces, while chemical adsorption involves a physical adsorption followed by the combining of the adsorbate with substrate atoms to form a new compound.

Surface adsorption is a dominant factor in determining the performance of a boundary lubrication. The function of a boundary lubricant is to modify the solid surfaces in some way by adsorption so as to provide a surface film which is more easily sheared than the bulk metal itself. Such lubricant films need often to be only one or two molecules in depth in order to lower the coefficient of friction between the two contact surfaces.

Oiliness

Oiliness of a lubricant is an important property, particularly for an extreme pressure lubrication. It reflects the capacity of a lubricant to attach firmly to the surface of moving engineering surfaces. No direct test method is available for measuring oiliness.

References

Ambrosio, J.A.C. 2001. Material Strain Rate Sensitivity. Crashworthiness, Vienna, 2001.

ASTM International. 2011. ASTM D7155-11, Standard Practice for Evaluating Compatibility of Mixtures of Turbine Lubricating Oils. West Conshohocken, PA.

ASTM International. 2013. ASTM D892-13e1, Standard Test Method for Foaming Characteristics of Lubricating Oils. West Conshohocken, PA.

ASTM International. 2015. ASTM D2896-15, Standard Test Method for Base Number of Petroleum Products by Potentiometric Perchloric Acid Titration. West Conshohocken, PA.

ASTM International. 2016. ASTM D2270, Standard Practice for Calculating Viscosity Index from Kinematic Viscosity at 40°C and 100°C. West Conshohocken, PA.

ASTM International. 2017. West Conshohocken, PA. ASTM E92, Standard Test Method for Brinell Hardness of Metallic Materials. ASTM International, West Conshohocken, PA.

ASTM International. 2017a. ASTM D97-17a, Standard Test Method for Pour Point of Petroleum Products. West Conshohocken, PA.

ASTM International. 2017b. ASTM D217-17, Standard Test Methods for Cone Penetration of Lubricating Grease. West Conshohocken, PA.

ASTM International. 2017c. ASTM D664-17, Standard Test Method for Acid Number of Petroleum Products by Potentiometric Titration. West Conshohocken, PA.

ASTM International. 2017d. ASTM E10, Standard Test Method for Brinell Hardness of Metallic Materials. West Conshohocken, PA.

ASTM International, 2017e. ASTM E18, Standard Test Method for Brinell Hardness of Metallic Materials. West Conshohocken, PA.

Bailey, A.I. 1961. Friction and adhesion of clean and contaminated mica surfaces. J. Appl. Phys. 32: 1407–1412.

Bayer, R.G. 2004. Mechanical Wear Fundamentals and Testing, Revised and Expanded: CRC Press.

Behm, R.J. and Hösler, W. 1986. Scanning tunneling microscopy. pp. 361–411. *In*: Vanselow, R. and Howe, R. (eds.). Chemistry and Physics of Solid Surfaces VI. Berlin, Heidelberg: Springer Berlin Heidelberg.

Bhushan, B. 1998. Contact mechanics of rough surfaces in tribology: Multiple asperity contact. Tribology Letters 4(1): 1–35. Doi: 10.1023/A:1019186601445.

Bhushan, B. 2001. Modern Tribology Handbook. 2nd ed. ed: CRC Press.

Bhushan, B. 2013. Principles and Applications of Tribology. 2nd edition ed. New York: John Wiley & Sons, Inc.

Bhushan, B. 2013. Introduction to Tribology. second edition ed. New York: John Wiley & Sons Inc.

Binnig, G., Quate, C.F. and Gerber, C.H. 1986. Atomic force microscope. Physical Review Letters 56(9): 930–933.

Bobilev, A.V. 1987. Mechanical and technological properties of metals: Manual: Moscow: Metallurgia.

Bradley, R.S. 1932. The cohesive force between solid surfaces and the surface energy of solids. Phil. Mag., 853–862.

Cameron, A. 1981. Basic Lubrication Theory. Ellis Horwood Limited.

de María, P.D. 2016. Industrial Biorenewables: A Practical Viewpoint. John Wiley & Sons.

Dean, E.W. and Davis, G.H.B. 1929. Viscosity variations of oils with temperature. Chemical and Metallurgical Engineering 36: 618–619.

Degarmo, E.P., Black, J.T. and Kohser, R.A. 2003. Materials and Processes in Manufacturing, 9th ed., John Wiley & Sons Inc.

Dowling, N.E. 2012. Mechanical Behavior of Materials, 4th edition. Pearson Education Limited, Harlow, United Kingdom.

Gellman, A.J. and Spencer, N.D. 2002. Surface Chemistry in Tribology. Edited by Carnegie Mellon University: Department of Chemical Engineering.

Greenwood, J.A. and Williamson, J.B.P. 1966. Contact of nominally flat surfaces. Proc. R. Soc. London, Ser. A 295(1442): 300–319.

Halling, J. 1979. Principles of Tribology. Macmillan Press Ltd., London, United Kingdom.

Hardy, W.B. and Bircumshaw, I. 1925. Boundary lubrication—plane surfaces and the limitations of Amontons' Law. Proc. R. Soc. London. A (108): 1–27.

Holm, R. 1946. Electric Contacts, Gerbers, H., ed. Stockholm, Sweden.

https://www.nature.com/articles/srep33730#supplementary-information.

International Organization for Standardization. 1999. ISO 12922, First edition, Lubricants, industrial oils and related products (class L)—Family H (Hydraulic systems) In Specifications for categories HFAE, HFAS, HFB, HFC, HFDR and HFDU: International Organization for Standardization.

ISO. The International Organization for Standardization. https://www.iso.org/.

Larsson, R. and Andersson, O. 2000. Lubricant thermal conductivity and heat capacity under high pressure. Proceedings of the Institution of Mechanical Engineers, Part J: Journal of Engineering Tribology 214(4): 337–342. doi: 10.1243/1350650001543223.

Ludema, K.C. 1996. Friction Wear and Lubrication: CRC Press.

Luyckx, S. and Love, A. 2004. The relationship between the abrasion resistance and the hardness of WC-Co alloys. Edited by The Journal of The South African Institute of and Mining and Metallurgy.

Mutton, P.J. and Watson, J.D. 1978. Some effects of microstructure on the abrasion resistance of metals. Wear 48(2): 385–398. Doi: https://doi.org/10.1016/0043-1648(78)90234-X.

O'Connor, J.J., Boyd, J. and Avallone, E.A. 1968. Standard Handbook of Lubrication Engineering. New York: McGraw-Hill Book Company.

Paris, P. and Erdogan, F. 1963. Closure to discussions of a critical analysis of crack propagation law (1963, ASME J. Basic Eng., 85: 533–534). Journal of Basic Engineering 85(4): 534–534. Doi: 10.1115/1.3656903.

Rabinowicz, E. 1995. Friction and Wear of Materials. Second edition ed. New York: John Wiley & Sons.

Rios, P.F., Dodiuk, H., Kenig, S., McCarthy, S. and Dotan, A. 2007. The effect of polymer surface on the wetting and adhesion of liquid systems. Journal of Adhesion Science and Technology 21(3-4): 227–241. Doi: 10.1163/156856107780684567.

Sarker, A.D. 1976. Wear of Metals. U.K: Pergamon Press.

Schijve, J. 2009. Fatigue as a phenomenon in the material. pp. 13–58. *In*: Schijve, J. (ed.). Fatigue of Structures and Materials. Dordrecht: Springer Netherlands. Doi: 10.1007/978-1-4020-6808-9_2.

Sinha, S.K. and Briscoe, B.J. 2009. Polymer Tribology: Imperial College Press.

Somorjai, G.A. and Li, Y. 2010. Introduction to Surface Chemistry and Catalysis. 2nd ed: Wiley, Hoboken, NJ.

Stachowiak, G.W. and Batchelor, A.W. 1993. 2 Physical properties of lubricants. pp. 11–57. *In*: Stachowiak, G.W. and Batchelor, A.W. (eds.). Tribology Series. Elsevier. Doi: 10.1016/S0167-8922(08)70576-5.

Stachowiak, G.W. 2005. Wear—Materials, Mechanisms and Practice. U.K: Wiley.

Thomas, T.R. 1999. Rough Surfaces. 2nd ed: Imperial College Press, London, U.K.

Tian, Pengyi, Dashuai Tao, Wei Yin, Xiangjun Zhang, Yonggang Meng and Yu Tian. 2016. Creep to inertia dominated stick-slip behavior in sliding friction modulated by tilted non-uniform loading. Scientific Reports 6: 33730. Doi: 10.1038/srep33730.

Uchida, Shin-nosuke, Hisayoshi Sato and Masanori O-hori. 1979. Two-Dimensional Measurement of Surface Roughness by the Light Sectioning Method. Vol. 28.

Uetz, H. and Föhl, J. 1978. Wear as an energy transformation process. Wear 49(2): 253–264. doi: https://doi.org/10.1016/0043-1648(78)90091-1.

Chapter 3
Friction

3.1 Introduction

Friction, as a phenomenon, is the force that resists sliding and is described in terms of a coefficient of friction. Two physically different forces of friction should be distinguished: The static friction force and the kinetic friction force. Static friction force is the minimal force needed to initiate sliding. Its value is related to the atomic structure of the sliding surfaces and the adhesion interactions. Kinetic friction force is the force needed to keep the motion. Moreover, it can be considered as the mechanism that converts the energy of motion into dissipated heat. Both forces are highly important. In various tribological applications, either a high or low value of friction may be required.

The microscopic mechanisms that generate friction are: Adhesion, mechanical interlocking of surface asperities (ploughing by surface asperities) deformation and fracture, plastic deformation by wear particles, and third bodies (Hsu, 1996).

We mentioned in Chapter One that there are two regimes of friction, namely, static friction between non-moving surfaces and kinetic (or dynamic) friction between moving surfaces. Accordingly, if two solid surfaces are loaded together and a tangential force is applied, then the static friction force is the value of the tangential force which is required to initiate sliding. Likewise, the kinetic friction force is the tangential force which is required to maintain sliding. From an engineering point of view, friction is a major cause of energy waste dissipated as heat and a major cause of failure in machinery components (Bayer, 2002).

Generally, friction depends very markedly on three main factors: The area of contact between the mating surfaces, the nature of the adhesion or junction at the regions of contact, and the way in which the formed junctions are sheared during sliding (Rothbart and Brown, 2006). For rough surfaces sliding over each other, the friction force is proportional to the real area of contact between both surfaces, which is smaller than the apparent (or nominal) contact area. In mechanical systems, friction may be increased or decreased depending on the contact between sliding surfaces and operating parameters such as roughness, degree of work hardening, and surface cleanliness. Understanding of friction mechanisms requires understanding of surface interaction and the mechanism involved between contacts.

3.2 Contact of Solid Surfaces

To discuss the contact of two solid surfaces, we should distinguish between two expressions: Apparent area and real area.

The apparent contact area A is the arithmetic (or projected) area of mating surfaces. This area has no effect in determining the friction and the overall interaction between mating surfaces.

The real contact area A_r determines the nature of the interaction between two surfaces. This area is proportional to the applied load and independent of the total interfacial area.

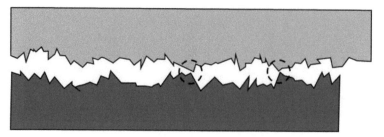

Figure 3.1: Illustration of microscopic contact between two surfaces.

Generally, we should consider that all solid surfaces have surface asperities which may be represented as very small sphere-shaped protuberances. Hence, it is fair to assume that all the contact between two surfaces takes place at those asperities at which there is atom-to-atom contact, as shown in Fig. 3.1. The asperity contact region is generally referred to as junction (or spot), and the sum of the areas of all junctions establishes the real area of contact which will obviously be much smaller than the total interfacial area. A great number of methods to measure the real area of contact have been attempted, but all methods have shortcomings.

The contact region of solid surfaces subjected to loads will initially connect at only a few junctions in order to support the applied load. As the load is increased, the surfaces move closer together, a larger amount of higher asperities on both surfaces come into contact. The existing contacts grow and deform to support the increasing load which is referred to as junction growth. Thus, surface deformation occurs in the region of the contact asperities, establishing stresses that oppose the applied load. The mode of surface deformation may be elastic, which is characterized by linear relation between stress and strain. On the other hand, plastic deformation is characterized by a more complex stress-strain relation. The applied load induces a generally elastic deformation of the solid surfaces at the tips of the asperities. Local plastic deformation may take place. Thus, in most contact situations, we find a mixture of both elastic and plastic deformations.

3.3 Theories of Friction

In considering the potential sources of friction, it is reasonable to consider the interaction between sliding surfaces and the mechanism of the energy loss. There are two main sources of surface interaction, which are adhesion between the contact surfaces and material deformation (or material displacement). According to these considerations, we will introduce various proposed theories of friction.

3.3.1 Adhesion Theory of Friction

The simple adhesion theory is most often attributed to Bowden and Tabor (Bowden and Tabor, 1950) assuming that the contact between two loaded solid surfaces occurs at the tip of the asperities. Consequently, the friction in dry sliding can be expressed as the minimum force required to shear the welded junction formed by adhesion bonds between contacting asperities.

In the elastic regime, the area of each asperity contact is approximately constant, while the number of contacts increases with the load. The pressure over the asperities is assumed to be high enough to cause plastic deformation, as illustrated in Fig. 3.2. Consequently, the real area should increase linearly until the external loading force is balanced by the counteracting contact pressure integrated over the real area. For ideal elastic-plastic materials, the relation between the normal load W and the real area of contact A_r is written as:

$$W = p \cdot A_r \tag{3.1}$$

where p is the yield pressure of the material.

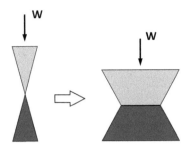

Figure 3.2: The pressure over the asperities is assumed to be high enough to cause plastic deformation.

For clean surfaces, no inter-diffusion or recrystallization of metal atoms takes place at the junction. The adhesion occurs in a cold condition, and the condition predominating at the interface of the junction is like "cold welding" of metals. Subsequently, strong junctions are formed at the interface.

When a tangential force is applied to slide one surface over the other, shear stresses develop over the junction interfaces to resist this force. At low shear stresses, the interatomic forces prevent atoms from sliding, and the material around the contact deforms elastically (Paulo Davim, 2011). When a critical shear stress is reached, the applied force is greater than the interatomic forces, and sliding starts. If it is assumed that all junctions have the same shear strength S, the adhesive friction force F_{ad} necessary to shear all the junctions and produce sliding will be:

$$F_{ad} = A_r S \qquad (3.2)$$

Ploughing Effect

Bowden and Tabor (Bowden and Tabor, 1950) defined ploughing as the part of friction resulting from the penetration of hard asperities in a softer surface and ploughing out a groove by plastic flow in the softer surface. The contribution to friction force from ploughing (or deformation) (Halling, 1979) is called the ploughing friction F_{plugh}.

Assuming a hard asperity of conical shape with semi-apex angle θ, rubbing upon a softer counterface, as shown in Fig. 3.3. The total projection area A_p in the direction of motion for a total number of asperities n is given by

$$A_p = n\frac{\pi r^2}{2} \qquad (3.3)$$

The ploughing friction force is obtained by considering A_p which is being displaced by plastic flow p, that is

$$F_{plough} = A_p p \qquad (3.4)$$

Thus, it is convenient to treat the adhesion and ploughing friction forces separately and express the total friction force F as:

$$F = F_{ad} + F_{plough} \qquad (3.5)$$

Then,

$$F = A_r S + A_p p \qquad (3.6)$$

In Equation 3.6, A_r and A_p represent the vertical and horizontal projected areas, respectively. For unlubricated surfaces, the ploughing term $A_p p$ is very small than the adhesion term $A_r S$ and could be neglected. So, Equations 3.1 and 3.6 altogether, could be written as:

$$F = \frac{WS}{p} \qquad (3.7)$$

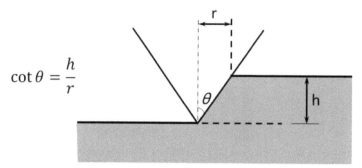

$$\cot \theta = \frac{h}{r}$$

Figure 3.3: The ploughing of a soft surface by hard conical asperity.

Regarding to the mathematical representation of the friction coefficient μ (see Equation 1.1), the above equation could be rewritten as:

$$\mu = \frac{S}{p} = \frac{F}{W} \tag{3.8}$$

Both S and p are material properties. The usual S/p ratio is between 0.17 and 0.2, which is comparable to the value of μ in practice for clean metals in dry sliding ($\mu \approx 0.2$) (Ludema, 1996). Although adhesion theory provides an explanation to the Amontons laws of friction (see Equation 3.6), there are enough exceptions to make this theory become subjected to criticism which can be summarized in the following points:

1) The theory didn't give an adequate explanation as to how the strong asperity junctions are formed, given that in many practical situations the sliding temperature is quite low.

2) The theory considered that friction was independent on surface roughness, but many researches realized that this is not quite true in dry and boundary lubricated sliding (Rabinowicz, 1995; Ghabrial and Zaghlool, 1974; Yamaguchi et al., 2014).

3) Experiments indicated that brittle non-metals showed friction properties similar to those of metals, although plastic deformation does not occur.

4) The theory was made on a ground that the normal force passing the sliding surfaces together results in strong adhesion between them. However, there was insufficient discussion about the formed junctions in case of removing this force.

However, it was the first theory to highlight the importance of the mechanical properties of sliding metals in friction.

Rubenstein (Rubenstein, 1956) introduced a general theory of friction based on the assumptions that the adhesion theory of friction is valid and that the asperities, at which contact occurs, deform elastically. The postulated model was for a rough surface on which spherical or cylindrical asperities are uniformly distributed. The deduced empirical equation relates friction force F to normal load W.

$$F = CW^\beta \tag{3.9}$$

where the parameter C is dependent on the physical properties of the sliding materials, the degree of surface roughness and the apparent area of contact. The parameter β is dependent on the physical properties of the sliding materials only and it lays somewhere in the range of 0.66 to 1. Obviously, when the first law of friction is obeyed, β is exactly 1.

In fact, according to Equation 3.8, the coefficient of friction should not exceed 0.2 for contacting metals with similar hardness or 0.3 for a hard metal sliding on a softer one. However, experimental friction values are often much higher, thus indicating that other mechanisms contribute to friction, such as junction growth and work hardening (Paulo Davim, 2011).

3.3.2 *Junction Growth Theory*

The fact that high friction forces are obtained in metal sliding in vacuum, motivated researchers to think that contact area might become enlarged under the additional shear force. Bowden and Tabor (Bowden and Tabor, 1950) proposed junction growth theory (or modified adhesion theory) taking into consideration that yielding must take place as a result of the combined normal and shear stresses (Halling, 1979). They considered two rough surfaces subjected to normal load W and friction force at the interface. To explain their hypothesis, they considered 2-dimensional stress system. In this case, maximum shear stress can be calculated using Mohr's circle construction. The maximum shear stress is represented by the radius R of the circle, as shown in Fig. 3.4.

Thus,

$$R^2 = S^2 + \left(\frac{p}{2}\right)^2 \tag{3.10}$$

Equation 3.7 demonstrated that yielding is dependent on a combination of normal and shear stresses. Now, if a tangential friction force is applied, plastic flow at asperity contact will take place. This flow causes an increase in the contact area, thus, junction growth is brought about by the superposition of the shear stress and the normal stress. As the area increases, the normal and shear forces must decrease, and junction growth continues until the combined stresses obey Equation 3.10. The solution of this 3-dimensional case was assumed as:

$$\left(\frac{W}{A}\right)^2 + \alpha\left(\frac{F}{A}\right)^2 = k^2 \tag{3.11}$$

where A represents the area of contact of the junction, α and k are constants. Now, if the shear S at the junction contact is zero, the pressure must equal the yield limit value p. Bowden and Tabor, finally, deduced the following equation:

$$A^2 = \left(\frac{W}{p}\right)^2 + \alpha\left(\frac{F}{p}\right)^2 \tag{3.12}$$

Notice that the term (W/p) is the total area of contact as considered in Equation 3.1, and the term $\alpha(F/p)^2$ represents the increase caused by the friction (or shear) force.

Now, according to Equation 3.12, at a constant normal load W, the area of contact A will increase with increasing friction force F till force reaches its limiting value. Therefore, on application of additional incremental tangential force, there will be further plastic flow at constant shear stress, resulting in an incremental contact area. Bowden and Tabor called this increase the junction growth.

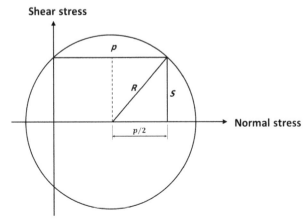

Figure 3.4: The Mohr's circle construction.

According to the junction growth theory, in order to reduce maintenance costs and increase bearing life, interface shear strength of contacting surfaces need to be as low as possible. Also, understanding this mechanism suggests applying a surface coating technology (coating the surface by a thin film of low shear strength materials) to enhance friction resistance.

3.3.3 Asperity Interlocking Theory

The basis of this theory is to consider the effect of asperity interactions on the sliding behavior of rough surfaces where asperity interlocking is to be one of the basic causes of friction. Coulomb (1736–1803) was the first to develop a model to understand friction due to the interlocking of asperities. He assumed that when two asperities run into each other or an asperity digs into the opposing surface, plastic deformations of softer asperities have to take place by the application of force, Fig. 3.5. This leads to frictional resistance.

Edwards and Halling (Edwards and Halling, 1968) introduced an analysis for the interlocking theory considering two wedge-shaped asperities interacting, as illustrated in Fig. 3.6.

In their model, the deformation is assumed to be plastic and the instantaneous shearing force and normal force over a complete horizontal displacement is possible to be calculated. They suggested that it was possible to extend the theory to cover surfaces consisting of any array of asperities with a given height distribution which interact elastically and plastically, depending on their height.

However, the interlocking theory is not very well recognized and there were some objections:

1) Actually, asperities "interfere" with each other's passage rather than interlock.

2) According to this, the friction force is directly proportional to the roughness of sliding surfaces. However, it was observed that the coefficient of friction increases when the surface becomes very smooth.

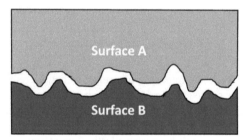

Figure 3.5: Asperity interlocking surfaces.

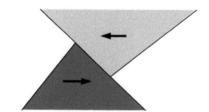

Figure 3.6: Two wedge-shaped asperities interaction.

3.3.4 Stick-Slip Theory

In practical applications, sliding of a surface over another may be designed to proceed at a constant velocity. However, under certain conditions, sliding velocity may fluctuate significantly. As a result, the friction force does not remain constant as a function of some other variables, such as time, distance, and any another cause of oscillation (Rabinowicz, 1995; Zuleeg, 2015). Consequently, the motion alternates

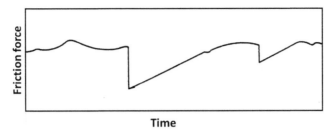

Figure 3.7: Friction force, of stainless steel sliding on nickel, as a function of time.

periodically between adhesion and sliding. This phenomenon of oscillating friction between two surfaces is referred to as stick-slip.

At microscopic scale, during sliding of the surface, an asperity can be in contact or it can be some distance apart from a flat counter surface. Therefore, these asperities are undergoing stick-slip and slip-stick transitions during motion of the surfaces. This theory is also considered as an alternative description of the adhesion theory.

In the actual stick-slip condition, the sliding bodies temporarily stop sliding and then slide for some short distance before becoming stationary again. So, if we plot the friction force as a function of time, the plot will be as shown in Fig. 3.7.

Prevention of Stick-slip

Practically, stick-slip in machinery can result in inaccurate operation or even significant damage to the mechanical parts. This phenomenon has been extensively studied, and it is recognized as a major source of problems, such as excessive bit wear, premature tool failures and poor drilling rate (Zuleeg, 2015; Kyllingstad and Nessjøen, 2009; Matsuzaki, 1970). The problems are closely related to the high peak speeds occurring during the slip phase. The high rotation speeds in turn lead to extreme accelerations and forces, both in axial and lateral directions. Hence, various methods have been adopted in order to minimize or prevent stick-slip motion (Matsuzaki, 1970; Kyllingstad and Nessjøen, 2009). For instance, a special lubrication oil is added, or a dopant-enhanced hydraulic fluid is used to minimize the chance of stick-slip. Also, changes can be made in lubrication methods; the stiffness of the driving system may be increased, or may change the finish of the sliding surface.

3.4 Laws of Sliding Friction

The mathematical discussion of the friction force depends on definite assumptions which are realized in the empirical laws of friction and are found to be in close agreement with practical experiments. As previously mentioned in Chapter One, there are three classic laws of dry sliding friction:

1. Friction force is directly proportional to the applied load, this can be expressed mathematically by Equation 1.1.
2. Friction force is independent on the apparent area of contact.
3. Once motion starts, the kinetic friction is nearly independent of the sliding velocity.

The first two laws are the pioneers and are often referred to as Amontons laws (Amontons, 1699) and they are reasonably well obeyed for sliding of metals. However, in the sliding of polymers (plastics), the laws are not so well obeyed. The third law was introduced by Coulomb (Coulomb, 1785). Other laws were introduced based on experimental studies (Bayer, 2002; Stachowiak, 2005):

1. The static friction coefficient is slightly greater than the kinetic friction coefficient, or simply, the dependence of friction force on sliding velocity is very small.
2. Friction depends upon the nature of the surfaces in contact.

3. Sliding friction is greater than rolling friction.
4. The direction of frictional force is always opposite to the direction of motion.

Thanks to the rapid development in measurement devices, it is fair to believe that in the future many new explanations for the friction phenomena will become available.

It is important to mentioned that, at the nano-scale regime, the above laws do not govern the friction between two surfaces as quantum mechanical effects, such as quantum confinement of electrons, and large surface area to volume ratio dominate and change the physical behavior of objects. This is because, at the molecular level, overall friction depends on that of the individual molecules, which is not predictable as statistical averages cannot be taken due to the very small number of atoms available at this scale. This friction of individual molecules depends on many factors, such as movement of individual atoms, number of atoms available and in contact, and direction and velocity of travel at the atomic level, with respect to the orientations of individual atoms, all of which contributes to making it unpredictable. The estimation of friction at macro-scale involves statistical averaging of friction due to a very large number of atoms, which is, therefore, predictable and independent of the area of contact and variations in the properties of individual atoms. However, at nano-scale, the number of atoms is very small, and the number of atoms available affects friction to a large extent. As the number of atoms in the contact increases with the area between surfaces in contact, friction at nano-scale becomes dependent on the area of contact, which results in breaking down of classical theories, such as Amontons' laws and Coulomb's law of friction (Sankar et al., 2019).

3.5 Laws of Rolling Friction

Rolling friction is a force resisting the motion which takes place when a ball, tire or wheel surface rolls on a counter surface. Nevertheless, a rolling body may be of an irregular outline, such as boulders or pebbles. Irreversibility in the deformation of contacting materials can also be considered as one of the sources of rolling friction (Ludema, 1996). In this section, the rolling friction of smooth surfaces of high geometric perfection will be introduced, while the other situations of irregular outline surfaces are out of our scope.

When we compare rolling friction and sliding friction, the first thing we notice is that the rolling friction is much more strongly connected to deformations than sliding friction. In other words, if the roller and surface are perfectly rigid, there will not be any rolling friction. Also, the energy loss in rolling is much less than in sliding (about 2–3 orders of magnitude less), provided the bodies considered are reasonably rigid.

To the best of our knowledge, the first significant work done to understand rolling friction was by Coulomb, in 1781, when he introduced the first two laws of rolling friction. Furthermore, some laws of sliding friction were generalized to rolling situations (Rabinowicz, 1995):

1. The friction force is directly proportional to the applied load.
2. The friction force varies inversely with the radius of curvature of the rolling elements.
3. The friction force is lower for smoother surfaces than for rough surfaces.
4. The static friction force is, generally, much greater than the kinetic friction force.
5. The kinetic friction force is lightly dependent on the rolling velocity.

Unlike the case of sliding friction, a formula for the force of rolling friction is not easily derivable. This is due to the fact that there are many physical mechanisms leading to rolling friction, including plastic deformation, elastic hysteresis and adhesion hysteresis. However, according to the illustration in Fig. 3.8, the rolling friction formula can be empirically expressed as

$$F_r = \mu_r W \tag{3.13}$$

where,

F_r is the rolling friction force, N
μ_r is the rolling friction coefficient
W is the applied normal load, N

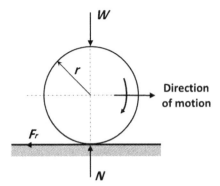

Figure 3.8: Rolling friction.

To account for the radius of rolling elements, the rolling resistance can be expressed as

$$F_r = \mu_{rl} W/r \tag{3.14}$$

where,

μ_{rl} is the rolling friction coefficient expressed in *mm*

r is the radius of the wheel, ball, cylinder, etc.

In most practical situations, we cannot consider the case of pure rolling due to the fact that the region of contact is elastically and plastically deformed. That is, the points of contact are lying in different planes. Consequently, it is not possible for pure rolling to take place, except at a very small number of points; at the other points, a combination of rolling, sliding and slip is found. The sliding (or slip) velocity v_s is generally less than 5% of the rolling velocity v_r. The total rolling friction F_R may be written in the form

$$F_R = \frac{v_s}{v_r} \mu_K \tag{3.15}$$

where,

μ_K is the kinetic coefficient of sliding friction.

Here, we may cite some causes of losses during rolling. First, an energy loss due to a lack of perfection of the rolling geometry, which will cause a roughness component of friction. Second, an energy loss caused by the plastic deformation of asperities on the rolling surfaces. Third, elastic hysteresis losses resulting from stress-release cycles occurring during rolling.

All these losses are in an order of magnitude of 10^{-4} and should add up to the rolling friction force (Tomlinson, 1929; Greenwood et al., 1961).

3.6 Factors Affecting Friction and Wear

Friction and wear are not material properties; they are system functions. Therefore, they are supposed to be governed by tribo-system parameters, which are considered as factors affecting the friction and wear behavior. Although it is convenient to consider these factors under different headings, it should be clear that these factors are interacting so that it is difficult to separate them one from another. These factors are normal load, geometry, relative surface motion, sliding speed, surface roughness of the rubbing surfaces, type of material, system rigidity, temperature, stick-slip, relative humidity, lubrication and vibration.

The role of the scale factor is illustrated schematically in Fig. 3.9 (Myshkin and Goryacheva, 2016). The scheme shows how different factors are interrelated depending on the scale of the process in question. For instance, friction and wear at the micro and nano levels occur on very smooth contact area, therefore, the role of adhesion and surface forces is negligible. Miniaturization of the friction contact models requires a transfer from the volume properties of the materials to their surface features evaluated by an atomic

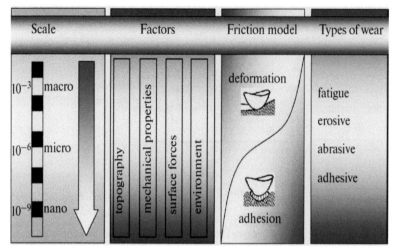

Figure 3.9: Illustration of the factors affecting friction and wear. After Myshkin et al. (2016), with permission from Elsevier publishing.

force microscopy, measurements of adhesion and micro and nano indentation. It is worth noticing that, in the course of the mutual sliding of contacting solids, the mechanical properties of their materials may change under the action of the friction induced heat. Nonetheless, Surface heating may also result from high loading, which has a drawback on surface film formation and material hardness.

On the other hand, friction and wear in rolling contacts is a much more complex phenomenon because of its dependence on further factors, including slip during rolling, and energy losses during mixed elastic and plastic deformations. This issue will not be considered in the current work.

The analysis of known wear modes proves that all of them are also related to the friction force components. At the same time, the fatigue wear is conditioned mainly by deformation of the material by friction. The adhesive wear is affected by the surface forces when the material undergoes breakage and is transferred to the rubbing bodies. A number of the factors influencing friction and wear behavior will be summarized below.

3.6.1 Applied Load

Generally, an increase in an applied load results in an increase in the friction force, and hence, a temperature rise which affects the hardness of the rubbing surfaces. Furthermore, the effect of the applied load on friction and wear behavior is attributed to the effect of the load on asperity interaction. The transition from mild to severe wear is related to the interaction of the plastic zone beneath the contacting asperities (Rabinowicz, 1995).

Experimental investigation (Chowdhury et al., 2011) of the effect of an applied normal load on friction and wear properties of an aluminum disc sliding against a stainless steel pin indicated that the value of friction coefficient decreases with the increase of the normal load. The wear rate, on the other hand, increases with the increase of the normal load. The friction at the time of starting is low and remains at its initial value for some time, and the reason for this low friction is the presence of a contaminating layer of moisture and oxide. During initial rubbing, the deposited layer breaks up and clean surfaces come in contact, this increases the bonding force between the contacting surfaces. At the same time, due to the inclusion of trapped wear particles and the roughening of the substrate, the friction force increases due to an increase of the ploughing effect. An increase of the surface temperature, a viscous damping of the friction surface and an increased adhesion due to micro-welding or deformation or hardening of the material might have some role on this increment of the friction coefficient as well, as shown in Fig. 3.10a. After a certain duration of rubbing, the increase of roughness and other parameters may reach a certain steady state value and, hence, the value of the friction coefficient remains constant for the rest of time. Figure 3.10b shows that the wear rate increases with an increase of the normal load. This is due to the fact that, as the normal load increases, frictional heat is generated at the contact surface and, hence, the strength of the materials decreases.

It is concluded that in many metal pairs, in the high-load regime, the coefficient of friction decreases with the load. Increased surface roughening and a large quantity of wear debris are believed to be responsible for the decrease in friction (Bhushan, 1996).

Likewise, studying the effect of light loads on the coefficient of friction and wear depth (Whitehead, 1950) suggested that the coefficient of friction remains essentially constant until some critical value. Figure 3.11 shows the coefficient of friction and wear depth as a function of the load for a sharp diamond tip sliding on three smooth materials. The coefficient of friction of $Si(111)$ and SiO_2 coating starts to increase above some critical load values for which the contact stresses correspond to their hardness. An increase in the wear depth also starts to take place above the critical load. Very little plastic deformation and ploughing contributions are responsible for low friction at loads below the critical load.

It can be drawn that, in the case of materials with surface films which are either deliberately applied or are produced by reaction with the environment, the coefficient of friction may not remain constant as a function of the load. In many metal pairs, in the high-load regime, the coefficient of friction decreases with the load. Increased surface roughening and a large quantity of wear debris are believed to be responsible for a decrease in friction (Bhushan, 1996).

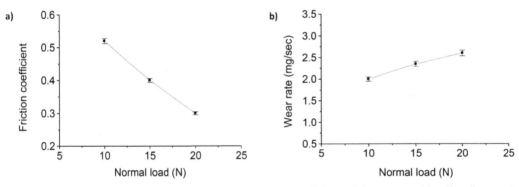

Figure 3.10: Effect of the variation of a normal load on (a) friction coefficient and (b) wear rate. After Chowdhury et al. (Chowdhury et al., 2011), with permission from IJENS.

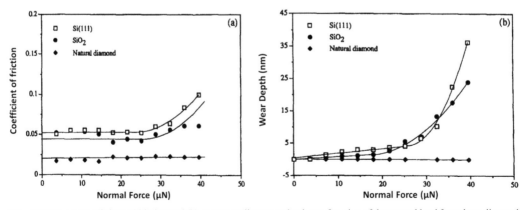

Figure 3.11: (a) Coefficient of friction and (b) corresponding wear depth as a function of the normal load for a sharp diamond tip sliding on Si (111), SiO_2 coating and natural diamond in air at a sliding velocity of 4 μm/s using friction force microscopy. After Bhushan et al. (1996) with permission form Elsevier.

3.6.2 Temperature

In general terms, as the temperature is increased, the coefficient of friction and wear rate are typically affected in three essential points:

a) mechanical properties of the rubbing materials.
b) the formation of a surface-contaminating film.
c) properties of the lubricant.

Generally, a higher temperature results in a decrease in the hardness of a material. As a result, the tendency for asperities to adhere and the wear rate increase with decreasing hardness. At high temperatures, softening of surfaces results in a greater flow, ductility and a larger real area of contact, which results in stronger adhesion. High temperatures can also result in diffusion across the interface. In a metal-metal contact, a high temperature may result in an increased solubility, and in a polymer-polymer contact, interdiffusion strengthens the contact, which results in stronger adhesion (Rabinowicz, 1995).

Metals commonly used at high temperature conditions include tool steels, and alloys with base composition of cobalt, chromium and molybdenum. Nonetheless, at a temperature above about 850°C, it is necessary to use ceramics, cermits or other advanced materials.

Another effect of high temperatures is the deterioration of the lubricant fluid, first by oxidation of the oil and then by thermal degradation. This indicates an operating limit beyond which organic fluids are ineffective in reducing friction and wear.

3.6.3 Sliding Speed

The sliding speed is a major factor that plays a significant role for the variation of friction and wear rate. The third law of friction, which states that friction is independent of the velocity, is not generally valid. The coefficient of kinetic friction as a function of the sliding velocity generally has a negative slope. Changes in the sliding velocity result in a change in the shear rate, which can influence mechanical properties of the mating materials.

Furthermore, the effect of speed could be arising from increased surface temperatures. Four of the most important consequences are:

a) High hot-spot temperature increases reactivity of the surfaces and the wear fragments with the environment.
b) Rapid heating and cooling of asperity contacts can lead to metallurgical changes which can change the wear process.
c) Surface melting, which, in some cases, resulted in reducing the friction and wear.

Usually, the friction coefficient at the beginning of sliding is low and remains at its initial value for some time. This low friction is due to the presence of a layer of foreign material. The surface, in general, comprises moisture, oxides of metals, deposited lubricating material, etc. Figure 3.12 shows the variation of the friction coefficient with the duration of rubbing at different sliding speeds for aluminum (Chowdhury et al., 2011).

Aluminum readily oxidizes in air, so that, at the start of rubbing, the oxide film easily separates both material surfaces and there is a little or no true metallic contact, also, the oxide film has a low shear strength. During initial rubbing, the film (deposited layer) breaks up and clean surfaces come in contact, increasing the bonding force between the contacting surfaces. At the same time, due to the inclusion of trapped wear particles and roughening of the substrate, the friction force increases due to an increase of the ploughing effect. Increase of the surface temperature, viscous damping of the friction surface, increased adhesion due to microwelding or deformation or hardening of the material might have some role on this increment of friction coefficient as well. After a certain duration of rubbing, the increase of roughness and other parameters may reach a certain steady state value and, hence, the value of the friction coefficient remains constant for the rest of the time.

In the curve of Fig. 3.13, it is seen that the value of the friction coefficient of aluminum decreases with the increase of the sliding speed. This may be due to a change in the shear rate which can influence mechanical properties of the mating materials.

The decrease of the friction coefficient of aluminum with the increase of the sliding speed may be due to a change in the shear rate which can influence mechanical properties of the mating materials. The strength of these materials is greater at higher shear strain rates, resulting in a lower real area of contact and a lower coefficient of friction in a dry contact condition (Bhushan and Jahsman, 1978). These findings are in agreement with the findings of Chowdhury et al. for mild steel, ebonite and GFRP sliding against mild steel (Chowdhury and Helali, 2008).

Figure 3.14 shows variation of the wear rate with the variation of speed. From this figure, it is observed that wear rate increases with an increase of the sliding speed. This is due to a fact that the duration of rubbing is the same for all sliding speeds, while the length of rubbing is more in the case of a higher speed. The mild steel-mild steel couples also show similar behavior, i.e., wear rate increases with the increase of the sliding speed (Chowdhury and Helali, 2007).

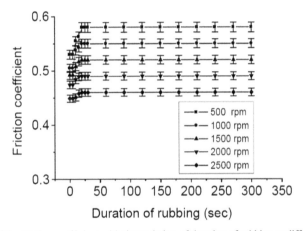

Figure 3.12: Variation of the friction coefficient with the variation of duration of rubbing at different sliding speed. After Chowdhury et al. (Chowdhury et al., 2011), with permission from IJENS.

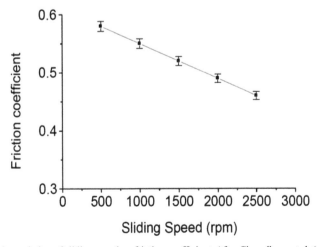

Figure 3.13: Effect of the variation of sliding speed on friction coefficient. After Chowdhury et al. (Chowdhury et al., 2011) with permission from IJENS.

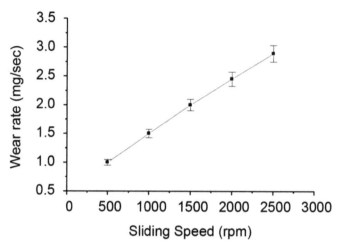

Figure 3.14: Effect of the variation of sliding speed on wear rate. After Chowdhury et al. (Chowdhury et al., 2011), with permission from IJENS.

3.6.4 Material Compatibility

Rabinowicz (Rabinowicz, 1966) has postulated that metal pairs with low 'metallurgical compatibility' will exhibit low friction and low wear. Metallurgically compatible metals are defined as those which show a high degree of mutual solubility. He attempted to find a correlation between adhesion, friction and wear results from several studies, with compatibility of the material pairs used. A positive correlation with friction and a great correlation with wear was deduced. However, he found zero correlation of compatibility with adhesion.

Based on the above, it can be drawn that the metal pairs chosen for boundary lubricated or unlubricated sliding applications should have low mutual solubility in order to reduce friction and wear.

3.6.5 Surface Roughness

Surface roughness is one of the important parameters that defines the tribological properties of sliding counterfaces. In dry sliding of metallic surfaces, surface roughness has a great impact on a variety of physical phenomena, including friction, wear, contact mechanics, adhesion and interfacial separation. Furthermore, the magnitude of friction forces is related to the surface roughness of contacting bodies and material properties of the friction pairs in contact. In general, high roughness means increasing the proportion of friction and adhesion, thereby increasing the rate of connection shearing, which leads to increased wear.

In the case of boundary lubrication, the friction coefficient increases as the rate of solid contacts increases due to greater surface roughness. On the other hand, wear grains trapped in the surface roughness can enhance the lubrication mechanism. Furthermore, it is reported that new surfaces produced by shearing chemically react with components of the additives in lubricants (Yamaguchi et al., 2014). Figure 3.15 shows the variation of friction coefficient and wear rate with a normal load applied at different surface roughness of Al-Si casting alloy under dry and lubricated conditions (Al-Samarai et al., 2012).

In elastomers, the friction behavior of sliding against a hard counterface could be described by the two-term model of friction, and it is implied that the total friction was the sum of two independent contributions due to adhesion and deformation. Studies have shown that the effect of the rough surface of a deformable soft material, an elastomer, on the friction was much less significant than the influence of a hard mating metal surface during sliding. Thus, it is important to know the effect of surface roughness of a harder mating metal on the friction coefficient of rubber/metal tribo-pairs during sliding. Feng (Feng et al., 2016) proposed that the friction coefficient of rubber is mainly dominated by the surface roughness of a hard counter surface when the soft rubber slides against the hard, rough surface under oil lubrication. The arithmetic roughness value *Ra* has a relatively greater influence on the magnitude of the friction coefficient, but this does not mean that lower surface roughness of the counterpart is better. The adhesion friction increases as a result. The lower friction value is attributed to the appropriate values of *Ra*, *Rq* and *Rz*, as presented in Fig. 3.16.

It is important to mention that the standard surface roughness parameters (e.g., *Ra*, *Rq*) normally used by designers do not sufficiently describe contact surfaces (Dzierwa, 2017). In addition, different standards use differing parameters. Although many experimental works have been carried out on the surface roughness and topography of contact surfaces, correlations between surface roughness and friction and wear are not yet clearly defined. Therefore, several experimental investigations were carried out in order to investigate the influence of surface topography parameters on the rate of wear and friction and correlations between roughness parameters and friction and wear (Reizer et al., 2011; Sedlaček et al., 2009). It was demonstrated that the basic amplitude parameters commonly used to describe tribological characteristics are not sufficient for determining the tribological properties of contact surfaces (Dzierwa, 2017). Thus, other surface topography parameters should be used to describe the tribological properties of friction pairs.

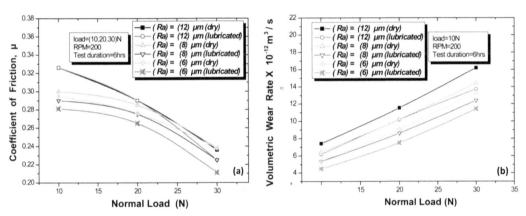

Figure 3.15: Variation of surface roughness Ra = 6, 8 and 12 μm and the normal load under dry and lubricated conditions with (a) Coefficient of friction and (b) Volumetric wear rate. After Al-Samarai et al. (Al-Samarai et al., 2012).

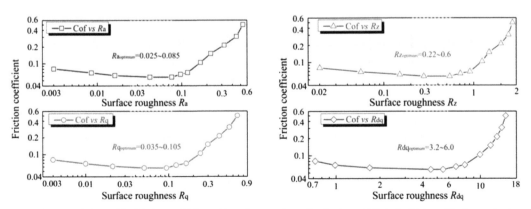

Figure 3.16: Variations of the friction coefficient with surface roughness *Ra*, *Rq*, *Rz* and *Rdq* in double logarithmic coordinates. After Feng et al. (Feng et al., 2016), with permission from Elsevier.

3.6.6 Environment and Surface Films

Occasionally, on metallic and nonmetallic surfaces, there will a greasy or oily film derived from the environment. Besides, due to chemical reactivity of the surfaces and the tendency of molecules to adsorb on it, an adsorbed film in the order of a monolayer could be also found.

In fact, even a small amount of contaminant films may be much more effective in reducing the adhesion of some metals than of others. For example, a very small amount of oxygen (perhaps enough to give a monolayer) can produce a remarkable reduction in the adhesion of iron, whereas far more oxygen is required to produce a comparable reduction in the adhesion of copper (Ludema, 1996; Bhushan, 2013). However, this is not always the case. If, for example, the metal oxide is hard and the conditions favor abrasive wear, the continuous formation of oxidized wear fragments may lead to a large increase in wear rate. For a soft metal covered by a brittle oxide, it has been found that there are three regimes of friction over a range of loads, as illustrated in Fig. 3.17: Regime A where the oxide film is intact, Regime B as a transition region, and regime C where the oxide film is fractured.

The surrounding atmosphere can have a remarkable effect on friction; in many, air or water vapor reduce the wear rate. Friction and wear are often influenced by the amount of handling of specimens with human hands, cleaning methods, the method of storage, the number of passages of slider, and many other factors. These invisible films on such surfaces vary in thickness up to 300 nm.

Adsorbed gas, water vapor layers, and greasy or soapy contaminant films usually reduce the severity of surface interaction often by one or more orders of magnitude. Their effects could be considered as those of lubrication, though to formalize concepts in this topic it would be necessary to characterize the thin films in terms of their thickness and viscosities. Nonetheless, in some cases, the films are worn out in the initial period of running and subsequently have no effect.

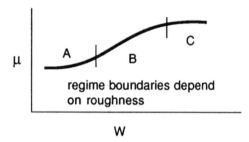

Figure 3.17: The impact of an applied load on friction for metals with brittle oxides.

3.7 Measurement of Friction

3.7.1 Sliding Friction Measurement

Friction measurement involves the quantification of force of contact generated in the interface that varies versus time and space. Commonly, the friction coefficient or its equivalent, as a dimensionless quantity, is used to represent the magnitude of the resistance to relative motion between bodies (Blau, 1992). Measurement techniques involve either the direct measurement of forces that resist relative motion between two or more bodies of matter or the indirect measurement of the effects of those forces. Some tribometers are designed to measure the force required to initiate relative motion (static friction force), and some are intended to measure the resisting force on objects that are already in motion (kinetic friction force). In lubrication conditions, tribometers are designed to focus on metering the effects of the interfacial fluid or grease.

Many friction measurement systems are designed to simulate specific applications, however, other measurement systems are designed to isolate and study frictional effects under highly controlled laboratory conditions that do not simulate any particular engineering application. As a result, there are literally hundreds, if not thousands of devices designed to quantify frictional resistance. Basic friction measurement methods are described here.

1. The inclined plane method, ASTM D4521.

This simple method to measure friction force (F_f) is based on a static balance of forces that requires only the angle (θ) of tilt to be measured at the instant when relative motion begins. According to Fig. 3.18, the static friction coefficient (μ_s) is given by:

$$\mu_s = \tan \theta \tag{3.16}$$

2. The horizontal plane method, ASTM D4521.

This method is directly utilized to measure the force required to begin motion and to continue relative motion, i.e., to measure static and kinetic friction forces between a specimen and counterface, as shown in Fig. 3.19. A typically 1000 gm dead weight is placed on the inclined slider. The inclined slider is raised until sliding begins. The coefficient of friction is equal to the tangent of the angle at which sliding begins. There are a number of factors that influence static friction, such as dwell time, operator positioning and condition of the sled. It should be noted that a higher degree of variability is generally reported for the static friction about 4–15% on paper materials. Kinetic (or dynamic) friction is the preferred measurement and generally has a variability of 2–8% on paper materials.

3. Indirect friction measurements.

Friction, both static and kinetic, can be measured indirectly by monitoring parameters, such as the current drawn by a motor that moves one of the sliding components against another (e.g., a sliding valve or a mechanical face seal). Eissenberg and Haynes (Eissenberg and Haynes, 1992) have reviewed this topic. Motor current measurements to infer the state of friction tend to be relative or comparative in nature, rather than the absolute. One compares the motor current from an operating device to a reference value from the same kind of device when it is operating properly. Electronic sensors can be used to shut down equipment when required and before more serious seizures or failures occur.

4. Friction measurement using torque data (Blau, 2013).

It is interesting to estimate the friction of mechanical assemblies from torque data. Examples include mechanical face seals, threaded bolts, brake pads on rotors, and rotating shafts.

We should emphasise that the uncertainty in the friction measurement from torque data is related to the application. For example, the measurement uncertainty in the case of a shaft in a bushing of known diameter is less than if the source of the frictional torque occupies a range of radial distances from the center of rotation, as in brake pads, face seals or clutches. According to Fig. 3.20, the friction force (F_f) that produces a torque (T) acts at a fixed radius from the center of rotation. Torque is the product of friction force times the radius from the center of rotation, therefore:

$$T = F_f r \tag{3.17}$$

and the friction coefficient (μ), in terms of the measured torque, is

$$\mu = T/(r\,P) \tag{3.18}$$

where (P) is the normal force on the block.

In friction tests, the friction force is usually measured and plotted as a function of time or sliding distance during the test. Figure 3.21 illustrates a general form where there is no difference between the static and the kinetic coefficients (i.e., there is no breakaway force), as well as no stick-slip. This system has a single value for the coefficient of friction, based on the average force to sustain motion. However, this is not the only example; in some cases, the static coefficient of friction is not the same as the kinetic coefficient of friction. Consequently, the force required to initiate sliding is higher than the average force needed to sustain motion. The static and kinetic coefficients are generally based on these two values, respectively. In addition, the stiffness of the test rig can result in a stick-slip behavior and differences between the static and kinetic coefficients. In both examples, the friction force will not be constant but it

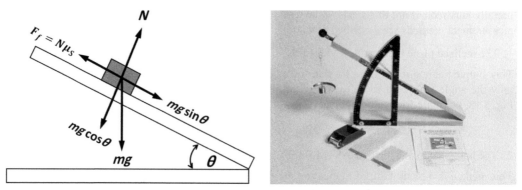

Figure 3.18: The inclined plane friction measurement method.

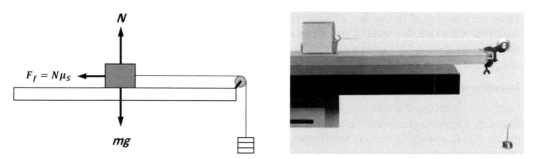

Figure 3.19: The horizontal plane friction measurement method.

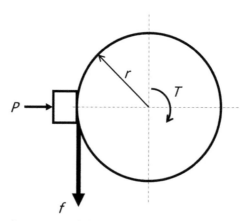

Figure 3.20: Friction measurement using torque data.

will fluctuate in the tests. It is possible that more than one type of curve could be observed for the same material with different instrumentation and apparatus stiffness (Bayer, 2004).

On the other hand, at nanoscale contacts, friction is usually measured experimentally by means of scanning probe microscopies (SPM), where the surface topology of the material is obtained using physical probes that scan the sample. The SPM has a cantilever which contains a tip that scans the surface of the sample. Piezoelectric scanners and actuators are attached to the cantilever in order to control and detect its motion, and computers are used to receive the signals from this tip movement, either as photons from laser or as tunneling current, depending on the type of SPM used. There are many different types of SPMs, but the most prominent among them are scanning tunneling microscopy (STM) which utilizes quantum tunneling of electrons to image a surface (Deng et al., 2013; Behm and Hösler, 1986; Sankar et al., 2019), and atomic force microscopy (AFM) which has the tip moving over the sample physically and signals received using reflection of laser light off the cantilever. AFM is the most advanced general purpose SPM technique used primarily in most experimental studies meant for studying the surface topography and nanofriction. AFM, either in intermittent contact (IC) mode, in which the tip vibrates over the surface without keeping constant contact, or in Frequency Modulation (FM) mode, is generally the most used method for experimental measurement of friction at the nanoscale (Salapaka, 2008). Depending on the materials studied, the AFM can also be Beetle style or Single Tube style. Figure 3.22 illustrates a schematic of the Beetle Style AFM with three hollow legs made of a piezoelectric ceramic, typically Lead Zirconium Titanate (PZT), on which a small disk with all sensing components attached to it is fixed. This is mainly used in studying hard and metallic samples.

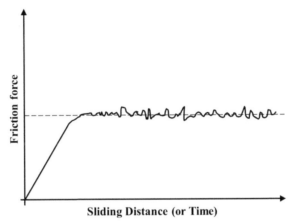

Figure 3.21: A typical friction curve.

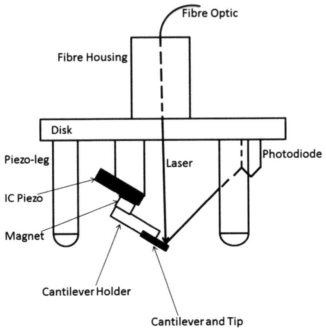

Figure 3.22: Beetle style AFM used for experimental study of friction at nanoscale.

3.7.2 Rolling Friction Measurement

Rolling friction, or rolling resistance, is the positive force resisting the motion of a rolling body (ball, tire or wheel) on a surface, as illustrated in Fig. 3.23. This force is mainly caused by non-elastic effects; that is, not all the energy needed for movement of the rolling element is recovered when the pressure is removed.

Another cause of rolling resistance lies in the slippage between the wheel and the surface, which dissipates energy. The main forms of this resistance are plastic deformation of the surface and hysteresis losses. In analogy with sliding friction, rolling resistance is often expressed as a coefficient times the normal force. This coefficient of rolling resistance is, generally, much smaller than the coefficient of sliding friction.

The coefficient of rolling friction (CRF) is defined by the following equation (Hersey, 1969):

$$F_r = C_r N \tag{3.19}$$

where,

F_r is the rolling friction force.
N is the normal force, the force perpendicular to the surface on which the wheel is rolling.
C_r is the dimensionless CRF.

The coefficient of rolling resistance for a slow rigid wheel on a perfectly elastic surface, not adjusted for velocity, can be calculated by:

$$C_r = \sqrt{z/d} \tag{3.20}$$

where,

z is the sinkage depth, as shown in Fig. 3.23.
d is the diameter of the rigid wheel.

Likewise, an empirical formula for the coefficient of rolling friction for cast iron mine car wheels on steel rails can be calculated as shown below (Hersey, 1970):

$$C_r = 0.0048(18/d)^{\frac{1}{2}}(100/W)^{\frac{1}{4}} \tag{3.21}$$

where

d is the wheel diameter (in).
W is the load on the wheel (lbs).

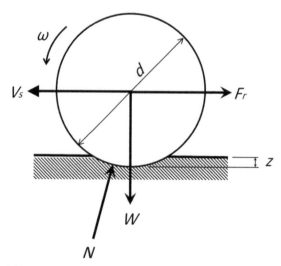

Figure 3.23: Illustration of a hard wheel rolling on and deforming a soft surface.

The driving torque (T) to overcome rolling friction (F_r) and maintain steady speed on level ground (with no air resistance) can be calculated as:

$$T = \frac{V_s}{\omega}C_r \tag{3.22}$$

where

V_s is the linear speed of the rotating body (at the axle).

ω is the rotating speed.

Typical ranges of coefficients of sliding friction are presented in Table 3.1 (Zayed, 1997; Zayed, 2004).

Table 3.1: Typical ranges of coefficients of sliding friction (Zayed, 1997; Zayed, 2004).

Tribo-combinations		sliding friction coefficient		static friction coefficient	
material	on material	dry	lubricated	dry	lubricated
Aluminum	Aluminum			1.05–1.35	0.3
	Steel			0.45	
	Mild Steel			0.61	
Brass	Steel			0.35	0.19
	Cast iron			0.3	
Bronze	Bronze	0.2	0.06		0.11
	Cast iron	0.18	0.08	0.22	
	Steel	0.18	0.07	0.19	0.1
Cast iron	Cast iron		0.1	0.1:0.15	0.07:0.16
	Steel	0.17:0.24	0.02:0.05	0.18:0.24	0.1
	Mild steel			0.23:0.4	0.13:0.21
	Oak			0.49	0.07
Copper	Copper			1	0.08
	Cast iron			0.3:1.05	
	Mild steel			0.36:0.53	0.18
Oak	Oak	0.2:0.4	0.05:0.15	0.4:0.6	0.18
Steel	PTFE	0.03:0.05	0.1		
	PA66	0.3:0.5			
	POM	0.35:0.45			
PTFE	PTFE	0.35:0.55			
	Steel			0.05–0.2	
POM	POM	0.4:0.5			
Glass	Glass			0.4:0.9	0.1:0.6
	Metal			0.5:0.7	0.2:0.3
	Nickel			0.78	0.56
Zinc	Zinc			0.6	0.04
	Cast iron			0.21:0.85	

PTFE   Polytetrafluoroethylene
POM   Polyoxymethylene
PA66   Polyamide (Nylon)

3.8 Frictional Heating

In any sliding operation, almost all the energy dissipated in friction appears in the form of heat at the interface. This process could be thought of as a transformation of energy, where mechanical energy is transformed into internal energy or heat, which causes the temperature of the sliding bodies to increase. The exact mechanism by which this energy transformation occurs may vary from one sliding situation to another, and the exact location of that transformation is usually not known for certain (Bhushan, 2001). It is widely accepted that nearly all of the energy dissipated in friction is, generally, used up in plastic deformation which is directly converted to heat in the material close to interface. The energy dissipation, called frictional heating, is responsible for the increase in temperature at the contact region of the sliding surfaces (Uetz and Föhl, 1978). Frictional heating has such an important impact on the tribological behavior of so many sliding systems that all tribo-systems must be designed with thermal considerations in mind (Floquet, 1983).

The frictional heating results in a surface layer which is physically different from the bulk of the material. The interfacial temperature of such a layer is higher than the environmental temperature. Historically, the observation of this Beilby layer, as it was named, was observed during metal polishing and first reported by Beilby (Beilby, 1903). Formation of such a surface layer was confirmed by Cochrane (Cochrane, 1938) and Bowden et al. (Bowden and Hughes, 1937).

Frictional heating and the resulting surface contact temperatures can have a significant impact on the tribological behavior and cause failure of many sliding tribo-systems. High temperature is responsible for changes in the friction and wear behavior of the material or the behavior of any lubricant present in the contact. Also, the temperature gradients around the contacts are very steep and can be responsible for softening and shear failure of the near-surface layer of the material in many situations.

Tribology of elastomers, polymers and their composites is significantly affected by interface temperatures. Frictional heating can cause surface temperatures to reach the melting or softening temperature of thermoplastics, resulting in a drastic change in their tribological behavior. Lancaster (Lancaster, 1971) introduced a critical surface temperature limit called the "*pv* limit" which is often used in the design of dry plastic bearings. This limit is a combination of contact pressure and sliding velocity and causes the surface temperature to reach the critical temperature of the material.

As we discussed in Chapter 2, friction is produced due to the microscopic asperity contacts. During sliding, junctions continue to be made and broken, and the hot spots on the surface shift their positions. The high temperature associated with the asperity contacts is often referred to as "flash temperature". That is, at any instant, there are usually several short-duration flash temperature rises at the various asperity contact spots within a nominal contact surface. In addition, the maximum surface temperature reached during sliding (i.e., the temperature component which added to the bulk temperature of the material) can also have an important influence on the tribological behavior of the contacting surfaces (Rabinowicz, 1995). For example, in dry sliding situations, high surface temperature rise may result in oxidation or even melting of contacting materials. In polymers, frictional heating is considered as one of the dominant factors influencing its physical and mechanical properties. Moreover, in lubricated sliding, surface temperature has a noticeable effect on the oil lubricant viscosity and on the durability of solid lubricant films (Dorinson, 1985).

3.9 Contact Surface Temperature

The ability to predict and measure the surface temperatures of actual contacting bodies is important if failure of tribological components is to be avoided. The temperature rise over large contact areas can be measured with some accuracy. Measurement of temperature rise over isolated micro-contacts is very difficult. Therefore, the transient temperature rise is generally calculated rather than measured. Furthermore, there is no single thermal analysis that will reasonably represent all the conditions of sliding.

There has been extensive work done on calculating contact surface temperature for moving or stationary bodies (Basu, 2011). Among several attempts made to predict the surface contact temperature, the most complete solution of temperature rise as a result of moving sources of heat on a semi-infinite

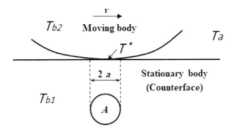

Ta Ambient temperature
Tb Bulk temperature
T* Flash temperature

Figure 3.24: Schematic diagram of temperature distribution around sliding contact.

body was presented by Jaeger (Jaeger, 1942b). The interface temperature can be calculated based on a concept of partition of heat developed by Blok (Blok, 1937b). This solution is adequate for a high contact-stress situation where real area of contact is approximately equal to the apparent area. The total contact temperature (T_C) produced during sliding is a combination of three separate heat sources, as illustrated in Fig. 3.24. These are the ambient temperature, the bulk temperature, and the asperity flash temperature.

The ambient temperature (T_a) is the temperature of the medium in which the body slides and, in most cases, may be considered as constant. The ambient (or nominal) temperature can be over 500°C for severe sliding cases, such as in brakes, but is usually much lower.

The bulk temperature (T_b) refers to the temperature of the entire body when run continuously, and after steady state conditions have been achieved. That temperature is, generally, less than 100°C.

The asperity flash temperature (T^*) occurs close to the area of true contact at which the energy is dissipated. This temperature can be very high (over 1000°C in some cases) but lasts only as long as the two asperities are in contact. It occurs in a very short period which could be less than 10 μs. At any instant, there are usually several short-duration flash temperature rises (ΔT^*) at various asperity contact spots within a nominal contact patch.

Thus, the total contact temperature (T_c) at a given point is given by:

$$T_C = T_b + \Delta T_a + \Delta T^* \tag{3.23}$$

In unlubricated sliding, the knowledge of these flash temperatures at the sliding interface is of fundamental importance for the tribological behavior, specially, that of thermoplastic polymers. Because of their short duration and because they occur only over small regions, direct measurement of flash temperatures is difficult and a theoretical estimate is usually thought of. Thus, calculating the temperature is the most convenient method for determining the flash temperature. In fact, there are many available models and it is not an easy task to decide which one to use (Jaeger, 1942a; Kuhlmann-Wilsdorf, 1987; Greenwood, 1991; Tian and Kennedy, 1994). Furthermore, other problems arise in the determination of the size and shape of the real contact area, and the geometric and thermal properties of the wear particles or any other third body within the contact. As a consequence, large discrepancies in the results can be obtained for the same contact situation.

The first significant contribution was introduced by Blok (Blok, 1937a), followed by another pioneering solution from Jaeger (Jaeger, 1942a), who obtained the flash temperature rise formula for the case of sliding between two semi-infinite planes of different geometrical configurations. The theoretical increase in bulk temperature ΔT_{max} calculated according to these models is given as:

Blok – Model:

$$\Delta T_{max} = 9.16 \times 10^{-4} \mu p \sqrt{2l} \tag{3.24}$$

Jaeger – Model:

$$\Delta T_{max} = 4.2 \times 10^{-4} \frac{\mu F_N \sqrt{v}}{b\sqrt{l}} \qquad (3.25)$$

where

μ Coefficient of friction

p Contact pressure (MPa)

F_N Normal load (N)

v Sliding speed (m/s)

l Semi-length of the sliding body parallel to sliding direction (m)

b Semi-width of the sliding body perpendicular to sliding direction (m)

Based on Equations (3.17) and (3.18), three models were developed:

1. Challen et al. (1976):

$$T^* = F_p \cdot F_N + F_{Tb} \cdot T_S \qquad (3.26)$$

where

$$F_P = \frac{1.65 F_S F_m \mu v^{\frac{1}{2}}}{F_S + F_m / v^{\frac{1}{2}}} \qquad (3.27)$$

and

$$F_{Tb} = \frac{F_S - 0.65 F_m / v^{1/2}}{F_S + F_m / v^{1/2}} \qquad (3.28)$$

and

$$F_S = \frac{1}{\pi a^2 k_S} \left[\left(\frac{Z_1}{Z_2}\right)^2 (Z_3 - Z_2) + \left(\frac{Z_1}{Z_2}\right)(Z_2 - Z_1) \right] \qquad (3.29)$$

and

$$F_m = \frac{\alpha_m^{1/2}}{3.25 k_m a^{3/2}} \qquad (3.30)$$

2. Samyn et al. (2008):

$$T^* = T_S + 4.2 * 10^{-4} \frac{\mu F_N v^{\frac{1}{2}}}{b\sqrt{l}} \qquad (3.31)$$

3. Zhang et al. (1998):

$$T^* = \frac{0.236 \mu F_N v}{2l(k_m + k_S)} \qquad (3.32)$$

where:

F_N Normal force

F_S Stationary body function

F_m Moving body function

T_a Ambient temperature

T_S Counterface temperature

T_b Bulk temperature

$Z_{1,2,3}$ Z coordinates of stationary body

a   Radius of contact area

b, l   Dimensions of stationary body

k_s Thermal conductivity of stationary body

k_m Thermal conductivity of moving body

v   Speed of moving body

α_S Thermal diffusivity of stationary body

α_m Thermal diffusivity of moving body

μ Coefficient of friction

The assumptions necessary for temperature calculation using previous models include various interfacial properties, which are usually unknown due to many difficulties in their exact determination. Crucial differences in the calculated flash temperature using the various models varied by almost 600°C using the same input parameters. Moreover, the agreement between the measured and predicted values of temperature rise at the sliding interface is generally lacking due to consideration of simplistic models to facilitate analytical solutions (Mitjan and Joze, 2001).

For polymers, mainly visco-elastic hysteresis losses in the near-surface region result in heat generation. The molecular mechanisms for conversion of mechanical energy into heat are dispersion and viscous flow. Another source of heat is the origination and breakdown of adhesive bonds between the polymer element and its counterface. These processes are energetically nonequivalent, and the energy difference causes the generation or absorption of heat (Ludema and Tabor, 1966).

3.10 Contact Temperature Measurement

In any tribosystem, most of the frictional energy is generally used up in plastic deformation which is directly converted to raise of the temperature at the interface. Although the transient temperature rise is generally calculated rather than measured, as discussed before. It is often necessary to measure interface (or surface) temperature experimentally. The interface temperature could be measured using one of the following techniques (Bhushan, 2013).

i) Contact temperature measurement using thermocouple sensors:

The thermocouple sensor involves wires of two dissimilar metals connected together so as to give rise to a thermal electromotive force (emf) potential which is a function of the difference in the temperature between the two junctions and is independent of the gradients in the wires. One junction is held at a known reference temperature (cold junction) and the temperature of the other measuring junction (hot junction) can be inferred by comparison of the measured total EMF with an empirically derived calibration table (Reed, 1982).

Generally, thermocouples are utilized to measure contact temperature in two techniques: Embedded and dynamic. In the embedded technique, a small hole is drilled through the stationary component of a sliding pair and extends just beneath its rubbing surface, as shown in Fig. 3.25. Also, thermocouples can be embedded in the moving component, but slip rings or a similar means will be required to gain access to the thermocouple output. Several thermocouples can be embedded at different depths and at various

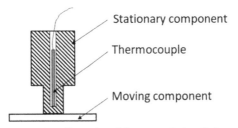

Figure 3.25: Illustration of thermocouple installation.

locations along the sliding path to monitor surface temperature distribution. A thermocouple can be made part of the sliding surface by placing it in a hole which extends to the surface and then grinding the thermocouple flush with the surface.

In the dynamic technique, a thermocouple junction is formed at the sliding interface by the contacting bodies themselves. Temperature rise at contact of dissimilar metals can be simply obtained from the measurement of the thermal emf potential produced at the interface. This technique was first developed to study contact temperatures at the interface between a cutting tool and workpiece during metal cutting. Currently, it is used to measure the surface temperature of bearing interfaces involving dissimilar metals. It is important to mention that the dynamic technique gives higher values of measured temperatures and faster transient response than embedded thermocouples. However, it can be used only in metallic tribosystems with dissimilar metals and also requires electrical contact with a moving body. Figure 3.26 is a schematic of the dynamic thermocouple technique used for the measurement of surface temperatures of two dissimilar metals.

ii) Non-contact measurement sensors:

Some of the most successful measurements of sliding contact temperatures have used techniques involving the detection of thermal radiation. If the temperature of the surface is high enough, radiation in the visible part of the spectrum can be detected (Bhushan, 2001). Several different radiation measurement techniques, including infrared thermocouples, pyrometry, photographic, metallographic, and photon detection, have been used with success in measuring surface temperatures.

Infrared Thermocouples are unpowered sensors that measure surface temperatures of materials without contact. They can be directly installed on conventional thermocouple controllers, transmitters and digital readout devices as if they were a replacement for thermocouples. Since the temperature is generally not constant over the field of view, the detector output is a function of the average temperature over the contact area. In order to improve the accuracy of the temperature measurement and to approach a point measurement, most modern detectors are equipped with optics which limit the field of view to a small spot size, around 100 to 500 μm diameter.

Pyrometers (or radiation thermometers) come in a variety of configurations. These devices are ideal for making point temperature measurements on circuit boards, bearings, motors, steam traps or any other device that can be reached with the probe.

Photographic technique utilizes infrared-sensitive film for studying sliding components. In these cases, the camera is focused on the moving component as it emerges from a sliding contact. The temperature distribution is best determined by measuring the optical density of the developed negative. The system must be calibrated in order to determine the density–temperature relationship of the film in the test configuration. Figure 3.27 shows a photograph of hot spots on a tool steel pin sliding at high speed on a sapphire disk. Temperatures of the spots were estimated to range from 950°C to 1200°C (Quinn and Winer, 1985).

In metallographic technique, an optical or scanning electron microscope is used to examine the microscopic changes in the surface temperature of the sliding component. This change in microstructure can be detected after metallurgical sectioning of the sliding body in a plane normal to the sliding direction. The technique gives only a crude estimate of the temperature rise and can be used for materials which

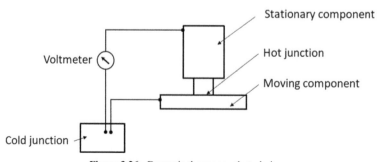

Figure 3.26: Dynamic thermocouple technique.

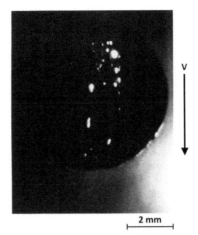

V

⊢ 2 mm ⊣

Figure 3.27: Photograph of hot spots on a steel pin sliding at 2 m/s on a sapphire disk with a load of 26 N. Photo taken after 25 min of sliding. Exposure time = 1 s. After Quinn and Winer (Quinn and Winer, 1985), with permission from Elsevier.

go through known changes in microstructure or micro-hardness at the temperatures expected in sliding (Wright and Trent, 1973).

References

Al-Samarai, R., Ahmad, H.K. and Al-Douri, Y. 2012. Evaluate the effects of various surface roughness on the tribological characteristics under dry and lubricated conditions for Al-Si alloy. Journal of Surface Engineered Materials and Advanced Technology 2(3): 167–173. Doi: 10.4236/jsemat.2012.23027.

Amontons, G. 1699. De la resistance causée dans les machines (About resistance and force in machines). Mem l'Academie R A: 257–282.

ASTM D4521-96, Standard Test Method for Coefficient of Static Friction of Corrugated and Solid Fiberboard (Withdrawn 2001), ASTM International, West Conshohocken, PA, 1996, www.astm.org.

Basu, B. and Kalin, M. 2011. Frictional heating and contact temperature. Tribology of Ceramics and Composites. Doi: 10.1002/9781118021668.ch6.

Bayer, R.G. 2002. Wear Analysis for Engineers. New York HNB Publishing.

Bayer, R.G. 2004. Mechanical Wear Fundamentals and Testing. New York: Marcel Dekker, Inc.

Behm, R.J. and Hösler, W. 1986. Scanning tunneling microscopy. pp. 361–411. *In*: Ralf Vanselow and Russell Howe (eds.). Chemistry and Physics of Solid Surfaces VI, Berlin, Heidelberg: Springer Berlin Heidelberg.

Beilby, G.T. 1903. Surface flow in crystalline solids under mechanical disturbance. Proceedings of the Royal Society of London 72: 218–225.

Bhushan, B. and Jahsman, W.E. 1978. Measurement of dynamic material behavior under nearly uniaxial strain condition. International Journal of Solids and Structures 14: 739–753.

Bhushan, B. 1996. Tribology and Mechanics of Magnetic Storage Devices. New York: Springer-Verlag.

Bhushan, B. and Ashok V. Kulkarni. 1996. Effect of normal load on microscale friction measurements. Thin Solid Films 278(1): 49–56. Doi: https://doi.org/10.1016/0040-6090(95)08138-0.

Bhushan, B. 2001. Modern Tribology Handbook. 2nd ed. ed: CRC Press.

Bhushan, B. 2013. Introduction to Tribology. Second edition ed. New York: John Wiley & Sons Inc.

Blau, P.J. 1992. Scale effects in sliding friction: An experimental study. pp. 523–534. *In*: Singer, I.L. and Pollock, H.M. (eds.). Fundamentals of Friction: Macroscopic and Microscopic Processes. Dordrecht: Springer Netherlands.

Blau, P.J. 2013. Friction measurement. pp. 1343–1347. *In*: Jane Wang, Q. and Yip-Wah Chung (eds.). Encyclopedia of Tribology, Boston, MA: Springer US.

Blok, H. 1937a. General discussion on lubrication. Proc. Inst. Me & Eng., London.

Blok, H. 1937b. Theoretical study of temperature rise at surface of actual contact under oiliness lubricating conditions. Gen. Disn. Lubn. Inst. Mech. Eng. 2: 222–235.

Bowden, F.P. and Hughes, T.P. 1937. Physical properties of surfaces. IV. Polishing, surface flow and the formation of the Beilby layer. Proceedings of the Royal Society of London. Series A, Mathematical and Physical Sciences 160(903): 575–587.

Bowden, F.P. and Tabor, D. 1950. The Friction and Lubrication of Solids. U.K.: Clarendon Press Oxford.

Challen, J.M. and Dowson, D. 1976. The calculation of interfacial temperature in pin-on-disc machine, The wear of non-metallic materials. The 3rd Leeds-Lyon Symposium on Tribology, MEP, The university of Leeds, London.

Chowdhury, M.A. and Maksud, H. 2007. The effect of frequency of vibration and humidity on the wear rate. Wear 262(1): 198–203. Doi: https://doi.org/10.1016/j.wear.2006.05.007.

Chowdhury, M.A. and Maksud, H. 2008. The effect of amplitude of vibration on the coefficient of friction for different materials. Tribology International 41(4): 307–314. Doi: https://doi.org/10.1016/j.triboint.2007.08.005.

Chowdhury, M.A., Khalil, M.K., Nuruzzaman, D.M. and Rahaman, M.L. 2011. The effect of sliding speed and normal load on friction and wear property of aluminum. International Journal of Mechanical & Mechatronics Engineering IJMME-IJENS 11(1): 45–49.

Cochrane, W. 1938. Polish on metals. Proceedings of the Royal Society of London. Series A, Mathematical and Physical Sciences 166(925): 228–238.

Coulomb, C.A. 1785. Théorie des Machines Simples, en ayant regard au Frottement de leurs Parties, et a la Roideur dews Cordages. Mem. Math. Phys., X, Paris: 161–342.

Davim, J.P. 2011. Tribology for Engineers. Cambridge, UK: Woodhead Publishing Limited.

Deng, Z., Klimov, N.N., Santiago, D.S., Li, T., Xu, H. and Cannara, R.J. 2013. Nanoscale interfacial friction and adhesion on supported versus suspended monolayer and multilayer graphene. Langmuir 29(1): 235–243. Doi: 10.1021/la304079a.

Dorinson, A. and Ludema, K.C. (eds.). 1985. Temperature effects in friction, wear and lubrication. Chapter 15. In Tribology Series, pp. 429–471. Elsevier. Doi.org/10.1016/S0167-8922(08)70854-X.

Dzierwa, A. 2017. Effects of surface preparation of friction and wear in dry sliding conditions. Tribologia 2: 25–31.

Edwards, C.M. and Halling, J. 1968. An analysis of the plastic interaction of surface asperities and its relevance to the value of coefficient of friction. Journal of Mechanical Engineering Science 10: 101.

Eissenberg, D.M. and Haynes, H.D. 1992. Motor-current signature analysis. pp. 313–318. *In*: Blau, P.J. (ed.). ASM Handbook, Friction, Lubrication, and Wear Technology. ASM International, Materials Park, OH.

Engineering ToolBox. 2004. Friction and friction coefficients for various materials. [online] Available at: https://www.engineeringtoolbox.com/friction-coefficients-d_778.html [11/10/2019].

Feng, D., Shen, M., Peng, X. and Meng, X. 2016. Surface roughness effect on the friction and wear behaviour of Acrylonitrile–Butadiene Rubber (NBR) under oil lubrication. Tribology Letters 65(1): 10. Doi: 10.1007/s11249-016-0793-5.

Floquet, A. 1983. Thermal considerations in the design of tribometers. Wear 88(1): 45–56. Doi: https://doi.org/10.1016/0043-1648(83)90311-3.

Ghabrial, S.R. and Zaghlool, S.A. 1974. The effect of surface roughness on static friction. International Journal of Machine Tool Design and Research 14(4): 299–309. Doi: https://doi.org/10.1016/0020-7357(74)90019-5.

Greenwood, J.A., Minshall, H. and Tabor, D. 1961. Hysteresis losses in rolling and sliding friction. Proc. R. Soc. Lond.

Greenwood, J.A. 1991. An interpolation formula for flash temperatures. Wear 150(1): 153–158. Doi: https://doi.org/10.1016/0043-1648(91)90312-I.

Halling, J. 1979. Principles of Tribology. London: Macmillan Press Ltd.

Hersey, M.D. 1969. Rolling friction, April 1969 pp. 260–275. Transactions of the ASME: 260–275.

Hersey, M.D. 1970. Rolling friction. Journal of Lubrication Technology: 83–88.

Hsu, S.M. 1996. Fundamental mechanisms of friction and lubrication of materials. Langmuir 12(19): 4482–4485. Doi: 10.1021/la9508856.

Jaeger, J.C. 1942a. Moving sources of heat and the temperature at sliding surfaces. Proc. R. Sot. NSW 66: 203–224.

Jaeger, J.C. 1942b. Moving sources of heat and the temperature at sliding contacts. Proc. Roy. Soc. NSW 76: 203–224.

Kalin, M. and Vizintin, J. 2001. Comparison of different theoretical models for flash temperature calculation under fretting conditions. Tribology International 34: 831–839.

Kuhlmann-Wilsdorf, D. 1987. Temperatures at interfacial contact spots: Dependence on velocity and on role reversal of two materials in sliding contact. Journal of Tribology 109(2): 321–329. Doi: 10.1115/1.3261361.

Kyllingstad, A. and Nessjøen, P.J. 2009. A New Stick-Slip Prevention System. SPE/IADC Drilling Conference and Exhibition, Amsterdam, The Netherlands, 2009/1/1.

Lancaster, J.K. 1971. Estimation of the limiting PV relationships for thermoplastic bearing materials. Tribology 4(2): 82–86. Doi: https://doi.org/10.1016/0041-2678(71)90136-9.

Ludema, K.C. and Tabor, D. 1966. The friction and visco-elastic properties of polymeric solids. Wear 9(5): 329–348. Doi: https://doi.org/10.1016/0043-1648(66)90018-4.

Ludema, K.C. 1996. Friction Wear and Lubrication. CRC Press.

Matsuzaki, A. 1970. Methods for preventing stick-slip. Bulletin of JSME 13(55): 34–42. Doi: 10.1299/jsme1958.13.34.

Myshkin, N. and Goryacheva, I. 2016. Tribology: Trends in the half-century development. Vol. 37.

Quinn, T.F.J. and Winer, W.O. 1985. The thermal aspects of oxidational wear. Wear 102(1): 67–80. Doi: https://doi.org/10.1016/0043-1648(85)90092-4.

Rabinowicz, E. 1966. Compatibility criteria for sliding metals. Friction and Lubrication in Deformation Processing, New York.

Rabinowicz, E. 1995. Friction and Wear of Materials. Second edition ed. New York: John Wiley and Sons.

Reed, R.P. 1982. Thermoelectric thermometry: A functional model. Temperature—its Measurement and Control in Science and Industry, New York.

Reizer, R., Galda, L., Dzierwa, A. and Pawlus, P. 2011. Simulation of textured surface topography during a low wear process. Tribology International 44(11): 1309–1319. Doi: https://doi.org/10.1016/j.triboint.2010.05.006.

Rothbart, H. and Brown, T.H. 2006. Mechanical Design Handbook, Measurement, Analysis, and Control of Dynamic Systems. New York: McGraw-Hill.

Rubenstein, C. 1956. A general theory of the surface friction of solids. Proceedings of the Physical Society. Section B 69(9): 921.

Salapaka, S.M. and Salapaka, M.V. 2008. Scanning probe microscopy. IEEE Control Systems Magazine 28(2): 65–83. Doi: 10.1109/MCS.2007.914688.

Samyn, P. and Schoukens, G. 2008. Calculation and significance of the maximum polymer surface temperature T* in reciprocating cylinder-on-plate sliding. Polymer Engineering & Science 48(4): 774–785. Doi: 10.1002/pen.21004.

Sankar, K.M., Kakkar, D., Dubey, S., Garimella, S.V., Goyat, M.S., Joshi, S.K. and Pandey, J.K. 2019. Theoretical and computational studies on nanofriction: A review. Part J: Journal of Engineering Tribology.

Sedlaček, M., Podgornik, B. and Vižintin, J. 2009. Influence of surface preparation on roughness parameters, friction and wear. Wear 266(3): 482–487. Doi: https://doi.org/10.1016/j.wear.2008.04.017.

Stachowiak, G.W. 2005. Wear—Materials, Mechanisms and Practice. U.K.: Wiley.

Tian, X. and Kennedy, F.E. Jr. 1994. Maximum and average flash temperatures in sliding contacts. Journal of Tribology 116(1): 167–174. Doi: 10.1115/1.2927035.

Tomlinson, G.A. 1929. A molecular theory of friction. Phil. Mag. 7th series (7): 905–930.

Uetz, H. and Föhl, J. 1978. Wear as an energy transformation process. Wear 49(2): 253–264. Doi: https://doi.org/10.1016/0043-1648(78)90091-1.

Whitehead, J.R. 1950. Surface deformation and friction of metals at light loads. Proceedings of the Royal Society of London. Series A. Mathematical and Physical Sciences 201(1064): 109.

Wright, P.K. and Trent, E.M. 1973. Metallographic methods of determining temperature gradients in cutting tools. Journal of Iron Steel Institute 211: 364–388.

Yamaguchi, K., Chiaki, S., Tsuboi, R., Atherton, M., Stolarski, T. and Sasaki, S. 2014. Effect of surface roughness on friction behaviour of steel under boundary lubrication. Proceedings of the Institution of Mechanical Engineers, Part J: Journal of Engineering Tribology 228(9): 1015–1019. Doi: 10.1177/1350650114540624.

Zayed, A. 1997. Summary for Engineers. Alexandria: Delta.

Zhang, M.Q., Song, L., Zeng, H.M., Friedrich, K. and Karger-Kocsis, J. 1998. Frictional surface temperature determination of high-temperature-resistant semicrystalline polymers by using their double melting features. Journal of Applied Polymer Science 63(5): 589–593. doi: 10.1002/(SICI)1097-4628(19970131)63:5<589::AID-APP6>3.0.CO;2-Q.

Zuleeg, J. 2015. How to Measure, Prevent, and Eliminate Stick-Slip and Noise Generation with Lubricants. Edited by SAE International.

Chapter 4

Wear

4.1 Introduction

The Committee of the Institution of Mechanical Engineers (IMechE) has defined wear as "the progressive loss of substance from the surface of a body brought about by mechanical action" (Halling, 1979). According to a definition provided by Kragelskii (Kragelskii, 1965), wear is defined as "the destruction of material produced as a result of repeated disturbances of the frictional bonds". Both definitions are satisfactory.

As a natural phenomenon, wear generally occurs when two surfaces with a relative motion interact with each other. Contrary to popular belief, wear is not necessarily a harmful phenomenon, i.e., a controlled wear is desirable in some situations. The evidence of beneficial wear can be clearly seen in the case of polishing, grinding and many other forming operations. Although the amount of material removed by wear is usually quite small, it is considered as one of the important reasons of failure in many mechanical parts.

Modern researches have recognized four main types of wear: Adhesive, Abrasive, Surface fatigue and Corrosive wear. Furthermore, erosion, cavitation, galling, thermal, fretting and impact wear can be considered as minor types of wear. Wear is generally a harmful phenomenon that cannot be eliminated, but can be controlled by careful design and material selection, or by proper lubrication. Wear is an integral part of tribology, thus, in order to achieve a consistent operation of machinery, an understanding of wear is important.

4.2 Wear Mechanisms

A general classification of wear mechanisms (or types) is still an open issue. An earlier research (Burwell, 1957) had established that the wear of metals can be divided into four main groups, in which several distinct and independent mechanisms are involved: Adhesive, abrasive, surface fatigue and corrosive wear. Each wear mechanism is governed by its own laws and, in many occasions, this mechanism may act in such a way as to affect the others. It is important to emphasize that it is not always easy to differentiate between these types of wear, as they are inter-related and rarely occur separately. Other types of wear, such as erosive, fretting, galling, thermal, impact, cavitation, impact, electro-arcing and biocorrosion wear are also considered as types of wear. According to some estimates, two-thirds of all wear encountered in industrial situations takes the form of adhesive and abrasive wear mechanisms (Bhushan, 2013). Wear by all mechanisms, except for the fatigue mechanism, occurs by the gradual removal of material.

4.2.1 Adhesive Wear

This mode of wear is considered to be the most common form of wear in mechanical systems. It occurs whenever two surfaces are in sliding contact or pressed together, whether lubricated or not. Usually, adhesion occurs at asperity junctions, followed by plastic shearing of the tips of the softer asperities. The removal of material takes the form of small particles which adhere to the other surface, as illustrated in Fig. 4.1. As the sliding continues, the transferred fragments may come off the surface to which they are adhered and be transferred back to the original surface, or separate as wear debris particles. Wear debris formed in this type of wear are characterized by relatively large particles composed of two surfaces (Sasada, 1979). Furthermore, wear rates from this adhesive mechanism should be very sensitive to the presence of contamination layers or a lubricant film (Paulo Davim, 2011).

Several mechanisms have been proposed for the detachment of a fragment of a material. Archard (Archard, 1953) suggested that if the force required to shear through the interface of the materials is larger than the force required to shear through one of the bulk materials, the shearing will occur at the original interface or in the weakest region in one of the two materials, as shown in Fig. 4.2.

In most cases of dissimilar material combinations, wear particles are formed from the softer material. However, fragments of the harder counterpart are also formed. This suggests that there are local regions of low strength within the harder material. Also, fragments of the harder material may be produced by the fatigue process which takes place after a number of repeated stresses. Wear debris particles produced by adhesive wear usually show a fragment pulled off one surface and adhered to the other, as shown in Fig. 4.3 (Dwivedi, 2010).

Experimental results for unlubricated sliding of similar and dissimilar metal combinations suggested the following quantitative laws of adhesive wear (Holm, 1946):

1. Wear volume V is generally directly proportional to the applied load W.
2. Wear volume V is generally directly proportional to the sliding distance x.
3. Wear volume V is generally inversely proportional to the hardness H of the worn surface.

Based on the above quantitative laws, wear equation for adhesive wear is given by the following expression:

$$V = k\frac{Wx}{H} \tag{4.1}$$

where k is a nondimensional wear coefficient dependent on the materials in contact and their cleanliness. In fact, the above equation is not always perfectly obeyed, however, in almost all cases, it reasonably represents the experimental data (Rabinowicz, 1995). Depending on the operating conditions, the value of k typically ranges from 10^{-8} to 10^{-4} for mild wear and from 10^{-4} to 10^{-2} for severe wear for most material combinations (Bhushan, 2013).

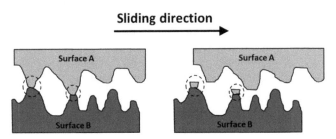

Figure 4.1: The mechanism of adhesive wear.

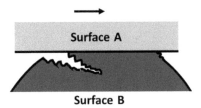

Figure 4.2: Shearing occurs in the weakest region of the two materials.

Figure 4.3: A micrograph showing typical morphology of an adhesive wear. After Dwivedi (Dwivedi, 2010) with permission from Elsevier.

A hypothetical model of generation of a hemispherical wear particle during a sliding contact, shown in Fig. 4.4, was introduced by Archard (Archard, 1953) in order to derive Equation 4.1.

He considered that the contact is made up of asperities with an average radius of a, thus, at each asperity junction the area of contact is πa^2 and supports a load of $H\pi a^2$, where H is the hardness of the softer material (Tabor, 2000). Therefore, for a total number n of junctions present at any instant during sliding, the total applied load will be given by

$$W = Hn\pi a^2 \qquad (4.2)$$

Note that, by definition, the term $n\pi a^2$ represents the real area of contact A.

Now, assume that the surface will completely pass over each asperity in a sliding distance of $2a$ and that the wear debris formed at each asperity interaction is of a hemispherical shape of volume $2/3\pi a^3$. Also, consider that all junctions are of the same size, then the total wear volume per unit distance of sliding will be

$$V = \frac{n\left(\dfrac{2}{3}\pi a^3\right)}{2a} = \frac{1}{3}n\pi a^2 \qquad (4.3)$$

From Equations 4.2 and 4.3, the total wear volume can be given by

$$V = \frac{W}{3H} \qquad (4.4)$$

Archard assumed that there was a constant probability k_c that any asperity junction will produce a wear fragment. According to Equation 4.4, the total volume of transferred fragments formed in sliding through a distance x can be given by the relation

$$V = \frac{k_c}{3}\frac{Wx}{H} \qquad (4.5)$$

The worn volume obtained from Equation 4.5 is identical to Equation 3.26, except that the wear coefficient k is replaced by the arbitrary constant $k_c/3$. In fact, the factor three, which somehow mysteriously appears in the denominator, is a shape factor applicable in Equation 4.5 to the assumption of hemispherical

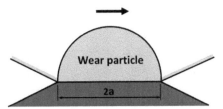

Figure 4.4: A hypothetical model of a hemispherical wear particle generated during a sliding contact.

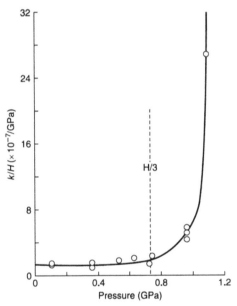

Figure 4.5: Wear coefficient/hardness ratio as a function of the average pressure for SAE 1095 steel. After Burwell (Burwell, 1957) with permission from Elsevier.

wear debris (Rabinowicz, 1995). For the assumption of a cubical wear debris, the factor would be one and so on. Nonetheless, Equation 4.5 agrees with the laws of adhesive wear, as mentioned before. Therefore, the physical meaning of the term $k_c/3$ can be considered as the wear volume fraction at the plastic contact zone, which is affected by material properties and the geometry of the asperity contact zone. Moreover, the interesting results obtained from Burwell and Strang (Burwell and Strang, 1952) while studying steel rubbing against steel postulated that k_c remains constant up to a pressure of about $H/3$. Figure 4.5 indicates that $H/3$ is the pressure at which the plastic zone under individual asperities begins to interact. Any increase in the pressure above $H/3$ causes the whole surface to become plastic so that the real area of contact is no longer proportional to the load. That is, the yield pressure equals $H/3$.

4.2.2 Abrasive Wear

Abrasive wear is probably the most distinct type of wear in daily life, in addition to most mechanical applications. This form of wear mostly occurs in tribo-systems of two rubbing surfaces which are different in hardness. The obvious surface topography feature of abrasive wear is that ploughing takes place during sliding, resulting in removal of surface material. Consequently, long abrasive grooves are formed on the softer (ductile) surface which usually runs in the sliding direction, as shown in Fig. 4.6.

In the case of brittle materials with low fracture toughness, wear occurs by brittle fracture (Bhushan, 2013). In these cases, the worn zone includes significant cracking.

Figure 4.6: Single abrasive wear track of Ultrahigh Strength Bainitic Steel. After Narayanaswamy et al. (Narayanaswamy, 2016), with permission from Elsevier.

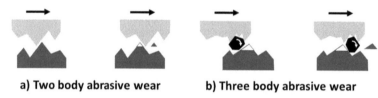

a) Two body abrasive wear   **b) Three body abrasive wear**

Figure 4.7: Schematic diagram for (a) Two body and (b) Three body abrasive wear.

The "abrasive wear" term refers to two distinct wear mechanisms: (1) Two-body abrasion, and (2) Three-body abrasion. Both situations are characterized by the ploughing-out of a softer material surface by a harder one, see Fig. 4.7.

(1) Tow-body abrasion wear

This occurs under two conditions; one of the rubbing surfaces must be harder than the other, and the harder surface must be rough. In such a case, asperities of the rough and hard surface slide against the softer one, dig into it, and damage the interface by a plastic deformation or a fracture. Examples of this wear mechanism are found in various machining operations, such as grinding and metal cutting.

(2) Three-body abrasion wear

This mechanism of abrasive wear is still of great significance. It arises when hard, abrasive particles are introduced between sliding surfaces and abrade either one of the mating surfaces or both of them. This form also includes abrasion by wear debris generated by other wear mechanisms. There are two conditions for this mechanism to take place. First, the abrasive particle must have considerable dimensions comparable to the gap between sliding surfaces. Second, the abrasive particle must be harder than at least one of the mating surfaces. Examples for this form of abrasive wear are found in free-abrasive lapping and polishing.

In order to deduce a quantitative law for abrasive wear, we assume a single contact point model where a hard conical asperity is indented against the flat surface forming a groove on it by ploughing, as illustrated in Fig. 4.8.

Assume that the abrasive asperity is subjected to a normal load W, has a projected area in the horizontal plane equal to πr^2, and H is the hardness of the softer surface. The change in normal load (dW) required to cause yield of the material would be expressed as:

$$dW = \frac{1}{2}\pi r^2 H \tag{4.6}$$

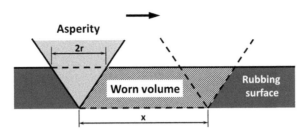

Figure 4.8: A simple model for abrasive wear showing a single contact of a hard conical asperity.

According to Fig. 4.8, the possible wear volume dV, which is ploughed by harder asperities after sliding a distance of x, is given by

$$dV = x\, r^2 \cdot \tan\theta \tag{4.7}$$

Equations 4.6 and 4.7 altogether give

$$dV = \frac{2dWx \cdot \tan\theta}{\pi H} \tag{4.8}$$

Integrating Equation 4.8 for contributions of all asperities, we will have

$$V = \frac{2\overline{\tan\theta}}{\pi} \cdot \frac{Wx}{H} \tag{4.9}$$

where roughness factor $\overline{\tan\theta}$ is an average of the slope values of all the individual asperities.

This relation is quite similar to Archard Equation 4.5 for adhesive wear, except that the factor $k_c/3$ is replaced by the term $\frac{2\overline{\tan\theta}}{\pi}$. In order to accommodate abrasive wear in terms of Equation 4.5, a parameter k_{abr} is introduced in Equation 4.9 as a nondimensional abrasive wear coefficient. This factor includes the geometry of the asperities, which depicts the effect of surface roughness on the abrasive wear mechanism. Finally, the volume removed by abrasive wear is given by

$$V = k_{abr} \frac{Wx}{H} \tag{4.10}$$

Typical experimental results suggested that the values of k_{abr} in two-body situations are in the range of 10^{-2} to 10^{-1}, whereas, in three-body situations, they are about 10^{-3} to 10^{-2} (Rabinowicz, 1995). Besides, the coefficient of friction during three-body abrasion is generally less than that in two-body abrasion, by as much as a factor of two. Based on this, two important remarks are drawn:

a) The abrasive grain in a three-body situation tends to roll rather than sliding and abrading the surface.

b) The rate of abrasive wear is generally two to three orders of magnitude larger than the adhesive wear.

4.2.3 Surface Fatigue Wear

This mechanism of wear is closely associated with the general fatigue phenomenon of materials where there is a correlation between the contact stress and the number of cycles required to failure. However, there are two differences between surface fatigue wear and ordinary fatigue; first, the high fluctuations in the life in fatigue wear applications compared to ordinary fatigue cases. Second, there is no fatigue limit stress in surface fatigue wear as in ordinary fatigue testing.

Practically, there are two common situations of wear by fatigue fracture: In rolling applications (e.g., rolling bearings, gears and cams), and in sliding on brittle surfaces (e.g., ceramic materials), see Fig. 4.9.

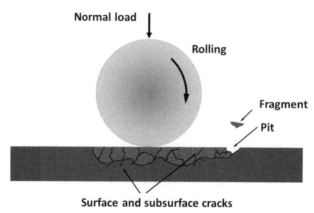

Figure 4.9: Schematic diagram for fatigue fracture in rolling motion.

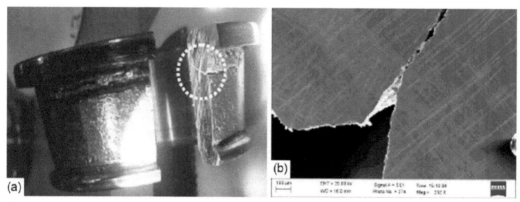

Figure 4.10: Images taken for a section of a bearing inner cone: (a) optical photograph of the cone section starting beneath the deep groove and propagating towards the bore surface encircled; (b) SEM image showing the origination of crack from the deep groove. After Gurumoorthy et al. (Gurumoorthy, 2013), with permission from Elsevier.

In such applications, during operation, under the applied normal loading, the contacting components are elastically deformed rather than plastically. Subsequently, stressing and unstressing the surface in the rolling or sliding track continues to take place. The repeated loading cycles may induce the formation of subsurface or surface cracks. After a critical number of cycles, these cracks will suddenly result in a breakup of the surface, forming large fragments and leaving large pits in the surface. Thereafter, deterioration of the surface by spalling or flaking becomes rapid, as shown in Fig. 4.10.

The occurrence of subsurface cracks is probably due to the fact that the point with maximum shear stress, and hence, the point with the maximum tendency for plastic yielding, is located a small distance below the surface, as shown in Fig. 4.11. Davies (Davies, 1949) reported that the higher the friction force between a rolling element and a mating surface, the closer to the surface the zone of maximum shear stress is. Therefore, at some point when the friction force is sufficiently high, the zone of maximum shear stress occurs at the mating surface. The location of the maximum stress is very important because it determines the formation of wear particles (Buckley, 1981). This would clarify the fact that surface fatigue wear can be generated at the surface as well as subsurface regions.

It is clear that surface fatigue wear in a rolling contact is distinguished by the formation of large fragments after a critical number of rolling cycles. Accordingly, typical experimental investigations of surface fatigue wear have considered running rolling contact devices until spalling occurs, and noting the rolling cycles (time) to spalling (Rabinowicz, 1995). Testing of a large number of rolling contact devices

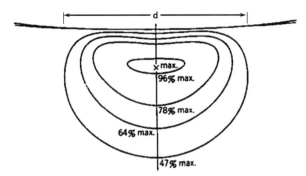

Figure 4.11: Position of maximum stress for elastic contact of a sphere and a flat surface (Davies, 1949).

has postulated that the critical value of high-cycle fatigue N_f, as defined above, is given by the following empirical relation (Lundberg and Palmgren, 1952):

$$N_f = bW^{-n} \tag{4.11}$$

where W is the normal load and b and n are experimental constants. The value of n depends on the shape of the rolling element. In the case of rolling ball bearings, the value of n is around 3.

Equation 4.11 has been broadly recognized in the design of rolling bearings. Its basic idea is that spalling or flaking can be treated as a statistical fracture phenomenon following the modified Weibull theory (Weibull, 1930).

In sliding, surface fatigue wear takes another form, probably resulting from repeated stress cycles applied to the surface with asperity interactions. During sliding, asperities can make contact without adhering or abrading and pass each other, leaving one or both of them plastically deformed. After a critical amount of such contact, the resultant cyclic plastic shear deformation leads to a subsurface crack initiation in the material surface. Further deformation may also lead to a crack propagation and coalescence parallel to the surface. Both surface and sub-surface cracks which open due to the repeated stressing will gradually grow, join, cross each other and meet the surface until wear debris are detached after a certain number of the stress cycles (Suh, 1980). The nature and number of crack initiation sites in the surface depend on the type of loading exerted on the surface as well as on the sliding. Although the mechanism of surface fatigue wear appears to be more convincing than that of the adhesion assumption, no satisfactory fatigue wear theory has been widely accepted yet.

As a final remark, we would like to draw the attention to a fact that, in some applications, corrosive conditions may seriously accelerate the surface fatigue wear of bearing. This may require the selection of a material that is less resistant to dry fatigue (Ludema, 1996).

4.2.4 Corrosive Wear

Corrosive wear (or Tribo-corrosion) is a dangerous form of corrosion because it leads to a reduction in a material's surface, so the failure can occur without a clear warning. It is characterized by a material degradation resulting from simultaneous action of wear and corrosion. Under these conditions, the material selection is a challenge, since the material has to effectively withstand wear, corrosion and their combined effect (Anaee and Abdulmajeed, 2016; Robert, 2007). Bio-tribocorrosion occurs in a biological environment where the presence of biological species creates further complexity in understanding the materials' degradation pathways and the dominant driving mechanisms generated by the interplay of wear and corrosion. On the positive side, the tribo-corrosion phenomena can be used as a manufacturing process, such as in the chemical–mechanical polishing of silicon wafers.

Corrosive wear occurs in situations in which the environment surrounding a rubbing surface chemically interact with it. Or simply, it is the case of wear in which sliding takes place in a corrosive medium, see Fig. 4.12. In atmospheric sliding, the most dominant corrosive medium is oxygen, and corrosive wear of metals in air is generally called "oxidative wear".

Figure 4.12: Corrosion wear in helical gear.

The mechanism of corrosive wear is extremely complex and not well understood. It involves the properties of contacting surfaces, the mechanics of the contact and the corrosion conditions. In the first stage, tribo-chemical reaction produces a film (e.g., oxides), typically less than a micrometer thick, on the material surfaces. This tends to slow down or even inhibit the corrosion. If such films strongly adhere to the surface and behave like the bulk material, the wear mechanism will be almost the same as that of the bulk material. However, practically, the sliding action wears away the chemical film resulting in a bare surface, so that the chemical attack can continue. Thus, chemical wear requires both corrosion and rubbing.

The corrosive wear rate is governed by the growth rate of the reaction film as well as the material removal rate. Therefore, models of the reaction layer growth and those of the layer removal become very important. In order to define the corrosive wear coefficient k_c, a model has been introduced by Quinn (Quinn, 1987) based on a case of an oxidative reaction between steel and normal atmospheric air. He assumed that an oxide film of steel is supposed to detach from the surface at a certain critical thickness, therefore

$$k_c = \frac{Ax}{\xi^2 \rho^2 v} \exp\left(-\frac{Q}{R_g T}\right) \tag{4.12}$$

where A is the Arrhenius constant, x is the distance along which a wearing contact is made, Q is the activation energy, R_g is the gas constant, T is the absolute temperature, ρ is the density of oxide, and v is the sliding velocity.

In general, it is supposed that the activation energy does not vary significantly between static and sliding conditions. Based on this assumption, the experimental wear results for the oxidation of steel during sliding gave the value of the Arrhenius constant in Equation (4.12) which is 10^3 to 10^{10} times larger than in static oxidation. The same assumption was made for the critical oxide film thickness for its self-delamination. It was successfully used to model the oxidation wear of steel, which was used to construct the wear map of unlubricated steel-steel sliding (Kato, 2002).

4.2.5 Fretting Wear

The word "fretting" was first introduced in 1927 by Tomlinson (Tomlinson, 1927) in order to refer to a small amplitude of oscillation ranging from a fraction of a microns to hundreds of microns. Fretting wear is a form of wear taking place at contacting surfaces when they undergo a low-amplitude oscillatory motion in the tangential direction. Since this form of wear is conjugated to vibration, it is commonly found in most machinery, bolt joints, shrink fits, splined couplings, transformer cores, between prosthetic hip joints and

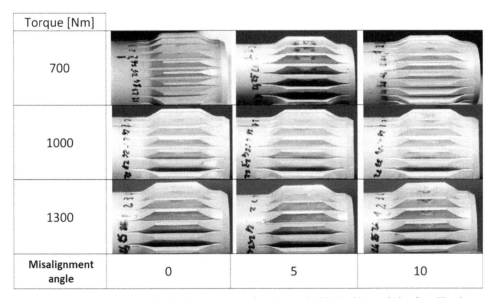

Torque [Nm]			
700			
1000			
1300			
Misalignment angle	0	5	10

Figure 4.13: Spline coupling after fretting wear test. After Curà et al. (2017) with permission from Elsevier.

bone, and in many other cases. Ultimately, fretting is a form of adhesive or abrasive wear where the applied normal load causes adhesion between asperities, while the low-amplitude oscillatory motion causes a rupture of the formed joints (Halling, 1979; Bhushan, 2013). However, fretting is considered as a special case of fatigue wear at the surface (Stachowiak, 2005). A reciprocating friction load produces surface stresses that can result in cracks and fretting fatigue. Consequently, fretting wear appears when cracks at the surface result in wear particles. Furthermore, fretting wear can cause a formation of surface stress raisers and, if the vibratory stresses are high enough, fatigue cracks propagate, leading to a complete failure.

Figure 3.20 shows fretting wear damage appearing on splined couplings because of the relative motion between teeth, mainly due to angular misalignments of these components (Curà and Mura, 2017). In particular, the right column of the figure shows the highest level of the wear damage with 10° misalignment. The central column shows specimens after tests performed with 5° misalignment and the left column represents tests performed without misalignment (in these last cases, no wear damage has been detected on the teeth surfaces).

In most cases, fretting is combined with corrosion; in such a case, the wear mode is known as fretting corrosion. For example, in the case of steel joints, chemical corrosion produces a fine reddish-brown oxidize powder, which is known as "cocoa". These oxide particles are abrasive. Because of the close fit of the surfaces and the oscillatory small amplitude motion, the surfaces are never brought out of contact. Therefore, there is a little opportunity for the products of the action to escape and a further oscillation causes abrasive wear and oxidation.

The fretting wear rate is directly proportional to the normal load for a given slip amplitude. In the low-frequency range, the frequency of oscillation has a minor effect on the wear rate per unit distance. On the other hand, the increase in the strain rate at high frequencies leads to an increased fatigue damage and an increased corrosion due to the rise in temperature. However, in a total-slip situation, the frequency has little effect (Waterhouse, 1992; Waterhouse, 1984).

There are various design changes which can be carried out to minimize fretting wear (Fridrici et al., 2003), such as:

1) Eliminating the oscillatory movement at the interface by increasing the normal load.
2) Modifying the frictional conditions by means of a solid lubricant coating, such as MoS_2, or a soft electroplated metal.

3) Copper-based thick coatings can be beneficial in fretting wear contacts.
4) Thermomechanical treatments can also be beneficial, especially if compressive residual stresses are induced.

4.2.6 Impact Wear

Impact wear (or percussive wear) is the wear of a solid surface due to percussion, which is a repetitive exposure to impact by another body. Repeated impacts result in a progressive loss of the solid material. Impact wear occurs by hybrid wear mechanisms which combine several of the following mechanisms: Adhesive, abrasive, surface fatigue, fracture and tribochemical wear.

Although impact wear is a source of failure in many industrial situations, it has generally not been studied as extensively as other wear mechanisms. As a result, actual impact wear data is quite scarce. This form of wear can be found, for example, in jaw crusher plates, railway points and excavator buckets. In fact, there are many other industrial situations in which impacting bodies are employed where failures due to impact wear can be costly. Moreover, safety, reliability and quality are greatly reduced. Excessive wear of impacting poppet valves in automotive engines can lead to a loss of the cylinder pressure and ultimately an engine failure. Failure of tools used for drilling rocks and other media raise cost concerns, not only associated with the need for a frequent replacement of parts, but also in the down time incurred. Potential impact wear problems are also being found in dental implants and heart valves, where health and well-being are at stake. Impact motion also widely exists in industrial machinery applied in the mining industry, metallurgy and electric power. Those impacts in the machines cause dynamic loads in the machine elements, such as the contact between a rolling bearing and its raceway, a gear transmission with clearance, and a cam-tappet mechanism. In the worst case scenario, the rapid variation of a load causes the occurrence of impact wear and a consequent failure, resulting in huge financial losses (Wang et al., 2015; Sarker, 1976; Bhushan, 2013).

In order to find some wear resistant materials or structures by adapting some special physical and chemical methods, several researches were introduced. Heat treatment and surface coating are likely the most effective methods that can enhance the impact and abrasion resistance with some considerations. In impact contact, a coated surface must possess high toughness in order to absorb the impacts that result in considerable instantaneous and repeated stress fields. The coating must also be sufficiently elastic to be able to accommodate any substrate deformation that may occur under the impact.

Many studies reflect the situation of the contact conditions in a dry contact and there have been fewer investigations on lubricated impact wear. In industry, lubricated shafts or rolling element bearings suffer failure because impact wear is very common. High-viscosity oil can effectively modify the impact wear and plastic deformation. However, if the oil supply is poor, local direct contact inside the dent takes place and, therefore, a deep and thin pit is formed. Under the same oil supply condition, the heavier the load, the deeper the local pit (Wang et al., 2015). In dry contact, spherical particles are formed inside the impact dent. On the other hand, in lubricated dents, especially in those with high viscosity oil, direct contact occurs and the spherical particles appear in clusters and in different sizes.

An expression identical to Archard's law (see Equation 4.5) can be obtained by recognizing that the term Wx represents sliding energy E_s. Consider n identical particles of mass m impacting a surface with the same velocity v. Their individual kinetic energy E_k is $1/2mv^2$ of the total kinetic energy of the particles. Only a fraction β results in material removal. Therefore, the impacting energy E_i resulting in wear is given by

$$E_i = \beta E_k = \beta \frac{1}{2} m v^2 n \tag{4.13}$$

Therefore, substituting this impact energy instead of the sliding energy into Archard's equation yields

$$V = k_i \frac{E_k}{H} \tag{4.14}$$

The impact wear coefficient k_i in this case can then be interpreted as a fraction of all impact events that results in material removal.

4.2.7 Erosive Wear

The term erosion may designate abrasive wear situations when only one surface is involved. The erosion process refers to the continuous loss of a solid surface due to mechanical interactions between the surface and a fluid. Abrasive erosion, slurry erosion, cavitation erosion, fluid erosion and spark erosion are the common types of erosive wear. They are caused by the impact of solid particles, liquid droplets, bubbles or electrical sparks (Stachowiak, 2005).

Abrasive erosion has been defined as the process of metal removal due to impingement of sharp solid particles on a surface, as illustrated in Fig. 4.14. As a result of the repeated impacts of sharp particles, wear debris are formed. It is clear that the erosive wear usually occurs if the particle hardness is greater than the material hardness. This is possibly the most important type of erosive wear and receives great attention, especially in the aerospace industry.

Erosion wear can be considered as a combination of impact and abrasive wear mechanisms. The main differences are that, first, the surface roughness produced during erosion may become relatively great, because an impinging particle may readily remove material from a low point (valley) on the surface. Second, the contact stress in erosion arises from the kinetic energy of particles flowing in an air or liquid stream as they encounter a surface. Although this type of wear is undesired in many situations, such as gas turbine blades and refractories in an electric arc melting furnace, it is deliberate in sand blasting, abrasive deburring and erosive drilling of hard materials.

The mechanism of erosion can be divided into three types (Anaee and Abdulmajeed, 2016):

a. Impingement corrosion, which occurs in two-phase or multiphase flow, as shown in Fig. 4.15a.
b. Turbulence erosion, which occurs in the regions of strong turbulence, such as the inlet end of heat exchanger tubes, as shown in Fig. 4.15b.
c. Removal of products leads to an increase in the corrosion by wearing the particles moving along the corroded surface, or by wearing between components in moving contacts with each other.

Since erosive wear is governed by the kinetic energy of hard particles impinging on surfaces, it is found that the erosion ratio, E (the mass of material removed divided by the mass of erosive particles striking the surface), depends on some factors contributing to the impact process (Divakar et al., 2005). These factors include impact angle, impact velocity and relative hardness of the material and incident particles, therefore

$$E = cv^n f(\alpha) \tag{4.15}$$

where v is the particle velocity, α is the impact angle with respect to the material surface and the coefficient n varies from ductile to brittle materials. In general, n ranges from 2 to 2.5 for metals and 2.5 to 3 for ceramics, while v usually ranges from 15 to 170 m/s (40 to 150 mph).

It was also suggested that the erosion loss rate increases approximately by (particle size)^m where $m \approx 3$. It has also been found that n is proportional to the particle size, perhaps due to fragmentation (Ludema, 1996; Paulo Davim, 2011). Although Equation (4.16) relates the erosion rate to the impact angle and particle velocity, it does not include the effect of particle hardness.

Hutchings (2017) derived a quantitative equation for determining the erosion ratio E based on the assumption that particles do not deform and the deformation of the surface is perfectly plastic with a constant indentation hardness, H. The erosion wear equation is then written as:

$$E = \frac{k\rho v^2}{2H} \tag{4.16}$$

where ρ is the density of the material being eroded, v is the particle velocity, and k is the proportion of the displaced material resulting as wear debris.

Similar to abrasive wear equations, the volume of erosive wear is inversely proportional to the hardness. The normal load in abrasive wear is replaced by kinetic energy (ρv^2) in erosive wear. However, Equation 3.36 does not include the effect of the impact angle and the shape and size of the particles. In fact, k depends on the impact angle, shape and size of the particles. The value of k typically ranges from

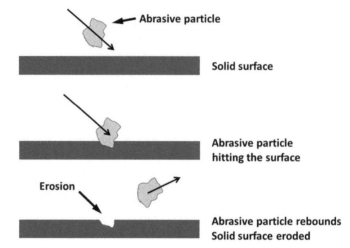

Figure 4.14: Schematic of erosive wear showing an abrasive particle hitting a surface.

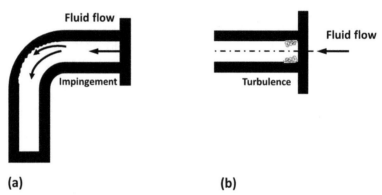

Figure 4.15: Erosion wear: (a) Impingement and (b) turbulence erosion.

10^{-5} to 10^{-1}. In case of brittle materials, erosive wear rate additionally depends on the fracture toughness of the material being eroded (Hutchings, 1992).

As mentioned above, erosive wear dependence on the impact angle is different for ductile and brittle materials. This is shown in Fig. 4.16 (Bitter, 1963).

In the case of a ductile solid, the material removal occurs by cutting and ploughing, similar to abrasion. Maximum erosion is found at an impact angle of about 20° to 30° and good erosion resistance to normal impact is found at low angles. However, at angles close to 90°, a fatigue mechanism is probably predominant (Halling, 1979).

On the other hand, in brittle solids, material will be removed by the formation and intersection of cracks that radiate out from the point of impact of the eroded particle. Brittle erosion is through flake fragmentation and removal of flakes. Brittle solids suffer severe erosion at normal impact angle ($\alpha = 90°$) and offer good erosion resistance at low angles (glancing impact). These relationships can be used to explain a number of erosive wear cases and are particularly useful in wear prediction.

Slurry erosion is generally defined as a mechanical interaction in which material is lost from a surface which is in contact with a moving particle-laden liquid. In fact, though there is no clear difference between abrasive erosion and slurry erosion, the terms often have different uses. Abrasive erosion may refer to low concentrations of solid in liquid, or it may refer to unknown concentrations. On the contrary, slurry erosion occurs when a solid-liquid mixture, specifically known as a "slurry", causes erosion. Generally,

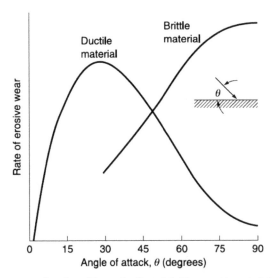

Figure 4.16: Rate of erosive wear as a function of the angle of attack (with respect to material plane) of impinging particles.

such a mixture is called a "slurry" when the solid phase is the focus of attention and the liquid is simply the carrier (Ludema, 1996).

Slurry erosion occurs in extruders, slurry pumps and pipes carrying a slurry of minerals and ores in mineral processing industries. The life of components used under slurry abrasion conditions is governed by the process parameters, properties of the abrasive particles in the slurry and the material properties. Pumps for moving slurries through pipelines may wear fast: Slurries pass through small gaps in the pumps at speeds of 100 m/s, sometimes even faster.

Cavitation erosion occurs in a situation when liquids (e.g., lubricating fluids) flow parallel to a metallic surface. Under a condition of tensile stress, the liquid may boil. Consequently, if this liquid is subjected to compressive stresses, i.e., to higher hydrostatic pressures, these bubbles will collapse very quickly, as shown in Fig. 4.17.

Furthermore, the formation of vapor bubbles in a lubricating fluid results in a formation of cavities on the metallic surface. Cavitation damage generally occurs in components such as marine propellers, dam slipways, steam control valves and centrifugal pumps, as shown in Fig. 4.18.

Fluid erosion (or liquid impingement erosion) is a process in which a material wears when small drops of liquid strike the surface of a solid at high speeds (300–1000 m/s). In such circumstances, a very high pressure, which exceeds the yield strength of most materials, is experienced. Therefore, a plastic deformation or a fracture can result from a single impact, and repeated impacts lead to pitting and erosive wear.

Fluid erosion can occur at locations where a working fluid attacks a pipe wall at a high flow velocity. The damage due to this action occurs downstream of elbows and T-tubes, and just behind orifices and valves. The most effective parameter for liquid impingement erosion is the flow velocity. Thus, it is theoretically possible to prevent the occurrence of liquid impingement erosion by operating power plants at low flow velocities. However, this is not realistic from the viewpoint of power generation efficiency.

It was postulated that the high-velocity impact of a liquid drop against a solid surface produces a high contact pressure in the impact region (Haymann, 1992). This is followed by a liquid jetting flow along the surface, radiating out from the impact area. In ductile materials, a single intense impact may produce a central depression, with a ring of plastic deformation around it where the jetting-out flow may remove the material by a tearing action. In brittle materials, circumferential cracks may form around the impact site caused by tensile stress waves propagating outward along the surface. Subsequent impacts can result in a material spall off the internal surface due to a compressive stress wave from the impact reflecting there as a tensile wave.

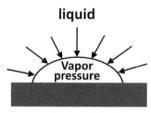

Figure 4.17: Schematic representation of a collapsing vapor bubble, as in cavitation.

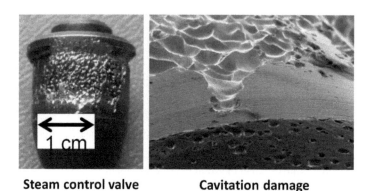

Steam control valve   **Cavitation damage**

Figure 4.18: Cavitation damage of steam control valve.

Spark erosion (or Electrical-Arc-Induced) is a well-known problem in the field of electrical contacts. It is found in situations when a high potential is present over a thin air film in a sliding process, where a dielectric breakdown occurs, leading to a sparking (arcing). When an electric spark occurs, a relatively high-power density (in the order of 1 kW/mm^2) occurs over a very short period of time (in the order of 100 μs) (Bhushan, 2013). The heating, generated from such an action, leads to considerable melting, hardness changes, damage of the material surface and finally catastrophic failures in electrical machinery. In a practical example, the case of copper slip rings against graphite-based brushes, the sparking damage on the copper slip ring can cause excessive wear of the brush by abrasion. There are a number of methods to eliminate the spark erosion, such as: (1) reducing the gap between the two surfaces, (2) eliminating the potential difference between the two surfaces, (3) introducing an insulator of a suitable dielectric strength between the two surfaces.

Spark erosion refers to the EDM (Electrical Discharge Machining) machining process which uses a shaped electrode to erode the metallic workpiece. The process involves both the workpiece and the tool being connected to a direct current; this provides a constant spark between the tool and the workpiece. The workpiece and the electrode are separated by a gap which is occupied by an insulating fluid. This fluid allows for the eroded debris to be washed away from the work area and is used to control the heating effect.

4.2.8 Thermal Wear

This type of wear is not considered as a typical mechanism of wear in general tribo-systems. However, it is generated by unexpected contact conditions, such as hard inclusions at the contact interface or a sudden overloading due to vibration. The mode of wear in such cases is governed by local surface melting caused by frictional heating or, in brittle materials, by surface cracking caused by thermal stresses. Diffusive wear is also included in the term "thermal wear", since it becomes noticeable only at high temperatures (Kato, 2002). A quantitative prediction of the wear volume may not be required. Even though the original material properties of a contact surface are degraded by losing significant chemical compositions, as a result of diffusion, the wear rate can still be increased by enhancing other wear modes, such as adhesive or abrasive wear (Lim and Ashby, 1987; Usui et al., 1984). Usui et al. (Usui et al., 1984) suggested that the wear rate increases linearly with $\exp(-\Delta E/kT)$, where k is the wear coefficient, T is the temperature, and E is the elastic modulus. However, models to predict the contribution of diffusive wear in such situations have not been found.

4.2.9 Other Types of Wear

Sometimes, in describing the mode of a surface failure, terms such as seizing, galling, scuffing and scoring are used. Many of these terms are old and apply principally to ductile metals. Each of them has several technical meanings, and describes different end results (Ludema, 1996).

Seizing is a severe damage of sliding surfaces that the driving system cannot provide a sufficient force to overcome friction; the sliding pairs cease to slide.

Galling is a severe form of adhesive wear that occurs when unlubricated, metallic or ceramic, surfaces are compressed against one another, at a slow speed. When the compressive forces in conjunction with the forces causing the sliding motion are sufficient, friction creates heat sufficient to weld the materials together. This unintentionally removes material from one material and places it onto the other one. In general, the main factors affecting galling are: design, applied load, contact area and degree of movement, lubrication and environment, and material properties (surface finish, hardness and steel microstructure).

Scuffing usually refers to a mode of failure of well lubricated metallic parts. Surfaces that are said to have scuffed become so rough that they no longer provide their expected function. Thus, scuffing is not a wearing mechanism, but primarily a surface roughening mechanism. Scuffed surfaces have a range of characteristic appearances depending on the chemical environment in which they operate. Some are shiny, some have grooves in them, and some are dull.

Scoring is a parallel phenomenon, sometimes established as a dull-appearing surface with no obvious roughening. Scored surfaces may only display an evidence of overheating of either the lubricant or the metal.

4.3 Wear-regime Maps

Wear is a complex function of the system which includes operating parameters, material properties, surface roughness, media and other properties. Consequently, for a given tribosystem, wear responses can vary significantly, depending on the conditions. Furthermore, no single wear mechanism applies over a wide range of conditions. This makes evaluation of materials in terms of wear resistance difficult. To this end, the approach of wear-regime maps has been introduced in order to provide guidance toward the proper selection of materials.

In principle, wear-regime maps are a graphical technique used to characterize various aspects of wear behavior in terms of the operating conditions of the tribosystem. Different forms of wear maps are typically used to identify ranges of operating parameters with wear mechanisms. Generally, two-dimensional or three-dimensional graphs were utilized where the axes are the operational parameters. Curves are plotted on these graphs to separate regions of different wear behaviors and to represent conditions of a constant wear rate. In addition to the generic name of wear-regime maps, such illustrations are also referred to as wear maps, wear mechanism maps, wear transition maps, or wear rate maps (Bayer, 2004).

The wear maps typically reveal different wear regimes according to the wear characteristics detected for the same material under different operating conditions. It can be divided into areas corresponding to different wear regimes. Prevailing wear mechanisms can give ultra-mild, mild, sever, or ultra-sever wear rates. Mild wear gives a smooth surface and severe wear produces a surface that is rough and deeply torn and the wear rate is usually high. The transition between mild and severe wear takes place over a wide range of sliding conditions.

To the best of our knowledge, the first wear map was developed by Lim and Ashby (Ashby and Lim, 1990; Lim and Ashby, 1987) for steel sliding against steel under a pin-on-disk configuration, as shown in Fig. 4.19.

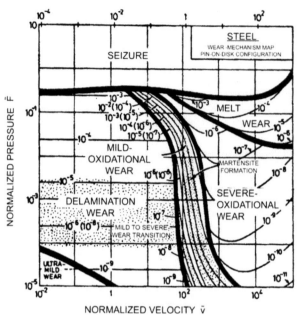

Figure 4.19: A wear map for low carbon steel based on physical modelling calibrated to experiments. After Ashby and Lim (Ashby, 1990) with permission from Elsevier Science Publishers.

The graph establishes pressure-velocity regions where different wear mechanisms prevail and, under relatively low pressures and velocities, material removal by delamination takes place. Mild wear occurs when oxidation due to frictional heating is relatively low, but under severe wear, wear particles are removed exclusively from the oxide layers. Melt wear or (Ultra-severe wear) occurs when flash temperatures reach the melting point.

Another example of two-dimensional wear maps has been published by Riahi and Alpas (Riahi and Alpas, 2003) as shown in Fig. 4.20. The measured wear rate, and the main wear mechanism operating in each wear regime are summarized in the wear map with log load versus log sliding speed axes.

The map depicts three major wear regimes, namely, ultra-mild, mild and severe wear regimes. The wear map indicates that the transition to ultra-mild wear occurs at loads of approximately 0.5 N, and it is not sensitive to the test speed. On transition to the mild wear regime, the wear rate sharply increases by approximately one order of magnitude, then increases steadily by increasing the load and the sliding speed. In the severe wear regime, the wear rates were up to three orders of magnitude greater than those in the mild wear regime. According to the wear map constructed, transition boundaries between the mild to severe wear, and the ultra-mild to mild wear regimes were linear on the log load versus log velocity scale. Subsequently, they developed an empirical relationship between the transition loads and the transition speeds.

Hsu and Shen (Hsu and Shen, 2000) suggested that, for a given material pair and a fixed discrete variable, there are five three-dimensional wear maps that can be used to describe wear systematically: Wear vs. speed and load, wear vs. speed and temperature, wear vs. speed and time, wear vs. load and temperature; and wear vs. load and time. Therefore, for a given material pair, a set of 20 wear maps will systematically define the wear behavior. These maps include: Five wear maps under dry sliding conditions, five wear maps under a nonreactive fluid, five wear maps under reactive lubricant conditions, and five wear maps under the same environmental and contamination conditions. In many instances, a complete set of wear maps is not needed in order to define the wear behavior. A selected set of maps will be sufficient to define the critical limits and operational boundaries for the materials pair in terms of acceptable wear behavior within those ranges.

An example of a three-dimensional wear map is shown in Fig. 4.21 for alumina under dry air, Purified Paraffin Oil (PPO), and water-lubricated conditions (Hsu and Shen, 1996). Figure 4.21a shows the

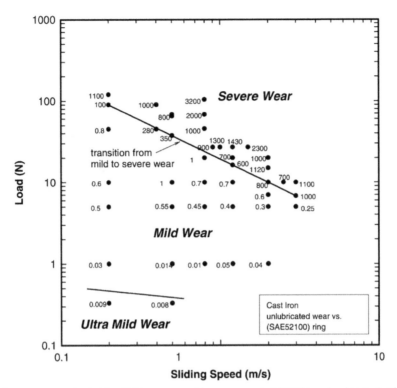

Figure 4.20: A wear map constructed for A30 type grey cast iron worn against 52100 type bearing steel. After Riahi et al. (2003) with permission from Elsevier Science Publishers.

alumina wear as a function of speed and load in room temperature and atmosphere conditions. The map simultaneously shows the functional dependence of wear with respect to speed and/or load. If both speed and load influence wear, their effect on wear can usually be seen in the topographical features as valleys and plateaus. For this case, it can be seen that alumina under this set of conditions has a very strong load dependence. Wear increases rapidly as load increases at a low speed. Under PPO lubrication, the speed effects are minimized. Figure 4.21b basically shows load dependence of wear with some speed influences. When water is used, as shown in Fig. 4.21c, the effect of speed is much more pronounced than in PPO's case. This is probably due to the tribochemical reactions that occur between alumina and water at certain speeds (Hsu et al., 1991).

Likewise, several kinds of wear maps have been introduced. Beerbower (Beerbower, 1972) proposed a conceptual wear mechanism diagram for steel under lubricated conditions as a function of the specific oil film thickness. While various mechanisms were reported in the literature, the diagram was constructed on a basis of inferences and isolated data. However, it illustrates the complex nature of the lubricated wear of steel.

deGee (deGee, 1989) proposed a simple system under well-lubricated conditions based on a large body of data generated under a set of standardized conditions. He referred to his maps as transition diagrams. He pointed out that as the severity of the wear test increases, as reflected by speed and load, the wear of steel under well-lubricated conditions progresses from no wear to mild wear, and then to scuffing. Furthermore, deGee found the concept of flash temperature useful in explaining the wear behavior.

In conclusion, the wear mapping technique can be used to present wear data systematically according to a hierarchy of parameters to define the wear system. The availability of the wear maps provides an opportunity to study different wear mechanisms in each region for a material in dry and lubricated environments. The diagram defines the number of wear mechanisms operating for a given system. The contour maps, lines of equal spacing and lack of curvature usually indicate the same dominant wear

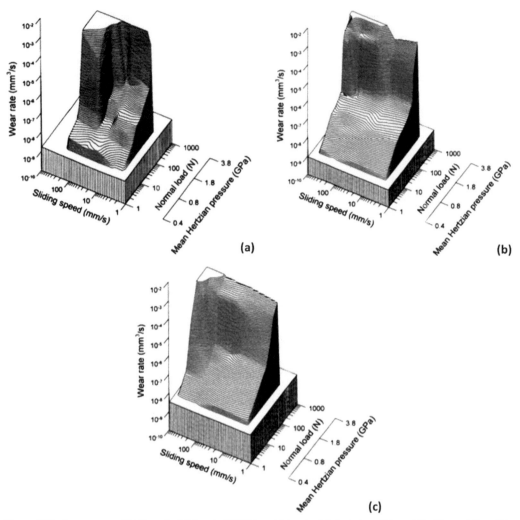

Figure 4.21: A wear maps of Al_2O_3 under (a) dry air, (b) PPO, and (c) water-lubricated conditions. After Hsu and Shen (Hsu and Shen, 1996) with permission from Elsevier Science Publishers.

mechanism. Valleys and plateaus usually suggest some change in the wear mode. In this way, regions with potentially different wear mechanisms can be identified. Critical experiments can then be conducted within those regions to identify the dominant wear mechanism, consequently constructing wear maps (Hsu et al., 1991).

4.4 Generation of Wear Debris

By definition, wear is a gradual removal of a material from contacting solid surfaces during their relative motion. In any tribosystem, the mechanism of wear involves generation of debris particles.

Since there are several mechanisms of wear, we should expect that there is no unique or typical form or size of wear debris. During dry, boundary and mixed lubricated sliding, wear particles are generally detached mechanically by micro-stresses resulting from the applied load and relative sliding. Wear particles are detached from sliding solids by micro-cutting of adhesive junctions between surfaces, mechanical failure of contacting asperities, surface spalling, plastic deformations of surfaces in a form of grooves and scratches, nucleation and propagation of surface and subsurface cracks and voids, oxidation, corrosion and chemical reactions (Zmitrowicz, 2005). Wear particles are also generated as a consequence of ploughing. Surfaces can be plowed by the wear particles, hard particles entrapped from the environment and by hard asperities of the counterface.

It was found that there is a clear difference in wear debris generated by soft and hard surfaces. Sliding behavior of materials with different harnesses (Pb, Zn, Al, Cu, Ni, Ti and AISI 1045 steel) showed that soft and ductile surfaces produced larger wear particles with a stronger tendency toward agglomeration, while hard surfaces produced smaller wear particles with weaker agglomeration tendency (Hwang et al., 1999).

Six basic particle types generated through the wear process including ferrous and non-ferrous particles where reported by Choudhury (Choudhury, 2014). They are detailed as follows:

1) Normal Rubbing Wear particles are generated during normal sliding wear in a machine and result from exploitation of particles of the mixed shear layer. Rubbing wear particles consist of flat platelets, generally 5 microns or smaller, although they might range up to 15 microns depending on equipment associations. There should be little or no visible texturing of the surface and the thickness should be 1 micron or less.

2) Cutting Wear particles are generated as a result of one surface penetrating another. There are two ways of generating this effect:

a) A relatively hard component can become misaligned or fractured resulting in hard, sharp edge penetrating a soft surface. The particle generated this way is coarse and large (2–5 microns wide and 25–100 microns long on average).

b) Hard abrasive particles in the lubricant, such as sand or wear debris from another part of the system, may become embedded in a softer surface (two body abrasion) such as Lead/Tin alloy in bearings. The abrasive particles protrude from the soft surface and penetrate the opposing wear surface. The maximum size of cutting wear particles generated in this way is proportional to the size of abrasive particles in the lubricant. Very fine wire-like particles with a thickness as low as 25 microns can be generated.

Under abrasive wear conditions, it is presumed that hard particles embedded in one surface will plow through a softer counterface and produce microchips. The attack angle, hardness and depth of penetration of abrading particles determine the shape and size of the debris. SEM of cutting-type debris from abrasive wear is shown in Fig. 4.22.

3) Spherical Particles are generated in the bearing cracks. Their presence gives an early warning of impending trouble as they are detectable before any spalling occurs. Rolling fatigue generates a few spheres over 5 microns in diameter, while the spheres generated by welding, grinding or corrosion are frequently over 10 microns in diameter.

A unique type of spherical wear debris is found in grinding swarf and when electric spark discharge damages metallic surfaces. The particles, as shown in Fig. 4.23, are formed from molten droplets caused by frictional heating from a grinding process (Rigney, 1992). They can also be produced by the hot arc developed in an electric discharge damage to a metal surface or from electric arc welding. Spherical debris can be found in oil samples from machinery operating in the vicinity of grinding operations.

4) Severe Sliding Wear Particles are identified by parallel marks on their surfaces. They are generally larger than 15 microns, with the length-to-width thickness ratio falling between 5–30 microns. Severe sliding wear particles sometimes show evidence of temper colors, which may change the appearance of the particle after heat treatment.

5) Bearing Wear Particles these distinct particle types have been associated with rolling bearing fatigues.

a) Fatigue spall particles constitute actual removal from the metal surface when a pit or a crack is propagated. These particles reach a maximum size of 100 microns during the micro spalling process. Fatigue spalls are generally flat with a major dimension-to-thickness ratio of 10:1. They have a smooth surface and an irregular circumferential profile.

b) Laminar (or flake) particles are very thin free metal particles with frequent occurrence of holes. They range between 20 to 50 microns in the major diameter, with a thickness ratio of 30:1. These particles

Figure 4.22: SEM micrograph of cutting-type debris from abrasive wear. After Rigney (Rigney, 1992) with permission from Elsevier Science Publishers.

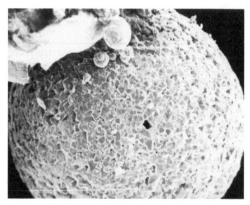

Figure 4.23: Spherical wear debris from grinding swarf. After Rigney (Rigney, 1992) with kind permission from Elsevier Science Publishers.

are formed by the passage of wear particles through a rolling contact. Laminar particles may be generated throughout the life of a bearing.

The initiation of cracks at the surface or in the material is the starting point of a process that may result in material detachment by delamination, debris generation and the formation of transfer layers. Also, wear debris can be formed by plastic deformation. Hard asperities plowing over a softer surface without cutting will produce ridges. Ridges can then be flattened by a further contact. Extrusions or lips are formed. These lips are broken off and become flat wear flakes (Glaeser, 2001), as shown in Fig. 4.24.

There are other ways to generate laminar flakes. For example, flakes are produced when a 310 stainless steel block slides against an M2 tool steel ring, as shown in Fig. 4.25. The high alloy content of this material assures its stable austenite (FCC) structure, i.e., it does not transform to martensite during plastic deformation. In this case, stainless steel transfers to the surface of the harder M2 ring and then it delaminates to form non-magnetic flake debris consisting only of stainless steel. Again, the flakes are not generated by direct delamination of base material.

6) Gear Wear: There are two types of particles associated with gear wear; pitch line fatigue and scuffing particles.

a) Pitch line fatigue particles from a gear pitch line have much in common with rolling-element bearing fatigue particles. They generally have a smooth surface and are frequently irregularly shaped. Depending upon the gear design, the particles usually have a major dimension-to-thickness ratio

Figure 4.24: Example of wear debris resulting from a plastic deformation in a block-on-ring wear test. After Rigney (Rigney, 1992) with permission from Elsevier Science Publishers.

Figure 4.25: Scanning electron image of Bake debris resulting when 310 stainless steel (stable austenite) transferred to an M2 steel ring and then delaminates. After Rigney (Rigney, 1992) with permission from Elsevier Science Publishers.

between 4:1 and 10:1. The chunkier particles result from tensile stresses on the gear surfaces which cause the fatigue cracks to propagate deeper into the gear tooth prior to spalling.

b) Scuffing or scoring particles are caused by too high load and/or speed. These particles tend to have a rough surface and jagged circumference. Even small particles may be discerned from rubbing wear by these characteristics. Some of the large particles have striations on their surface, indicating a sliding contact. Because of the thermal nature of scuffing, quantities of oxides are usually present and some particles may show evidence of partial oxidation that is tan or blue temper in color.

The former classification is not the only one, in fact, there are various classifications of shapes for wear particles (Zmitrowicz, 2005):

a) A rough division of shapes of wear debris into two categories, flake and non-flake, can be made. Flake-type debris refers to a particle that has a relatively uniform thickness, whereas non-flake type includes such debris types as powder, plates, ribbons, cylinders, spheres, irregular chunks and loose clusters.

b) Debris can roughly be divided into two groups: Long fibrous debris and very thin platelet debris. Both types of particles appeared in lubricant samples.

c)   Other types of wear debris are: Sheet-like, roll-like, long ropelike, aggregated and granular particles.

d)   Wear particles can be classified according to their origin, such as: Fatigue chunk particles, severe sliding particles, or laminar particles.

Wear debris can be defined with the aid of the following dimensions: Length, width and thickness. The flake debris can be characterized by their outline, e.g., straight edges/irregular edges or by curvature radii of arcs and by apex angles of corners, etc.

Table 4.1 is a summary of observations relating worn surface appearance to associated wear particle type (Lansdown and Price, 1986):

Table 4.1: Worn surface appearance and associated debris (Lansdown and Price, 1986).

Type of wear	Worn surface appearance	Wear debris
Mild wear	Fairly uniform loss of material; slight surface roughening	Fine, free metal particles < 10 μm
Adhesive	Tearing and transfer of materials from one surface to another	Large irregular particles > 10 μm, containing unoxidized metal
Two-body abrasive	Harder surface–little or no damage. Softer surface exhibits scores, grooves or scratches corresponding with rough asperities on harder surface. Harder surface scored or grooved, may appear to be machined, with grooves corresponding with hard embedded particles in counterface	Consists mainly of softer material (fine swarf), unoxidized material Contains swarf-like, unoxidized material from harder surface. May also contain softer material in lump form
Three-body abrasive	Surfaces have deep scratches or grooves	Fine, may contain some unoxidized metal, but mainly loose abrasive material
Fretting	Surfaces heavily pitted: Pits may be small, or larger, producing roughened surface area; oxidized appearance	Fine, fully oxidized, if from ferrous metal. Will contain Fe_2O_3 (rust coloured). Spherical particles, sometimes

In addition to the types detailed in Table 4.1, severe sliding wear in metals is associated with relatively large particles of metallic debris, while in mild wear, the debris is finer and formed of oxide particles. In the case of ceramics, the "severe" wear regime is associated with brittle fracture, whereas 'mild' wear results from the removal of reacted surface material. Laminate debris is a characteristic of severe wear.

A number of authors have noted crack nucleation at second-phase particles in a plastically flowing matrix, propagation of cracks parallel to the sliding direction and delamination leading to the formation of a debris, the debris thickness being related to the depth of the plastic zone.

The size, shape, structural and chemical details of particles are analyzed using various techniques, including optical microscopy, scanning electron microscopy (SEM), transmission and scanning transmission electron microscopy (TEM/STEM), energy dispersive and wavelength dispersive spectroscopy (EDS and WDS), Auger electron spectroscopy (AES), X-ray photoelectron spectroscopy (XPS), X-ray and electron diffraction. Based on such a comprehensive debris analysis, the condition of a system can be monitored and predictive maintenance can be applied.

4.5 Measurement of Wear (Tribo-measurements)

A valuable goal of many tribologists is to increase the reliability of tribosystems. Design optimization, material choice, lubrication and surface treatment are often proposed as solutions to this issue. Preliminary validation of machines and machine parts by laboratory testing are also required before integrating a new solution into the system. The choice of a tribological test method representative of the real system thus becomes a stage of first magnitude in the success of the whole project. Furthermore, laboratory investigation may also be applied to study undesired behavior of tribosystems. The term "tribometery" is defined as a set of technologies that allows for the measuring of the friction and wear behavior of a tribosystems (Richard

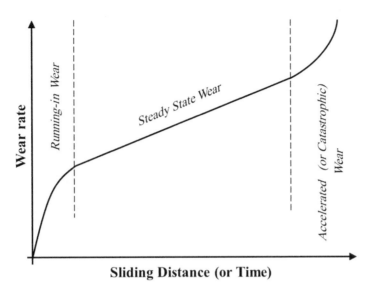

Figure 4.26: A typical wear curve.

et al., 2015). Historically, the term "tribometer" was first introduced in 1700s by Goldsmith in order to refer to a device to measure friction, but in current usage, the scope of the tribometer has been broadened to mean an instrument used to measure friction, wear, interface temperature, and other triboparameters.

Usually, volume is the fundamental measured for wear when wear is equated with loss or displacement of material. Nevertheless, in many material investigation studies, mass loss is used instead of volume loss. However, in engineering applications, we are more concerned with the change of a dimension due to volume or mass loss of the bulk material. Loss of a dimension frequently results in an increase in clearance or a change in contour. Beside the former measures for wear, there is a variety of other operational ones that may be used such as crazing, vibration level, noise level, life, and appearance of a surface.

Wear is generally described by "wear rate" which is defined as the volume (or mass) of material removed per unit sliding distance or per unit time. Other forms could be dimensionless, such as the depth of material per unit sliding distance, or the volume removed per apparent area of contact and per unit sliding distance. Wear rate is generally not constant, in general, it is a complex function of time (Bhushan, 2013).

A typical wear curve is shown in Fig. 4.26. At the initial stage, wear usually starts low and then rises. After a certain time, the wear rate remains constant for a period and may change if a transition from one mechanism to another occurs during the wear test. The initial stage during which the wear rate changes is known as the running-in wear. During this stage, wear depends on the initial material structure and properties and on surface conditions, such as surface finish and the nature of any films present. Following this transient stage, the wear goes through a steady state wear stage characterized by a constant wear rate. In some cases, after a certain sliding distance, an accelerated (or catastrophic) wear stage starts due to surface fatigue.

The wear rate for a material pair is normally presented in terms of a wear factor (or specific wear rate). Wear factor (K) is defined as the wear volume per unit applied normal load (F) and per unit sliding distance (x), and can be calculated using Equation 4.18.

$$K = \frac{V}{F \cdot x} \tag{4.17}$$

There are many types of wear, including sliding, scratch, abrasion and other mechanisms. Most laboratories associated with wear testing possess a variety of tribometers and procedures that are used to address each specific wear type. The choice of the proper tribometer is very important. Thus, in order to select a suitable wear test, some intrinsic points should be considered (Kennedy and Hashmi, 1998).

i)   The selected test should measure the desired properties.

ii) Whether the contact between the components is rolling, sliding, impact or erosion only, or a combination of these, the surface roughness of the rubbing faces should be similar to those of the actual applications.

iii) The loads and stresses in the test and actual application should be in the same range.

iv) Environmental conditions (temperature, humidity, sliding media, etc.) should simulate the actual working environment.

v)   The duration of the test and the materials used in testing should be the same as the actual materials used in the machine parts.

In the current section we will introduce a number of testing methods associated with different types of wear.

4.5.1 Sliding Wear Test

Sliding wear test is the most common testing method. A review of the apparatus used and their features indicate that these tribometers have the same general feature of the applications (Abdelbary, 2014). The test rigs accommodate a test specimen and provide different types of wear paths: Unidirectional, oscillating, spiral and others. The nature of wear situations has typically resulted in the development of unique apparatus and test methods in order to simulate these situations and to provide the needed data. ASTM G99 (ASTM, 2004b) and ASTM G133 (ASTM, 2005) are the standards for these tests, which do not specify specific values for the parameters, but allows those to be selected by the researchers to provide simulation model. Wear results are usually obtained by conducting a test for a selected sliding distance, and for selected values of load and speed. In some cases, results may be reported as plots of wear volume versus sliding distance using different specimens.

The most famous sliding wear tests are:

a) Unidirectional sliding: In unidirectional sliding applications, the pin-on-disc, Fig. 4.27, has been used extensively in wear test process. It has been also employed in many other applications, including material wear and friction properties at elevated temperatures, controlled atmospheres and with lubrication.

The basic configuration of this apparatus consists of a pin sliding against a rotating counterface disk. The motion is generally unidirectional at a constant speed. The test parameters that have been used include size and shape of the pin, load, speed and material pairs.

To model some aspects of the sliding of typical applications, block-on-ring, ASTM G137 (ASTM, 2017), pin-on-drum, ASTM G65 (ASTM, 2004a), ball-on-flat and other configurations were also used in many studies (Kelley et al., 1988; Roshon, 1974), as shown in Fig. 4.28.

b) Reciprocating sliding: Unlike unidirectional sliding wear tests, the pin-on-plate (or ball-on-plane) wear tests are used to estimate wear and friction during a reciprocating (oscillating) sliding. This test method is designed to simulate the geometry and motion experienced in many types of rubbing components, whose normal operation results in periodic reversals in the direction of the relative sliding. The resulting data from this type of movement may differ from that experienced by the same materials in a unidirectional sliding. Unlike material combinations, wear may occur at different rates depending on which material is the pin (or ball) and which is the flat plate. The generic configuration of a pin-on-plate tribotester is shown in Fig. 4.29. Normally, the rubbing mode is simulated by a radius-tipped, or flat-ended, test pin pressed against a flat counterpart which is moving backward and forward. An alternate configuration in which the specimen moves while the flat plate is fixed may be used.

The generic feature of the change in the direction of motion is that each cycle contains acceleration and deceleration portions. The sliding velocity profile tends to vary with different rigs and depends on the drive mechanism used. A reciprocating sliding test can be used for the simulation of fretting situations, by reducing the stroke of the motion to the range associated with fretting (Wayson, 1964). Amplitude, frequency and contact pressure have critical influence on the fretting wear of polymer tribosystems.

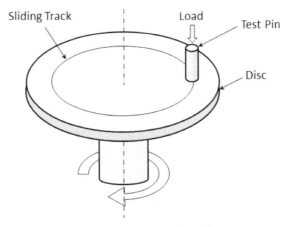

Figure 4.27: Schematic of a pin-on-disc sliding wear test.

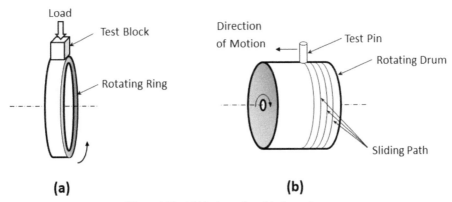

(a)   **(b)**

Figure 4.28: (a) block-on-ring, (b) pin-on-drum.

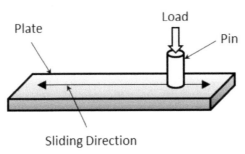

Figure 4.29: Diagram of the pin-on-plate test.

4.5.2 Rolling Wear Test

Generally, the rolling motion is a complex interaction between two contacting surfaces. However, pure rolling means that there is no relative slip, i.e., both mating surfaces are at the same velocity. It only occurs in a fraction of the total footprint of a revolute shape (ball, roller, wheel, etc.) rolling on another surface. In reality, often no pure rolling motion occurs between contacting partners; instead, a superposition of rolling and sliding motions takes place. In order to address wear properties in a rolling contact, several configurations have been developed (Harrass et al., 2010). A four ball (or three ball-on-plate) test rig is used to determine the wear characteristics in both dry and lubricated conditions. Basically, it consists of three steel balls of a specified diameter, which are clamped together and loaded against a flat disc specimen. These balls, driven by an upper fourth ball, are pressed into a cavity formed by the three clamped balls. The balls are housed in a steel cup and are free to roll over the surface of the disc located at the bottom of the cup, Fig. 4.30.

The motion of steel balls on the specimen plate is similar to the motion in a thrust bearing. As there are three lower balls, any given point on the test piece surface is loaded/unloaded three times per one full revolution. In this way, a cyclic loading acting on the surface of the disc is generated (Stolarski et al., 1998).

A twin disc apparatus can also be used to investigate the influence of wear in a rolling contact. It gives an opportunity to test polymers at different velocities corresponding to different contact points along the line of action. In fact, for modeling a gear with a slip ratio (close to the pitch point), a twin disc set-up is the appropriate solution (Sukumaran et al., 2012). Basically, it consists of two driven cylinders or discs pressed against one another, see Fig. 4.31. Two motors drive the shafts so that different rotational speeds can be applied. Thus, a great advantage of this test rig is the possibility to apply a relative slip of one surface on another. Hence, the critical elements of this test are control of the surface velocities of the two cylinders, alignment of the cylinders and geometrical tolerance of the cylinders. Another approach is to use slightly curved or crowned cylinders in order to simulate conditions of pure rolling, in which case the surface velocities of the two cylinders must be identical.

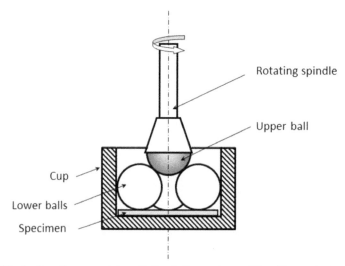

Figure 4.30: A schematic representation of a four ball apparatus used for rolling contact experiments.

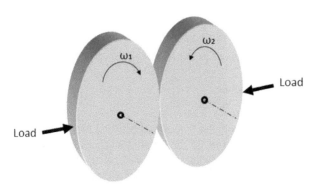

Figure 4.31: Basic configuration of a twin disc apparatus used in rolling wear tests.

4.5.3 Scratch Wear Test

Since an approximately linear correlation exists between wear resistance and hardness, a scratch test, ASTM G 171 (International, 2003), has often been used to give a guide to the abrasive wear resistance. This applied to metals, ceramics, polymers and coated surfaces. The basic principle for all scratch tests is nearly the same. It involves producing a scratch in a solid surface by moving a hard, sharp stylus of a specified geometry along a specified path. The normal force magnitude and the motion speed are constant, as shown in Fig. 4.32.

In addition, scratch speed, scratch load, loading rates, number of scratches and scratch length can be changed to give an enough flexibility to define a desired test. Usually, the scratch test is conducted under dry conditions and at room temperature; however, it is possible to conduct it under lubricated and elevated temperature conditions.

In scratch tests, there are two shapes of stylus indenters; circular cross-sections, such as cones, spheres and square-base pyramid shapes, Fig. 4.33. The main criterion is that the scratching process produces a measurable scratch in the tested surface without causing fracture, spalling or delamination. This test can be used to predict the amount of wear that takes place in a polymer under specified conditions. The micro mechanism that actually takes place during wear under each asperity of a hard surface can be reproduced on a macro scale using a scratch test. In other words, the wear mechanism, which takes place under each asperity in a nano-scale, can be magnified to a micrometer and millimeter scale and reproduced. This gives researchers a good idea about what goes on under each asperity. These values can be applied for multiple asperities using simulations, and can recreate the whole wear mechanism taking place on the surface of a polymer (Sinha et al., 2007).

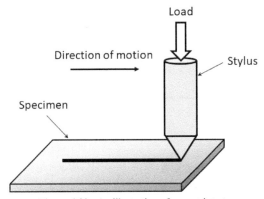

Figure 4.32: An illustration of a scratch test.

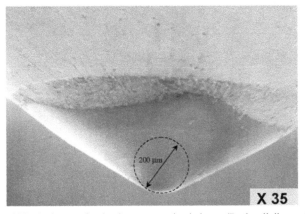

Figure 4.33: An image of a circular cross-section indenter (Rockwell diamond tip).

4.5.4 Abrasion Wear Test

In general, the main testing methods of abrasion wear are two-body and three-body abrasion. Two-body abrasion can be simulated by pins, of the specimen material, that are loaded and rotated against a spinning rough counterface, Fig. 4.34. The high contact pressure produces indentations and scratching of the wearing surfaces, and fractures and pulverizes the abrasive particles. It has been speculated that high-stress grinding abrasion facilitates material removal by a combination of cutting, plastic deformation and surface fracture on a microscopic scale, as well as tearing and fatigue or spalling on the macroscopic scale (Unal et al., 2005).

On the other hand, the three-body abrasion test was developed to simulate wear situations in which low-stress scratching abrasion is the primary mode of wear. Scratching abrasion occurs when lightly loaded abrasive particles impinge upon and move across the wearing surface, cutting and ploughing on the microscopic scale. In liquid environments, or slurry, corrosion may also contribute to the overall wear rate, in which case erosion-corrosion is the operative wear mechanism. A dry sand/rubber wheel test ASTM G65 is usually utilized to investigate the influence of various parameters on this mode of wear, such as abrasive particle size and shape and material parameters (Harsha et al., 2003; Bijwe et al., 2001). The test configuration is shown in Fig. 4.35.

During the test, the abrasive particles are introduced between the test specimen and a rotating wheel. The rim of the wheel is coated with rubber to avoid crushing the grains in the nip between the specimen and the wheel. The test specimen is pressed against the rotating wheel by means of a lever arm, while a controlled flow of grit abrades the test surface.

Wet sand/rubber wheel abrasion test ASTM G105 (ASTM, 2002) also simulates three-body abrasion and is very similar to the previous test. The wet abrasion test involves abrading a test specimen with a slurry containing grit of a controlled size and composition, Fig. 4.36.

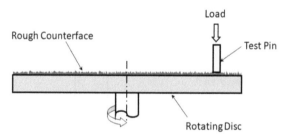

Figure 4.34: Schematic diagram of the two-body abrasion wear test apparatus.

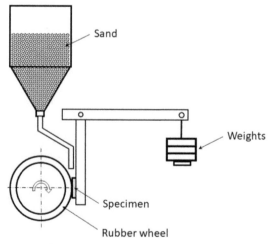

Figure 4.35: Schematic illustration of the sand/rubber wear test apparatus.

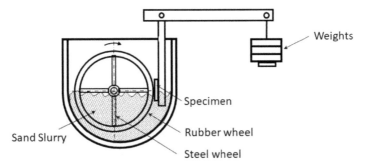

Figure 4.36: Schematic diagram of the wet sand/rubber wheel abrasion wear test apparatus.

The composition of the slurry is defined in terms of abrasive size, type and source, and the type and amount of water. The wear in both tests is determined by weight loss and converted to volume loss for material rankings. Wet abrasion tests have been also extensively used to investigate scratching abrasion of polymers and have been found to correlate well with many practical applications (Kennedy and Hashmi, 1998).

In addition to the former typical methods, there are many other methods to address the abrasion resistance in some special cases. For example, ASTM C241 is a standard test method for the abrasion resistance of a stone subjected to foot traffic. This test method covers the determination of the abrasion resistance of all types of stones for floors and steps, as well as similar uses where the wear is caused by the abrasion of foot traffic. It is suitable for indicating the differences in abrasion resistance between various building stones. It also provides one element in comparing stones of the same type.

4.5.5 Erosion Wear Test

Erosion is a mechanical degradation of a surface, resulting from a solid particle impingement causing a local damage combined with material removal. As a good simulation to this mode of wear, gas jet apparatuses have been employed in order to investigate solid particle erosion, as well as to rank materials in terms of the resistance to this mode of wear. This test covers the determination of the material loss by gas-entrained solid particle impingement erosion. The test rig is capable of eroding material from a test specimen under well controlled exposure conditions. The basic principle of the erosion test method involves a small nozzle delivering a stream of a gas containing abrasive particles, which impacts the surface of a test specimen, Fig. 4.37.

The apparatus may also include a means of controlling and adjusting the particle impact velocity, and the specimen location and orientation relative to the impinging stream. The standard conditions for the test are presented in ASTM standard G76-05 (International, 2005a), which lists the significant parameters to be

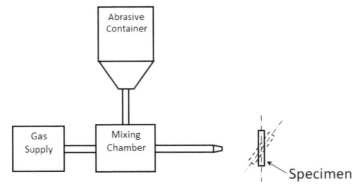

Figure 4.37: Schematic diagram of the erosion test apparatus.

controlled in the test and their tolerances. The determined amount of a material volume loss is normalized by the abrasive flow rate, to provide the erosion value which is defined as a wear volume per gram of the abraded material (Bijwe et al., 2001).

4.5.6 Impact Wear Test

Impact wear has been identified as one of the dominant types of wear in many mechanical applications, particularly in valves and seats. Although impact wear is the most likely to occur in these situations, there are no broadly used tests for this type of wear. However, several methods and mechanisms to determine impact wear and to compare the resistance of materials to this mode of wear have been suggested (Bayer et al., 1972; Mahoney et al., 1984). Where there is a high impact velocity and, therefore, a high impact energy, a plastic deformation occurs in the soft material in contact with a ductile extrusion away from the contact. On the other hand, if impacts occur on a brittle specimen, there is a little plastic deformation and fracture occurs (Slatter et al., 2011). Generally, apparatus used in impact wear tests can be grouped into two categories: Pure impact and compound impact. A hammer type impact wear test rig is usually applied for a pure or normal impact. In this case, there is no sliding or adhesion involved while the flat specimen is stationary. For a compound impact, that is combined impact and sliding, the flat member rotates or oscillates beneath the hammer or projectile. It is important to mention that there is no fixed number of impacts to be used in this method of evaluating materials for engineering applications; however, in some cases, somewhere between 10^3 and 10^4 impacts were selected. Similar to testing methods and techniques used for sliding wear, resistance of materials to impact wear can be characterized using weight (or volume) loss after a fixed number of repeated impacts.

References

Abdelbary, A. 2014. 7—Methodology of testing in wear. pp. 159–183. *In*: Abdelbary, A. (ed.). Wear of Polymers and Composites. Oxford: Woodhead Publishing.

Anaee, R.A.M. and Abdulmajeed, M.H. 2016. Tribocorrosion. *In*: Pranav, Darji, H. (ed.). Advances in Tribology. IntechOpen.

Archard, J.F. 1953. Contact and rubbing of flat surfaces. Journal of Applied Physics 24: 981–988.

Ashby, M.F. and Lim, S.C. 1990. Wear-mechanism maps. Scripta Metallurgica et Materialia 24(5): 805–810. Doi: https://doi.org/10.1016/0956-716X(90)90116-X.

ASTM International. 2002. ASTM Standard G105 In: Standard Test Method for Conducting Wet Sand/Rubber Wheel Abrasion Tests. West Conshohocken, PA.

ASTM International. 2003. ASTM Standard G 171-03. In Standard Test Method for Scratch Hardness of Materials Using a Diamond Stylus. West Conshohocken, PA.

ASTM International. 2004a. ASTM Standard G65. In: Standard Test Method for Measuring Abrasion Using the Dry Sand/Rubber Wheel Apparatus. West Conshohocken, PA.

ASTM International. 2004b. ASTM Standard G99. In: Standard Test Method for Wear Testing with a Pin-on-Disk Apparatus. West Conshohocken, PA.

ASTM International. 2005a. ASTM Standard G76 In: Standard Test Method for Conducting Erosion Tests by Solid Particle Impingement Using Gas Jets. West Conshohocken, PA.

ASTM International. 2005b. ASTM Standard G133. In: Standard Test Method for Linearly Reciprocating Ball-on-Flat Sliding Wear. West Conshohocken, PA.

ASTM International. 2017. ASTM G137-97. In: Standard Test Method for Ranking Resistance of Plastic Materials to Sliding Wear Using a Block-On-Ring Configuration. West Conshohocken, PA.

Bayer, R.G., Engel, P.A. and Sirico, J.L. 1972. Impact wear testing machine. Wear 19(3): 343–354. Doi: https://doi.org/10.1016/0043-1648(72)90125-1.

Bayer, R.G. 2004. Mechanical Wear Fundamentals and Testing, Revised and Expanded: CRC Press.

Beerbower, A. 1972. Boundary Lubrication. US Army, Office of the Chief of Research and Development, Contract No., DAHC19-69-C-0033.

Bhushan, B. 2013. Principles and Applications of Tribology. 2nd edition ed. New York: John Wiley & Sons, Inc.

Bhushan, B. 2013. Introduction to Tribology. Second edition ed. New York: John Wiley & Sons Inc.

Bijwe, J., Indumathi, J., John Rajesh, J. and Fahim, M. 2001. Friction and wear behavior of polyetherimide composites in various wear modes. Wear 249(8): 715–726. Doi: https://doi.org/10.1016/S0043-1648(01)00696-2.

Bitter, J.G.A. 1963. A study of erosion phenomena part I. Wear 6(1): 5–21. Doi: https://doi.org/10.1016/0043-1648(63)90003-6.

Buckley, D.H. 1981. Surface effects in adhesion, friction, wear and lubrication, Tribology series 5, New York: Elsevier.

Burwell, J.T. and Strang, C.D. 1952. Metallic wear. Proc. R. Soc.

Burwell, J.T. 1957. Survey of possible wear mechanisms. Wear 1(2): 119–141. Doi: https://doi.org/10.1016/0043-1648(57)90005-4.

Choudhury, S. 2014. Wear Debris Analysis.

Curà, F. and Mura, A. 2017. Evaluation of the fretting wear damage on crowned splined couplings. Procedia Structural Integrity 5: 1393–1400. Doi: https://doi.org/10.1016/j.prostr.2017.07.203.

Davies, R.M. 1949. The determination of static and dynamic yield stresses using steel ball. Pros. Roy. Soc.

Davim, J.P. 2011. Tribology for Engineers. Cambridge, UK: Woodhead Publishing Limited.

deGee, A.W.J. 1989. Wear Research for Industry—Examples of Application of the IRG Transition DiagramTechnique. Wear of Materials, New York.

Divakar, M., Agarwal, V.K. and Singh, S.N. 2005. Effect of the material surface hardness on the erosion of AISI316. Wear 259(1): 110–117. Doi: https://doi.org/10.1016/j.wear.2005.02.004.

Dwivedi, D.K. 2010. Adhesive wear behaviour of cast aluminium–silicon alloys: Overview. Materials & Design (1980–2015) 31(5): 2517–2531. Doi: https://doi.org/10.1016/j.matdes.2009.11.038.

Fridrici, V., Fouvry, S. and Kapsa, P. 2003. Fretting wear behavior of a Cu Ni in plasma coating. Surface and Coatings Technology 163-164: 429–434. Doi: 10.1016/s0257-8972(02)00639-4.

Glaeser, W.A. 2001. Wear debris classification. *In*: Bhushan, B. (ed.). Modern Tribology Handbook, New York: CRC Press.

Gurumoorthy, K. and Ghosh, A. 2013. Failure investigation of a taper roller bearing: A case study. Case Studies in Engineering Failure Analysis 1(2): 110–114. Doi: https://doi.org/10.1016/j.csefa.2013.05.002.

Halling, J. 1979. Principles of Tribology. London: Macmillan Press Ltd.

Harrass, M., Friedrich, K. and Almajid, A.A. 2010. Tribological behavior of selected engineering polymers under rolling contact. Tribology International 43(3): 635–646. Doi: https://doi.org/10.1016/j.triboint.2009.10.003.

Harsha, A.P., Tewari, U.S. and Venkatraman, B. 2003. Three-body abrasive wear behaviour of polyaryletherketone composites. Wear 254(7): 680–692. Doi: https://doi.org/10.1016/S0043-1648(03)00142-X.

Haymann, F.J. 1992. Liquid Impact Erosion. *In*: ASM Handbook: Friction, Lubrication and Wear Technology, 221–232. Metals Park, Ohio: ASM International.

Holm, R. 1946. Electric Contacts. Edited by H. Gerbers. Stockholm, Sweden.

Hsu, S.M., Lacey, P.I., Wang, Y.S. and Lee, S.W. 1991. Wear mechanism maps of ceramics. *In*: Chung, Y.W. and Cheng, H.S. (eds.). Advances in Engineering Tribology. Chicago, 123: Society of Tribologists and Lubrication Engineers.

Hsu, S.M. and Shen, M.C. 1996. Ceramic wear maps. Wear 200(1): 154–175. Doi: https://doi.org/10.1016/S0043-1648(96)07326-7.

Hsu, S.M. and Shen, M.C. 2000. Wear map. *In*: Bhushan, B. (ed.). Modern Tribology Handbook, New York: CRC Press.

Hutchings, I.M. 1992. Tribology: Friction and Wear of Engineering Materials. Boca Raton, Florida: CRC Press.

Hutchings, I. and Shipway, P. 2017. 6—Wear by hard particles. pp. 165–236. *In*: Hutchings, I. and Shipway, P. (eds.). Tribology (Second Edition). Butterworth-Heinemann.

Hwang, D.H., Kim, D.E. and Lee, S.J. 1999. Influence of wear particle interaction in the sliding interface on friction of metals. Wear 225-229: 427–439. Doi: https://doi.org/10.1016/S0043-1648(98)00371-8.

Kato, K. 2002. Classification of wear mechanisms/models. Proceedings of the Institution of Mechanical Engineers, Part J: Journal of Engineering Tribology 216(6): 349–355. Doi: 10.1243/135065002762355280.

Kelley, J.E., Stiglich, J.J. and Sheldon, G.L. 1988. Methods of characterization of tribological properties of coatings. *In*: Surface Modification Technologies. 1st International Conference of Surface Modification, 25–28, January 1988, Phoenix, AZ, USA.

Kennedy, D.M. and Hashmi, M.S.J. 1998. Methods of wear testing for advanced surface coatings and bulk materials. Journal of Materials Processing Technology 77(1): 246–253. Doi: https://doi.org/10.1016/S0924-0136(97)00424-X.

Kragelskii, I.V. 1965. Friction and Wear. London: Butterworth.

Lansdown, A.R. and Price, A.L. 1986. Materials to Resist Wear. 1st ed, Pergamon Materials Engineering Practice Series. New York: Pergamon Press.

Lim, S.C. and Ashby, M.F. 1987. Wear mechanism Maps. Acta Metallurgica 35: 1–24.

Ludema, K.C. 1996. Friction Wear and Lubrication: CRC Press.

Lundberg, G. and Palmgren, A. 1952. Dynamic capacity of roller bearings. Ingenicorsrenten-skapsakademiens Nr. 210.

Mahoney, N.J., Grieve, R.J. and Ellis, T. 1984. A simple experimental method for studying the impact wear of materials. Wear 98: 79–87. Doi: https://doi.org/10.1016/0043-1648(84)90218-7.

Narayanaswamy, B., Hodgson, P., Timokhina, I. and Beladi, H. 2016. The impact of retained austenite characteristics on the two-body abrasive wear behavior of ultrahigh strength bainitic steels. Metallurgical and Materials Transactions A 47(10): 4883–4895. Doi: 10.1007/s11661-016-3690-5.

Quinn, T.F.J. 1987. Friction, wear and lubrication, institution of mechanical engineers conference series. International Conference on Tribology, London.

Rabinowicz, E. 1995. Friction and Wear of Materials. Second edition ed. New York: John Wiley and Sons.

Riahi, A.R. and Alpas, A.T. 2003. Wear map for grey cast iron. Wear 255(1): 401–409. Doi: https://doi.org/10.1016/S0043-1648(03)00100-5.

Richard, C., Manivasagam, G. and Chen, Y.-M. 2015. Chapter 8: Measurement of Wear and Friction Resistance of Bulk and Coated Materials.

Rigney, D.A. 1992. Paper X (i) The role of characterization in understanding debris generation. pp. 405–412. *In*: Dowson, D., Taylor, C.M., Childs, T.H.C., Godet, M. and Dalmaz, G. (eds.). Tribology Series. Elsevier.

Robert, J.K.W. 2007. Tribo-corrosion of coatings: A review. Journal of Physics D: Applied Physics 40(18): 5502.

Roshon, D.D. 1974. Testing machine for evaluating wear by paper. Wear 30(1): 93–103. Doi: https://doi.org/10.1016/0043-1648(74)90059-3.

Sarker, A.D. 1976. Wear of Metals. U.K: Pergamon Press.

Sasada, T. 1979. The role of adhesion in wear of materials. Journal of Japan Society of Lubrication Engineering 24(11): 700–705.

Sinha, S.K., Chong, W.L.M. and Lim, S.-C. 2007. Scratching of polymers—Modeling abrasive wear. Wear 262(9): 1038–1047. Doi: https://doi.org/10.1016/j.wear.2006.10.017.

Slatter, T., Lewis, R. and Jones, A.H. 2011. The influence of cryogenic processing on wear on the impact wear resistance of low carbon steel and lamellar graphite cast iron. Wear 271(9): 1481–1489. Doi: https://doi.org/10.1016/j.wear.2011.01.041.

Stachowiak, G.W. 2005. Wear—Materials, Mechanisms and Practice. U.K.: WILEY.

Stolarski, T.A., Hosseini, S.M. and Tobe, S. 1998. Surface fatigue of polymers in rolling contact. Wear 214(2): 271–278. Doi: https://doi.org/10.1016/S0043-1648(97)00206-8.

Suh, N.P. 1980. Update on the delamination theory of wear, Fundamentals of friction and wear of materials. ASM Materials Science Seminar.

Sukumaran, J., Ando, M., De Baets, P., Rodriguez, V., Szabadi, L., Kalacska, G. and Paepegem, V. 2012. Modelling gear contact with twin-disc setup. Tribology International 49: 1–7. Doi: https://doi.org/10.1016/j.triboint.2011.12.007.

Tabor, D. 2000. The Hardness of Metals. New York: OUP Oxford.

Tomlinson, G.A. 1927. The rusting of steel surfaces in contact. Proc. R. Soc., London.

Unal, H., Sen, U. and Mimaroglu, A. 2005. Abrasive wear behaviour of polymeric materials. Materials & Design 26(8): 705–710. Doi: https://doi.org/10.1016/j.matdes.2004.09.004.

Usui, E., Shirakashi, T. and Kitagawa, T. 1984. Analytical prediction of cutting tool wear. Wear 100(1): 129–151. Doi: https://doi.org/10.1016/0043-1648(84)90010-3.

Wang, Jing, Leilei Lu and Zhikuan Ji. 2015. A study on the impact wear behaviors of 40Cr steel. Tribology Online 10(4): 273–281. Doi: 10.2474/trol.10.273.

Waterhouse, R.B. 1984. Fretting wear. Wear 100(1): 107–118. Doi: https://doi.org/10.1016/0043-1648(84)90008-5.

Waterhouse, R.B. 1992. Fretting wear. pp. 242–256. *In*: ASM Handbook, Vol. 18: Friction, Lubrication and Wear Technology. Ohio: ASM International, Metals Park.

Wayson, A.R. 1964. A study of fretting on steel. Wear 7(5): 435–450. Doi: https://doi.org/10.1016/0043-1648(64)90136-X.

Weibull, W. 1930. A statistical theory of the strength of materials.

Zmitrowicz, Alfred. 2005. Wear debris: A review of properties and constitutive models. Journal of Theoretical and Applied Mechanics 43(1): 3–35.

Chapter 5

Lubricants and Lubrication

5.1 Introduction

Typically, Tribological systems consist of four elements: (i) two matting surfaces, (ii) an interface between them, (iii) a medium in the interface, and (iv) the environment.

Generally, when introducing the term "lubricant", it simply suggests oil or grease introduced between the matting surfaces because they are the most common lubricants used in tribological systems. In fact, any substance capable of reducing friction, heat, or wear when deliberately introduced as a film between solid surfaces can be considered as a lubricant. In modern tribological applications, lubricants, according to their physical state, may be solid (e.g., graphite, molybdenum disulfide, hexagonal boron nitride, polytetrafluoroethylene), liquid (automotive and other machine oils) or gaseous (carbon dioxide, nitrogen, inert gases).

There are a number of definitions based on what constitutes a lubricant, but the most general and probably best definition is that "a lubricant is a substance which is capable of altering the nature of the surface interaction between contacting solids" (Rabinowicz, 1984). The functionality of lubricants is defined by their chemical structure and physical properties.

Most lubricants are used to reduce the friction force, the amount of wear, the degree of surface adhesion, or the interfacial temperature. However, in some applications, such as in metal cutting, the task of the lubricant may be to influence the way that chips are formed, or the nature of the surface finish. Also, it is possible that a lubricant may decrease friction while increasing wear. An example of this is in wearing-in situations where a special lubricant is used to produce fast wear during a running-in regime. In contrast, there is another example of a tribosystem in which a lubricant can reduce wear but increase friction, as in rolling bearings.

The term "lubrication" can be defined as the application of some substance(s) between two objects moving relative to each other in order to allow as much smooth operation as necessary. This is applied to solid film lubrication and fluid (liquid or gaseous) film lubrication. Fluid lubrication occurs when a thick, or thin, film of some fluid completely separates the matting surfaces. Solid lubrication arises when a soft solid film is interposed between the contacting surfaces. The situation in which the solid film arises as a result of a chemical reaction between the mating surfaces and the environment is called extreme pressure (EP) lubrication. Furthermore, a combination of solid and liquid lubrication is also feasible and may have a beneficial synergistic effect on the friction and wear performance of sliding surfaces.

However, in addition to the former definition, lubrication is also used to reduce oxidation and prevent rust, provide insulation in transformer applications, transmit mechanical power in hydraulic fluid power applications, or to seal against dust, dirt and water.

5.2 Types of Lubricants

5.2.1 Solid Lubricants

A solid lubricant is a solid material that provides lubrication between two surfaces moving in relation to one another (Campbell, 1972). Lansdown (Lansdown, 1996) introduced another definition stating that "a solid lubricant is basically any solid material, such as a thin film or a powder, that can be placed between two bearing surfaces and will shear more easily under a given load than the bearing materials themselves."

Solid lubricants are mostly required for lubrication under extreme conditions where the contacting surfaces in a tribological system must be kept effectively separated. Solid lubricants are essentially applied in nuclear reactors, high vacuum applications, aggressive environments, or any severe service conditions where lubricating oils or greases cannot be tolerated.

The crystal structures of most of these solids are unique. The atoms lying on the same layer are closely packed and strongly bonded to each other, while the layers themselves are relatively far apart, and the forces that bond them (e.g., van der Waals) are weak. When present between sliding surfaces, these layers can align themselves parallel to the direction of the relative motion and slide over one another with relative ease, thus providing low friction. In addition, strong interatomic bonding and packing in each layer is thought to help reduce wear damage. On the other hand, soft metallic lubricants have crystal structures with multiple slip planes and do not work-harden appreciably during a sliding contact. Dislocations and point defects generated during shear deformation are rapidly nullified by the frictional heat produced during the sliding contact (Erdemir, 2001).

The purpose of solid lubricants is to build up a continuous adherent hard or soft film on the rubbing surfaces. These films can be applied by mechanical, chemical, electrochemical or physical processes (2007). The main requirements of solid lubricants properties are:

i)   Low shear strength in the sliding direction.

ii) High compression strength in the direction of the applied load.

iii) Good adhesion of the solid lubricant to the substrate surface.

Solid lubricants are presented in tribological systems in any of these forms (Kopeliovich, 2012a):

i)   Coating (film) of a solid lubricant applied on the sliding surface.

ii) Composite coating consisting of particles of a solid lubricant dispersed throughout a matrix.

iii) Particles of a solid lubricant dispersed throughout the bulk of the surface.

iv) Powder of a solid lubricant delivered to the rubbing area (dry lubrication).

v)   Additives in lubricating oils or greases.

Classification of solid lubricants

The available solid lubricants can be divided into five groups:

1) Lamellar structure solids (inorganic)

In these, the atoms are bonded together in hexagonal rings forming thin parallel planes, as shown in Fig. 5.1. The planes are bonded to each other by weak Van der Waals forces. The layered structure allows a sliding movement of the parallel planes. Weak bonding between the planes determines the low shear strength and lubricating properties of the materials. Graphite, Molybdenum disulphide (MoS_2) and Boron nitride (BN) are the best known and most widely-used materials. Likewise, Cadmium chloride, Sulphides, Selenides molybdenum, Tungsten, Niobium, Tantalum, and Titanium are other examples of such materials. Additionally, new forms of carbon – fullerenes or Buckyballs (C_{60}), CaF_2 and $CaF_2 - BaF_2$ eutectic based coatings are also proposed as solid lubricants (Bhushan and Gupta, 1997; Bhushan et al., 1993).

The most important property of these materials is the ability to form a strongly adherent transferred film on to the tribological surfaces. Also, the materials at both surfaces develop a preferred orientation in order to reduce the mechanical interaction between the surfaces.

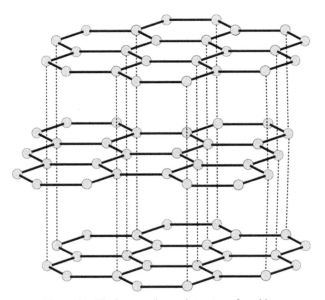

Figure 5.1: The hexagonal crystal structure of graphite.

Most lamellar solids have good wetting capability or chemical affinity for ferrous surfaces. On a rough or porous sliding surface, they fill in the valleys between asperities and/or pores, thus providing a smoother surface finish and a better support. When applied properly, these solids can also withstand extreme contact pressures without being squeezed out of the load-bearing surfaces (Erdemir, 2001).

Lamellar solids can be used in several forms, the most common ones are (Halling, 1979):

i)   Solid blocks, usually graphite and graphite carbons used in manufacturing of bearing. Carbon bearings made from amorphous carbons have higher strength and are commonly used for thrust bearings. On the other hand, highly crystalline electrographites are utilized in low load applications, such as electrical generator brushes.

ii) Dry powder or dispersion in fluids. Since Molybdenum disulphide (MoS_2) films tend to adhere better than graphite, it is often applied as a dispersion or dry powder to facilitate the assembly of close fitting parts and lubrication of metal working components.

iii) Bonded films. MoS_2 films are frequently bonded to metallic surfaces using resin binders. Such binders have maximum decomposition temperature limits, so there are limits for applied load and sliding speed.

iv) Grease additions, e.g., MoS_2 used as an additive to greases and oils in order to increase the effectiveness of lubricant under heavy loads. Additives will be extensively discussed later in the present chapter.

2) Soft metals (inorganic)

Low shear strength and high plasticity of some soft metals impose their lubrication properties. Examples of soft metals are: Lead (Pb), tin (Sn), bismuth (Bi), indium (In), cadmium (Cd) and silver (Ag). They can be used in a pure form, alloys, or in the form of coatings. Simple electroplating and vacuum evaporation can be used to deposit most of these metals as self-lubricating films, but dense and highly adherent films are produced by ion plating, sputtering, or ion-beam-assisted deposition techniques (Erdemir, 2001).

The thickness of the soft metallic film plays a major role in both friction and wear. Thin films (i.e., 0.5 to 1 μm thick) usually obtain low friction coefficients and wear rates, as shown in Fig. 5.2. However, too thin film tends to wear out quickly. Also, the friction coefficients of most soft metals tend to decrease as the ambient temperature increases, mainly because of additional softening and rapid recovery from strain hardening. Thick films result in large contact areas and hence high friction.

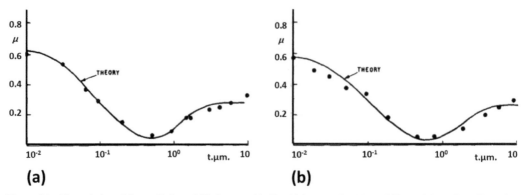

Figure 5.2: The variation of the coefficient of friction, μ, with film thickness, t, for (a) Lead film and (b) Indium film. After Sherbiney et al. with permission from Elsevier.

3) Polymers (Organic)

The molecular structure of polymeric materials consists of long-chain molecules parallel to each other. The bonding strength between the molecules is weak, therefore they may slide past one another at low shear stresses. The strength of the molecules along the chains is high due to the strong bonding between the atoms within a molecule. Such anisotropy of mechanical properties provides good lubrication properties of the materials. Other important advantages for using plastics are their lightweight, relatively inexpensive costs, and ease of fabrication. They can also be easily blended with other solids to make self-lubricating composite structures.

In addition to the former advantages, polymeric bearings have the following advantages:

i)   Vibrations absorption and quiet operation.

ii) Easy deformation to conform to mating parts.

iii) Easy machining or moulding.

iv) Low coast.

Polymers can be used in bulk and thin-film forms, or as binders for other solid lubricants. Polymer coatings can be produced on a tribological surface by first spraying or sprinkling the powders, then consolidating and curing them at high temperatures. The most widely used polymers are Polytetrafluoroethylene (PTFE), Polyimide (PI), Polyamide (PA), and Ultra High Molecular Weight Polyethylene (UHMWPE).

4) Oxides

Some oxides (fluorides and sulfates) become soft and lose thier shear strength at high temperatures and can therefore be used as lubricants. They can be used as thin or thick coating films, or it can be mixed with other solid lubricants to obtain lubrication over much wider temperature ranges. Potential applications for lubricious oxides include high-temperature seals, bearings, gears, valves and valve seats, variable stator vanes, and foil bearings (Erdemir, 2001).

Examples of oxides are (Re_2O_7, MoO_3, PbO, B_2O_3, NiO), of fluorides (CaF_2, BaF_2, SrF_2, LiF, MgF_2), and of sulfates ($CaSO_4$, $BaSO_4$, $SrSO_4$). Recent researches have demonstrated that the oxides of Ti, Ni, W, Mo, Zn, V, and B, become highly lubricious and can provide fairly low friction at elevated temperatures. Likewise, mixed oxides (CuO-Re_2O_7, CuO-MoO_3, PbO-B_2O_3, PbO-MoO_3, CoO-MoO_3, Cs_2O-MoO_3, NiO-MoO_3) can also provide wider operational ranges and can be prepared as alloys or composite structures to provide longer durability (Kanakia and Peterson, 1987).

The main disadvantages of oxide-based lubricants are that they are inherently brittle and thus may fracture easily and wear out quickly. Also, most of them do not provide proper lubrication down to room temperature.

5) New composite coatings

A new physical vapor deposition (PVD) technology is used for depositing a special barrier underlayer before the deposition of a conventional coating by chemical vapor deposition (CVD). These technologies have led to the development of a new generation of self-lubricating nanocomposite films and multilayer coatings. The hardness of these coatings could be as high as 20 GPa, while their friction coefficients are generally low even in open-air environments. Because of their hardness and low friction, they can be used in both sliding and machining applications (Gilmore et al., 1998).

Practical examples of such new coatings are composite architectures based on layers of a self-lubricating solid (MoS_2 or WS_2) and a metal, ceramic, or hard metal nitride or carbide. PbO/MoS_2 and ZnO/WS_2 nanocomposite films were also produced in recent years and tested for their lubricity and durability in a variety of environments (Walck et al., 1994). It was found that composite coating films perform significantly better during tribotesting than films composed entirely of MoS_2 or PbO and ZnO.

Electroless nickel, chromium, nickel-phosphorus coatings containing small amounts of graphite, MoS_2, PTFE, and diamond particles were also developed in recent years and used to achieve relatively thick films with self-lubricating properties (Moonir-Vaghefi et al., 1997).

6) Soaps

Soaps are metals (Lithium, Calcium, Sodium, Potassium) salts of the higher saturated and unsaturated fatty acids. They are prepared by chemical treating of oils and fats by strong alkaline solutions. A soap molecule is composed of a long non-polar hydrocarbon tale, which is hydrophobic (repelled by water) and a salt polar end, which is hydrophilic (water soluble). They often give the lowest coefficients of friction obtainable with solid lubricants but in general cannot be used above their melting points or at high loads (Mang, 2007).

The main function of soaps in lubrication technology is in the preparation of greases. The soap molecules attached to a substrate surface provide good adhesion of the soap lubricant and low coefficient of friction (Kopeliovich, 2012b).

5.2.2 Liquid Lubricants

The primary way by which a lubricant influences friction and wear is by reducing adhesion. This could be obtained by three mechanisms: absorption on the contact surfaces, chemical modification of the surface, or physical separation of the surfaces. For liquid lubricants, mechanical separation results from the response of the lubricant to being trapped between two surfaces under relative motion. Under such conditions, a liquid can support a normal load, thus providing separation between the two surfaces (Bayer, 2004). Liquid lubricants have the tendency to flow back into the region of contact, replenishing any of the lubricants that is displaced during the wearing action.

Liquid lubricants can evaporate and spread over available surfaces so that, with time, the amount of a lubricant available to the contact interface can decrease. Hence, an adequate supply of liquid lubricant should be maintained to avoid dry surfaces. They may also degrade with time as a result of oxidation, polymerization, or some other mechanism, with the consequence that the ability of the fluid to lubricate the contact may degrade.

Classification of liquid lubricants

Liquid lubricants may be classified according to their origin into three groups:

Mineral oils

Mineral lubricating oils are refined from petroleum and have a variety of physical and chemical properties. More highly refined mineral oils such as oxidation inhibitors are used for applications where higher temperatures or longer service periods require better ageing stabilities.

Synthetic oils

In applications with extremely high or extremely low temperatures or very heavy loads, synthetic lubricants may overcome the challenges better than mineral lubricants. Some types of synthetic oils are suitable for use in admixture with mineral oils (the so-called "semi-synthetic" oils) as well as additives for mineral oils. The majority of synthetic oils can be improved by additives. Examples of synthetic oils are Synthetic hydrocarbon, Chlorofluorocarbons, Esters, Silicones (polysiloxanes), Silanes, Polyphenyl ethers (PPEs), and Perfluoropolyether (PFPE).

Natural biodegradable lubricants

Biodegradable products of vegetable or animal origin are also considered for liquid lubrication, e.g., the effects of sunflower oil added to a base oil on the performance of journal bearings. The use of vegetable oils as lubricants is likely to increase due to environmental and governmental requirements and is becoming increasingly important (Ahmed and Nassar, 2013). Petroleum oils are also excellent lubricants and can be used up to a maximum temperature of 130–200°C (Bhushan, 2000). Additionally, water, refrigerants, and other liquids can provide some lubrication.

5.2.3 Gaseous Lubricants

Lubrication with a gas is analogous in many respects to lubrication with a liquid, since the same principles of fluid-film lubrication apply. Gaseous lubricants are applied in aerodynamic and aerostatic bearings where liquid lubricants would freeze or decompose. They offer several advantages over liquid lubricants. First, they are the lubricants with lowest known viscosity. The chemical and physical properties of most gases remain unchanged over a wide range of temperature. Clean lubricants are primarily applied in food, pharmaceutical and electronic industries. Another useful property of gases is that, in contrast to liquid lubricants, their viscosities increase with temperature. However, the main drawback of a gaseous lubricant is its relatively low viscosity which generally limits the load-carrying capacity of self-acting, aerodynamic bearings.

Examples of gaseous lubricants are: *Air* (which is suitable up to 650°C), CO (to 650°C), He and N (to 1000°C and more), *Hydrogen* (higher cooling capacity, but flammable) or *Methane* for gas turbines, working fluids like steam for steam turbines, and *fuel* or *exhaust gases* for internal combustion turbines. Reactive compounds such as CF_2Br_2 and SF_6 are sometimes added to the gas in closed circuit systems to improve its performance functions in critical and transition regimes. The use of gas lubrication is increasing in high-speed equipment: precision optical instruments, dental drilling, measuring instruments, electronic computers, precision grinding spindles, and in the pharmaceutical, chemical, food, textile and nuclear industries where contamination must be avoided and the loads are low. Externally pressurized gas bearing working in the aerostatic region can carry higher loads, are less critical on clearances and tolerances and can be used even at lower speeds (Václav, 1992).

5.3 Fluid Film Lubrication

Indeed, the modern period of our understanding of lubrication began with the work of Petroff (Petroff, 1883), Beauchamp Tower (Tower, 1884), and Osborne Reynolds (Osborne, 1886). Since then, developments in hydrodynamic bearing theory and practice have been extremely rapid in meeting the demand for reliable bearings in new machinery.

Till now, the *Reynolds equation* introduced in 1886 still remains the main tool for the calculation and design of rubbing machine elements. The equation is instrumental in designing deformable surfaces as in the case of rolling contact bearings, mechanical seals, gears, jet engines, internal combustion engines, artificial human joints, and many other areas. The *theory of lubrication* is concerned with the solution of the Reynolds equation, sometimes in combination with the equation of energy, under various lubricating conditions. The theory studied the case of introducing a fluid lubricant to a rotating shaft (journal) in a bearing, as shown in Fig. 5.3.

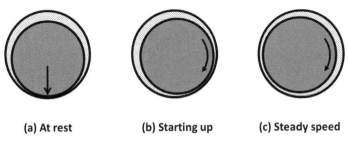

| (a) At rest | (b) Starting up | (c) Steady speed |

Figure 5.3: The three positions of a shaft rotating in a bearing.

5.3.1 *Laminar and Turbulent Flow*

In general, the consideration of a laminar viscous flow is the base for the analysis of the fluid flow. Laminar flow exists in bearings and machine elements at low relative velocities, in which the fluid flows in a series of parallel or concentric surfaces or layers, with relative velocities and without mixing between the layers. In contrast, at high velocities, turbulence occurs in the fluid films. The critical flow velocity at which turbulence is initiated is based on the non-dimensional Reynolds number (Re) which is the ratio of inertial to viscous forces, as given in Equation (5.1). Generally, a Reynolds number of about 2000 is the critical value above which turbulence occurs (Bhushan, 2013).

$$\mathrm{Re} = \frac{\rho v d}{\eta} = \frac{vd}{v} \qquad (5.1)$$

where

v	linear velocity,
d	diameter of a tube or the film thickness,
η	dynamic viscosity,
v	kinematic viscosity.

5.3.2 *Derivation of Reynolds Equation*

The Reynolds equation establishes a relation between the geometry of the surfaces, relative sliding velocity, the property of the fluid and the magnitude of the normal load the bearing can support.

There are several common methods for deriving the Reynolds Equation (Cameron and Robertson, 1962; Bhushan, 2000; Bhushan, 2013; Halling, 1979). These differ basically in the form in which the continuity condition is applied.

The principles of the theory are based on the assumption that the lubricant can be treated as isoviscous and laminar and the fluid film is of negligible curvature. The Navier-Stokes equations of motion and the continuity equation were combined into a single equation in lubricant pressure, the so called Reynolds equation, assuming:

1) constant viscosity, Newtonian flow
2) thin film geometry
3) negligible body force
4) no-slip boundary conditions

Consider two surfaces separated by the variable distance h, as illustrated in Fig. 5.4. For simplicity, the upper surface is assumed to be curved and the lower surface a plane. Also, assume that between the surfaces a fluid with density ρ and local viscosity is η. The viscosity is constant with respect to y. The local fluid velocity components in the directions of the positive x, y and z axes are u, v and w, respectively. The surface velocity components of the upper and lower surfaces in the directions of the positive x, y and z axes are denoted by U_1, V_1, W_1 and U_2, V_2, W_2, respectively.

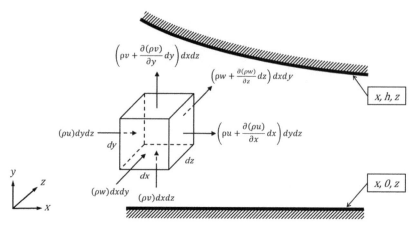

Figure 5.4: Schematic of mass flow between two surfaces and stresses acting on a fluid element, where X, Y, Z are the components of body forces in coordinate directions.

Considering an elementary volume of geometrical space whose sides have lengths dx, dy, dz, as shown in Fig. 5.4.

The rate of flow of fluid into the element is

$$(\rho u)\,dydz + (\rho v)\,dxdz + (\rho w)\,dxdy$$

while the rate of flow out of the element is

$$\left(\rho u + \frac{\partial(\rho u)}{\partial x}dx\right)dydz + \left(\rho v + \frac{\partial(\rho v)}{\partial y}dy\right)dxdz + \left(\rho w + \frac{\partial(\rho w)}{\partial z}dz\right)dxdy$$

Since we assume the fluid to be incompressible, these must be equal and, hence, we obtain the form of the continuity equation for incompressible fluid and steady flow

$$\frac{\partial(\rho u)}{\partial x} + \frac{\partial(\rho v)}{\partial y} + \frac{\partial(\rho w)}{\partial z} = 0 \tag{5.2}$$

The equation can be written as

$$\frac{\partial u}{\partial x} + \frac{\partial v}{\partial y} + \frac{\partial w}{\partial z} = 0 \tag{5.3}$$

If we consider the flat lower surface to lie in the plane $y = 0$ then the integrating the continuity Equation (5.3) across the lubricant film thickness h (from $y = 0$ to $y = h$) will be

$$\int_0^h \frac{\partial u}{\partial x}dy + \int_0^h \frac{\partial v}{\partial y}dy + \int_0^h \frac{\partial w}{\partial z}dy = 0 \tag{5.4}$$

Expressions for the fluid velocity components u, v, and w are obtained by consideration of the pressure on an elementary volume of fluid. The equations of motion under the assumptions stated above will take the form

$$\frac{\partial p}{\partial x} = \eta \frac{\partial^2 u}{\partial y^2} \tag{5.5}$$

$$\frac{\partial p}{\partial z} = \eta \frac{\partial^2 w}{\partial y^2} \tag{5.6}$$

$$\frac{\partial p}{\partial y} = 0 \tag{5.7}$$

Equation (5.7) indicated that the pressure is invariant in the y direction, i.e., across the film thickness. Integrating Equations (5.5) and (5.6) with respect to y

$$u = \frac{1}{2\eta} \frac{\partial p}{\partial x} y^2 + Ay + B \qquad (5.8)$$

$$w = \frac{1}{2\eta} \frac{\partial p}{\partial z} y^2 + Cy + D \qquad (5.9)$$

where A, B, C, D are the integration constant and can be evaluated according to the assumption of thin film geometry, the boundary conditions should be applied

$$u = U_1, w = 0 \text{ at } y = 0$$

$$u = U_2, w = 0 \text{ at } y = h$$

where U_1 and U_2 represent the velocity of the bearing surfaces.

Evaluation of the integration constants leads to the following velocity distribution for a fluid flow, under laminar conditions through a small clearance h as a result of the pressure gradient $\partial p / \partial z$, gives

$$u = \frac{1}{2\eta} \frac{\partial p}{\partial x} (y^2 - yh) + \left(1 - \frac{y}{h}\right) U_1 + \frac{y}{h} U_2 \qquad (5.10)$$

$$w = \frac{1}{2\eta} \frac{\partial p}{\partial z} (y^2 - yh) + \left(1 - \frac{y}{h}\right) W_1 + \frac{y}{h} W_2 \qquad (5.11)$$

Note that, according to boundary conditions, $W_1 = W_2 = 0$

Substituting u and w from Equations (5.10), (5.11) into Equation (5.4) and performing the integrations gives on rearrangement

$$[v]_0^h = -\frac{\partial}{\partial x}\left[\frac{1}{2\eta}\frac{\partial p}{\partial x}\int_0^h (y^2 - yh)dy\right] - \frac{\partial}{\partial z}\left[\frac{1}{2\eta}\frac{\partial}{\partial z}\int_0^h (y^2 - yh)dy\right] - \frac{\partial}{\partial x}\int_0^h\left[\left(1 - \frac{y}{h}\right)U_1 + \frac{y}{h}U_2\right]dy + U_2\frac{\partial h}{\partial x} \qquad (5.12)$$

Recognizing that

$$[v]_0^h = -(V_2 - V_1) \qquad (5.13)$$

The integration yields to the following Reynolds equation for lubrication pressure

$$\frac{\partial}{\partial x}\left(\frac{h^3}{\eta}\frac{\partial p}{\partial x}\right) + \frac{\partial}{\partial z}\left(\frac{h^3}{\eta}\frac{\partial p}{\partial z}\right) = 6(U_1 + U_2)\frac{\partial h}{\partial x} + 12(V_2 - V_1) \qquad (5.14)$$

There is the expected symmetry between the U and V terms on the right hand side. We note that if a positive pressure is to be developed in the film, it is necessary that these terms should have a resultant negative value. Recognize that the first term on the right hand side depends upon the surface velocity components in the x-direction and upon the slope of the upper surface. It is known as 'wedge' or 'velocity variation' term (Cameron and Robertson, 1962). The second term, being the difference of the vertical velocity components of the surfaces, has been called the "squeeze" term and it is not difficult to picture how the outward flow in a squeezed oil film contributes to the development of pressure.

In his original paper, published in 1886, Osborne Reynolds (Reynolds, 1886) put forward his equation in the form

$$\frac{\partial}{\partial x}\left(h^3\frac{\partial p}{\partial x}\right) + \frac{\partial}{\partial z}\left(h^3\frac{\partial p}{\partial z}\right) = 6\eta\left\{(U_1 + U_2)\frac{\partial h}{\partial x}\right\} + 2V_1 \qquad (5.15)$$

where an appropriate change of symbols from those actually used by Reynolds has been used to confirm with the symbolism of this book. Furthermore, Reynolds considered a case where there was no motion of the surfaces in the z direction, so that there are no W terms and the lower surface was taken to have no vertical component of velocity so that there is no V_2.

If we neglect pressure variation in the z direction, assume constant viscosity and have pure sliding ($V_1 = V_2 = 0$), the Reynolds equation becomes

$$\frac{d}{dx}\left(h^3 \frac{dp}{dx} \right) = 6\eta \, (U_1 + U_2)\frac{\partial h}{\partial x} \tag{5.16}$$

Liquids can be assumed to be incompressible, i.e., their density remains constant during flow. For an incompressible fluid and unidirectional tangential (rolling or sliding) motion, the Reynolds equation is given as

$$\frac{\partial}{\partial x}\left(\frac{h^3}{\eta}\frac{\partial p}{\partial x} \right) + \frac{\partial}{\partial y}\left(\frac{h^3}{\eta}\frac{\partial p}{\partial y} \right) = 12\bar{u}\frac{\partial(h)}{\partial x} \tag{5.17}$$

For a gas-lubricated bearing with perfect gas,

$$p = \rho RT \tag{5.18}$$

where R is gas constant and T is the absolute temperature. Therefore, ρ is replaced by p in the Reynolds equation. For unidirectional tangential (rolling or sliding) motion (Bhushan, 2013),

$$\frac{\partial}{\partial x}\left(\frac{ph^3}{\eta}\frac{\partial p}{\partial x} \right) + \frac{\partial}{\partial y}\left(\frac{ph^3}{\eta}\frac{\partial p}{\partial y} \right) = 12\bar{u}\frac{\partial(ph)}{\partial x} \tag{5.19}$$

In general, the Reynolds equation has to be solved using numerical methods such as finite difference, or finite element. Depending on the boundary conditions and the considered geometry, however, analytical solutions can be obtained under certain assumptions. Corresponding MATLAB code can also be applied upon solving of 1-D Reynolds equation using Finite Difference Method. A Matlab code for calculation of a semi-analytical solution of Reynolds equation using Grubin's approximation (Stachowiak, 1993; Morales, 2007). At first, the Reynolds equation is integrated analytically using Grubin's assumptions with respect to spatial coordinate and further numerically integrated with respect to temporal coordinate. The solution is then compared to the analytical Grubin's solution for the central film thickness.

Various generalized Reynolds equations were derived to weaken the assumptions used to derive the classical form. For example, compressible, non-Newtonian lubricant behavior can be considered. In tribology, the Reynolds equation is used to predict the thickness of the lubricant film, but also to predict the friction developed by the lubricant on the surfaces. Since many tribological contacts operate in highly loaded regime and thin films, the shear rates can be very high (in the order of 10^7–10^9). Many of the typical lubricants start to behave non-Newtonian in the contact conditions and, therefore, the Reynolds equation was generalized to the case of non-Newtonian lubricant (Tribonet, 2007).

5.3.3 Lubrication Regimes

Liquid lubricants can be brought into a converging contact due to rotation and pressure generated between the bodies.

Regarding the lubricant film thickness between mating surfaces in relative motion, four different forms of lubrication can be identified for self-pressure generating lubricated contacts: (i) Hydrodynamic, (ii) Elastohydrodynamic, (iii) Mixed, (iv) Boundary.

The ratio of the minimal film thickness h to the asperity height Ra is the parameter usually used for characterizing these regimes. The Stribeck curve, Fig. 5.5, is a fundamental concept in the field of tribology that can be applied to illustrate these regimes. The curve represents the dependence of the friction coefficient on the so-called Hersey number, bearing number, or bearing ratio

$$\text{Hersey number} = \frac{\eta u}{p} \tag{5.20}$$

where η is the kinematic viscosity, u the velocity of the moving part, and p the mean pressure in the fluid film.

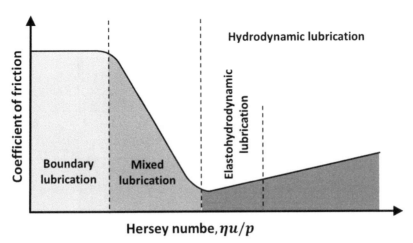

Figure 5.5: The Stribeck curve which summarizes the lubrication regimes by describing the relationship between friction, speed, load, oil viscosity, and oil film thickness.

i) Hydrodynamic lubrication

Hydrodynamic lubrication (HL or HD) or full film lubrication is a stable regime of lubrication and metal-to-metal contact does not occur during the steady-state operation of the bearing, as illustrated in Fig. 5.6.

It is the condition when the load carrying surfaces are completely separated by a relatively thick film of lubricant and is mostly associated with film thicknesses near or more than 5 µm in oil bearings (i.e., $h \gg R_a$) (Bhushan, 2013).

There are four essential elements in hydrodynamic lubrication: A liquid, a relative motion, the viscous properties of the liquid, and the geometry of the surfaces between which the convergent wedge of the fluid is produced (Mang, 2007). If the bearing is of a convergent shape in the direction of motion, the fluid adhering to the moving surface will be dragged into the narrowing clearance (wedge). This generates a fluid pressure sufficient to separate the surfaces and carry the applied load. This is the mechanism function of the hydrodynamic bearings widely used in modern industry. The flow of the lubricant film is laminar, but at thicknesses above 20 µm it becomes turbulent, leading to undesirable friction losses (Paulo Davim, 2011). However, a fluid film can also be generated solely by a reciprocating or oscillating motion in the normal direction towards each other (squeeze) which may be fixed or variable in magnitude.

Hydrodynamic lubrication is an excellent method of lubrication since it is possible to achieve coefficients of friction as low as 0.001, and there is no wear between the moving parts. On the other hand, adhesive wear may occur during start–stop operations.

The most famous hydrodynamic bearings are journal bearings and thrust bearings; these bearings might also be called "thick film" bearings. The operating speed of such bearings must be sufficient to allow the formation and maintenance of the fluid film.

Journal bearings are designed to support radial loads on rotating shafts, as illustrated in Fig. 5.7. It operates under average pressures of order 10 to 10^7 N/m^2 and speeds of revolution N of the order of 100 r/min. The loading conditions on the shaft can be characterized by a single dimensionless group, the Sommerfeld number S, defined by

$$S = \frac{\eta N}{P}\left(\frac{R}{C}\right)^2 \tag{5.21}$$

where R is the radius of the rotating shaft, C is the radial clearance between the bearing and the shaft, η is the viscosity of the fluid, P is the specific load given by

$$P = \frac{W}{LD} \tag{5.22}$$

where W is the applied normal load, L and D are the length and the diameter of the bearing, respectively.

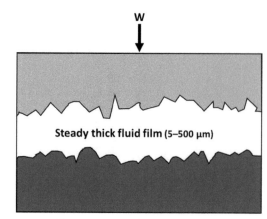

Figure 5.6: Hydrodynamic lubrication.

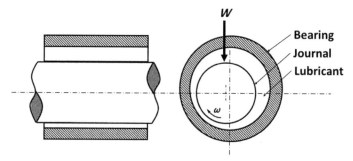

Figure 5.7: Schematic of a journal bearing.

The torque G to overcome the viscous resistance of the lubricant in a full bearing may be calculated fairly reliably, simply by assuming that the journal and bearing run concentrically

$$G = \eta N \left(\frac{4\pi^2 R^3 L}{60c} \right)$$

(5.23)

where c is the radial clearance.

Thrust bearings are a particular type of rotary bearings and, they are designed to support axial or thrust loads, as shown in Fig. 5.8. A typical thrust bearing consists of two inclined plane surfaces in relative motion to one another. The geometry of the bearing surfaces is commonly rectangular, to accommodate a linear motion, or sector shaped, to support a rotation, but other geometries are possible.

In both types of bearings, a special attention must be paid to two important issues:

a) Corrosive wear of the bearing surfaces can occur as a result of interaction with the lubricant. One of the most effective ways to minimize corrosive wear is by the participation of the lubricant and bearing surface in the formation of a relatively complete and inert film on the bearing surface (Bhushan, 2013).

b) The heating of the lubricant by the frictional force, since viscosity is temperature dependent. One method of accomplishing this is to maintain the desired viscosity of the fluid by cycling the lubricant through a cooling reservoir. Additives (viscosity index improvers) can be also used to decrease the viscosity's temperature dependence.

ii) Elastohydrodynamic lubrication

Elastohydrodynamic lubrication (which is commonly abbreviated as EHL or EHD) is a special case of hydrodynamic lubrication attributed to the high hydrodynamic pressure within the contact. It occurs in

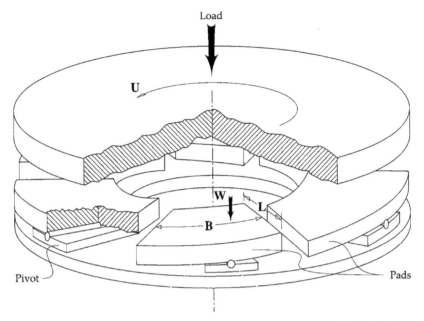

Figure 5.8: Schematic diagram of thrust bearing. After Stachowiak and Batchelor (Stachowiak, 2006) with permission from Elsevier publishing.

rolling and line contacts, such as rolling elements, gear teeth, cams and cylindrical roller bearings. In such cases, there is some interaction between the asperities. The applied load is high enough for the surfaces to elastically deform during the hydrodynamic action. Typically, in this lubrication regime, film thicknesses usually range from 0.5–5 μm, while the pressure is in the order of 1 GPa (Bhushan, 2000). Therefore, EHL is governed largely by surface finish, alignment and lubricant viscosity.

Figure 5.9 shows the distribution of the film pressure and film thickness between two steel cylinders in a rolling contact.

The general features of the figure are:

a) the film thickness h_c is nearly uniform over the contact zone, but displays a sudden decrease h_{min} just upstream of the trailing edge.

b) the pressure distribution curve follows the Hertzian ellipse over most of the contact zone, but a sharp second pressure maximum manifests itself at high speeds and light loads.

It is postulated that the EHL contact starts with a slowly converging inlet region where the lubricant is entrained and a hydrodynamic pressure is generated. The film pressure gradually increases in the inlet region until it reaches the leading edge of the Hertzian region where the pressure quickly builds to values that are essentially equal to the Hertzian contact stress. Under high pressure, the lubricant viscosity increases exponentially to the extent that the lubricant cannot escape because its viscosity is too high. Within the Hertzian region, the bodies are separated by a constant film thickness. At the end of the Hertzian region there is a constriction near the outlet that forms the minimum film thickness. Within the Hertzian region, the film pressure follows the Hertzian pressure, except for a sharp spike in pressure just upstream of the constriction at the outlet. Within the constriction, the film pressure drops rapidly to the atmospheric pressure (Errichello, 2015; Ren et al., 2002).

The development of an EHL film thickness formula requires a numerical solution of a coupled Reynolds equation and the elastic deformation of bodies. There are various formulas to estimate the EHL film thickness in a line contact configuration in terms of material properties, geometry, viscosity and pressure-viscosity coefficient of the lubricant in the inlet region (Anuradha and Kumar, 2012; Conry et al., 1987; Kumar et al., 2008; Dowson and Higginson, 1961; Lubrecht et al., 2009). Based on numerical

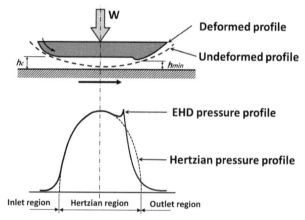

Figure 5.9: Distribution of film pressure and film thickness between two steel cylinders in rolling contact.

simulation results, Dowson and Higginson (Dowson and Higginson, 1961; Lubrecht et al., 2009) established a pioneer empirical formula to determine the minimum film thickness at the outlet of the EHL conjunction for line contact problems under fully flooded, isothermal conditions. The formula was revised by Dowson (Dowson, 1968) to make it compatible with the law of dimensionless analysis. In dimensional form, the equation becomes

$$h_{min} = 2.65 \underbrace{(\alpha^{0.54}\eta^{0.7}E'^{-0.03})}_{\text{material properties}} \underbrace{(R_e^{0.43}L^{0.13})}_{\text{geometry}} \underbrace{(u_e^{0.7}W^{-0.13})}_{\text{operation conditions}} \tag{5.24}$$

where

η viscosity,

α pressure-viscosity coefficient of the lubricant,

E' effective Young's modulus,

R_e geometrical average radius,

L contact length,

u_e rotation speed of bearing,

W applied load.

For most of the EHL regimes, the lubricant film thickness in the centre is approximately 4/3 of the minimum film thickness (i.e., $h_c \approx 4/3\ h_{min}$).

Recently, Anuradha et al. (Anuradha and Kumar, 2012) introduced a new formula for estimating the minimum film thickness in EHL rolling/sliding line contacts considering a shear thinning behavior. The proposed formula uses the reciprocal asymptotic iso-viscous pressure coefficient (α^*), which accommodates the variation of a piezo-viscous response throughout the entire range of pressure. The simulation results obtained are subjected to a regression analysis to yield the following minimum film thickness equation

$$\frac{h_{min}}{R} = 3.8 U^{0.702} G^{*0.516} W^{-0.123} \bar{R}_m \bar{S}_m \tag{5.25}$$

$$\bar{R}_m = \left(1 + 0.096 \frac{U^{0.748} W^{0.294}}{G^{*0.014} \bar{G}_{cr}^{1.833}}\right)^{-0.65(1-n)^{1.37}} \tag{5.26}$$

$$\bar{S}_m = \left(1 + 9.384 \frac{S^{3.07} U^{0.021} W^{-0.018}}{G^{*0.434} \bar{G}_{cr}^{0.134}}\right)^{-0.58(1-n)^{1.58}} \tag{5.27}$$

where

G^* dimensionless material parameter ($G^* = \alpha^* E'$),

R equivalent radius of contact,

S slide to roll ratio,

U dimensionless speed parameter ($U = \dfrac{\eta_o u_o}{E'R}$),

W applied load,

G_{cr}, n shear-thinning parameters (Bair, 2004),

u_o average rolling speed $u_o = (u_1 + u_2)/2$,

u_1, u_2 velocities of the bearing upper and lower surfaces,

α^* piezo-viscous coefficient,

η_o viscosity at $p = 0$

note that $\bar{R}_m$ and $\bar{S}_m$ reduced to 1 for $n = 1$ (Anuradha and Kumar, 2012).

It is important to mention one of the most significant parameters associated with EHL, which is called lambda ratio (λ). This ratio indicates the condition of the lubricant film formation and reflects the severity of asperity contact between mating surfaces [modern tribology handbook]. For ($3 > \lambda > 1$) the regime is called mixed or partial EHL, where the local lubrication film can be interrupted at the tip of tall asperities. The majority of rolling bearings are running in this regime. While ($1 > \lambda$) indicates that the contact operates in the boundary lubrication regime, in which severe surface distress is expected.

iii) Mixed lubrication

Mixed lubrication or partial lubrication regime is an intermediate stage between boundary lubrication and either hydrodynamic or elastohydrodynamic lubrication. As with all lubrication regimes, the transition into and out of mixed lubrication is not clearly defined. However, the transition between lubrication regimes depends on the shape of the bodies in contact, conformal or non-conformal. In a conformal contact the regime will move from hydrodynamic lubrication into mixed and then boundary lubrication as the load increases. A non-conformal contact will move from elastohydrodynamic lubrication into mixed lubrication and finally into boundary lubrication.

As mentioned before, mixed lubrication occurs when $\lambda = 1 \sim 3$ (most authors believe it takes place at about $\lambda = 3$) and the fluid film thickness h is less than 1 μm but greater than 0.01 μm. This indicates a strong dependence of the mixed lubrication on the surface roughness of contacting surfaces. However, it is difficult to ascertain the lubrication regime with any certainty using the concept of specific film thickness (the ratio of the lubricating oil film thickness to the composite surface roughness of the two contacting surfaces). The specific film thickness, though it is a very important parameter to identify the regime of operation of a journal bearing, does not conclusively establish the operative lubrication regime (Hirani, 2016).

A mixed regime occurs in many mechanical components, including gears, cams and valves at the operating conditions of very heavy loads, slow relative speed, insufficient surface area or if temperature is sufficiently high as to significantly reduce the lubricant viscosity. In such cases, the formation of the thick film necessary for hydrodynamic lubrication becomes difficult and interacting surfaces contact each other at several locations and tribo-surfaces operate in a mixed lubrication. Furthermore, the key damage processes that limit the performance of these components, such as wear, micro-pitting and scuffing, take place during this regime.

Figure 5.10 depicts a model of mixed lubrication regime where the applied load is supported by both the fluid film and the asperity contacts. On the figure, two regions could be distinguished: First, where a fluid film separates the tribo-surfaces, and second, where the asperities are in contact. Contacting surfaces and increased wear rate could lead to a cycle of adhesion, metal transfer, wear particle formation and eventual seizure. The worn out particles may also scratch the contacting surfaces and increase the wear rate.

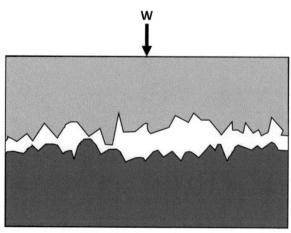

Figure 5.10: Mixed lubrication.

Therefore, the friction force in a conformal bearing comes from two parts, the contributions from viscous shear and the asperity contact. The friction due to a fluid viscous shear, also known as fluid friction, is the friction force between the fluid layers that are moving relative to each other. The friction due to the asperity contact, or dry friction, is the resisting force which prevents the relative motion of two sliding surfaces in contact (Wang, 2013). The friction force F_f in the mixed lubrication regime is the sum of the friction due to viscous shear τ and the friction due to asperity contact and sliding specified in the boundary lubrication, therefore,

$$F_f = F_{f \cdot fl} + F_{f \cdot C} \tag{5.28}$$

where $F_{f,fl}$ is a hydrodynamic friction force given by

$$F_{f \cdot fl} = \iint_{Afl} \tau dA_{fl} \tag{5.29}$$

The parameter $F_{f,C}$ represents the asperity interacting friction force given by

$$F_{f,C} = \iint_{Ac_i} f_c P_{c_i} dA_{c_i} \tag{5.30}$$

The overall coefficient of friction can be expressed as

$$\mu = \frac{F_f}{F_T} = \frac{F_{f \cdot fl} + F_{f \cdot C}}{F_T} \tag{5.31}$$

where the total normal load F_T is shared by the hydrodynamic force F_{fl} and the asperity interaction force F_C.

On the other hand, Echávarri Otero et al. (Echávarri Otero et al., 2017) presented a mixed lubrication model for highly-loaded non-conformal contacts. This model is based on a simplified analytical procedure that considers the thermal effects in the contact and the rheology of the lubricant. According to their procedure, the friction coefficient could be estimated with reasonable precision for different lubricants and operating conditions, both for point and line contacts. A specific film thickness and an average contact temperature could be also predicted.

iv) Boundary lubrication (BL)

Boundary lubrication (BL) regime is the left-hand section of the mixed lubrication region, see Fig. 5.5. This regime of lubrication occurs under high load and low speed conditions, which result in a decrease in the fluid viscosity. Under such severe conditions, the solid surfaces are so close together that appreciable contact between opposing asperities is possible. The coefficient of friction in boundary lubrication can

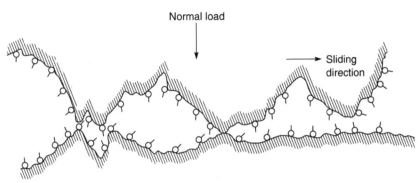

Figure 5.11: A boundary film.

increase sharply and approach high levels (about 0.1 or much higher), therefore, this regime governs the life of the components subject to wear.

While fluid-film lubrication is the desired mode of operation, a boundary lubrication regime cannot be avoided. It usually occurs in bearings, cam and tappet interfaces, piston rings and liner interfaces, pumps, transmissions, etc. Boundary lubrication also occurs during equipment startup and shutdown, when the bearing may operate in boundary rather than in unbroken fluid film conditions. Also, it can be found in toothed gear contacts, or in reciprocating motion (e.g., car valve on valve seat).

The boundary films are formed by physical adsorption, chemical adsorption and chemical reaction. The concept of boundary film is represented in Fig. 5.11, which shows a microscopic cross section of films on two surfaces and areas of asperity contact. A good boundary lubricant should have a high degree of interaction between its molecules and the sliding surface. As a general rule, liquids are good lubricants when they are polar and, thus, able to be adsorbed on solid surfaces (Bhushan, 2000). In addition to the polarity of liquid lubricants, the shape of their molecules governs the effectiveness, which determines whether they can form a dense and thick layer on the solid surface.

On the other hand, the solid surfaces should have a high wetting and a high surface energy, so there will be a strong tendency for the lubricant to wet as well as to adsorb on the surface easily. Furthermore, the surfaces should be hydrophobic, i.e., highly functional with polar groups and dangling bonds (unpaired electrons) so that they can react with lubricant molecules and adsorb them.

The effectiveness of boundary lubrication is strongly dependent on the contact temperature. As a result, a breakdown or failure of the lubricant film occurs by melting of a solid film and degradation of liquid films. It was reported that friction and wear with lubricants containing physically or chemically adsorbed additives deteriorate rapidly when the contact temperature reaches a critical value at which the additive molecules adsorbed (Fein et al., 1959). Moreover, failure in boundary lubrication occurs also by adhesive and chemical (corrosive) wear.

In order to enhance the formation of a chemically reacted boundary film, additives can be added to liquid lubricants. Additives are adsorbed at the surface to form a condensed film or react with the surface to form a metallic soap. It is then found that the addition of a small percent of a fatty acid, alcohol or ester may significantly improve the lubrication, even if the thickness of the lubricant film is no larger than around 10^{-2} μm. The role of additives will be covered in detail in this chapter.

Due to the different chemical composition of the lubricants, the appearance and morphology of the boundary films can be patchy, continuous or discrete, and have different colors, from green to brown to black. Several boundary lubricating films are illustrated in Fig. 5.12. These are photomicrographs taken of the wear scars tests conducted on silicon nitride at 2 GPa mean pressures (Gates and Hsu, 1995; Hsu and Gates, 2005). The films range from fluid-like, observed for 2-ethyl hexyl ZDP, Fig. 5.12(A) and calcium phenate, Fig. 5.12(B), to the more solid-like films of tricresyl phosphate, Fig. 5.12(C), and magnesium sulfonate, Fig. 5.12(D).

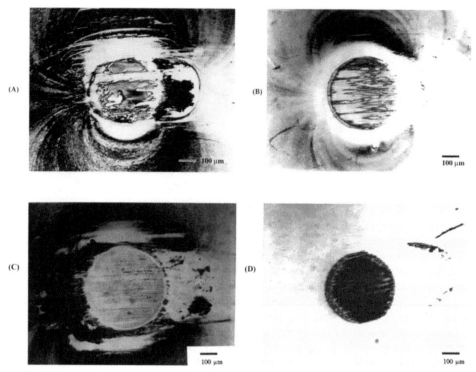

Figure 5.12: Optical micrographs of wear scars from wear tests on silicon nitride using paraffin oil. After Gates et al. (Hsu and Gates, 2005).

Equation (5.32) was introduced as an expression to estimate the fluid film thickness h based on modeling the lubricant rheology in the contact (Dowson and Higginson, 1959; Bhushan, 2000)

$$h = 1.63 \left\{ \frac{\eta_0^{0.7} \alpha_0^{0.54} V^{0.7} R^{0.43}}{L^{0.13} E'^{0.03}} \right\} \tag{5.32}$$

where

E' effective elastic modulus
R radius of bearing curvature
L bearing length
V surface velocity
α viscosity-pressure coefficient
η viscosity

5.3.4 Hydrostatic Lubrication

In contrast to the former types of lubrication, hydrostatic lubrication is a different type in which the contacting surfaces are completely separated by a thick film of liquid or gas fluid. The fluid lubricants are forced between the friction surfaces by an external pressure source, a pump, which feeds the pressurized fluid to the film. Subsequently, the friction surfaces can be separated from each other even when they are still at rest (at zero sliding speed). Thickness of hydrostatic films usually has values up to 100 μm and, therefore, they prevent contact between the asperities of even the roughest surfaces.

Hydrostatic lubrication differs from hydrodynamic lubrication in that the contact pressure is generated by an external pump instead of by viscous drag. Hydrostatic lubrication can be used instead of hydrodynamic lubrication when the latter proves to be not very effective. The main advantages of this type of lubrication are low friction, negligible wear, high load capacity, high degree of stiffness and an ability to dampen vibrations. However, high equipment costs, high installation expenses and complicated design are the main disadvantages.

There are several bearing applications of hydrostatic lubrication, including radial, axial, spindle, rotary table and turbine bearing. Furthermore, externally pressurized lubrication is used in the entire field of mechanical engineering, for large machines, where speed is in general low, such as in large telescopes and radar tracking units. High stiffness and damping of these bearings also provide high positioning accuracy in small high-velocity machinery, such as bearings in machine tools, high-speed dental drills, gyroscopes and ultracentrifuges.

Hydrostatic bearings

In machinery, hydrostatic bearings are important parts for lubrication and are essential for a reliable motion performance. They are designed for use with both incompressible and compressible fluids. In such bearings, the load-carrying capacity is achieved by fluid pads. The principle of these types of bearings is that an external pressure supply is used to continuously force a fluid lubricant through inlet channels into chambers between the bearing surfaces, see Fig. 5.13. That means these bearing surfaces are always separated by a thin lubricant film which prevents any friction between the bearing surfaces. This allows highly precise position control in the sub-micrometre range (Bhushan, 2013; Stachowiak, 1993).

As shown in the figure, the pump is used to draw the lubricant fluid from a reservoir tank to the bearing through a line filter. The pressurized fluid is supplied to the bearing before entering the central recess or pocket, then it passes through a compensating element (or flow restrictor) in which its pressure is dropped to some low value. The fluid then passes out of the bearing through the narrow gap between the bearing land and the opposing bearing surface, also known as the slider or runner. The compensating element is a simple device where a small change in flow rate results in a large pressure drop across the device. Therefore, when increasing (or decreasing) the supply pressure, the recess pressure will simply adjust itself through increasing (or decreasing) the film thickness so that the generated pressure force always balances the external load. Three common types of compensating elements for hydrostatic bearings include capillary tubes, sharp-edge orifice and constant-flow-valve compensation.

Unlike hydrodynamic bearings, hydrostatic bearings avoid the slip resistance caused by mixed friction during start-up and shut-down, which causes increased bearing wear. Hydrostatic bearings may be classified on the basis of the direction of the load that may be carried into thrust bearings, journal bearings and multidirectional bearings. An example of the essential features of a hydrostatic multi-section thrust bearing and the pressure distribution are illustrated in Fig. 5.14, and Fig. 5.15, respectively.

In Figs. 5.14 and 5.15, W is bearing load capacity, ω is the angular velocity of the rotating bearing ring, p is the hydrostatic pressure distribution, P_p is the pocket pressure, p_{en} is the feed pressure (pump

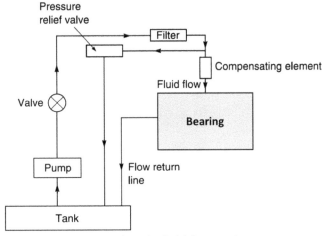

Figure 5.13: Schematic of a lubricant supply system.

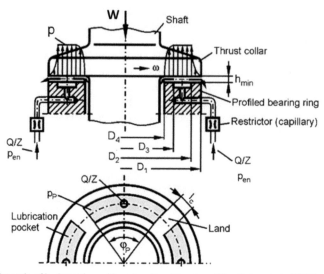

Figure 5.14: Schematic of hydrostatic multi-section thrust bearing. After Deters et al. (2014) with permission.

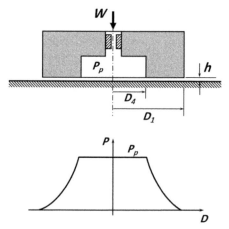

Figure 5.15: Hydrostatic pressure distribution for a circular hydrostatic thrust bearing.

pressure), φ_p is the circumferential angle of pocket, Z is the number of pockets, Q is the lubricant flow rate of bearing, D_1 is the outer diameter of profiled bearing ring, D_2 is the outer diameter of pocket, D_3 is the inner diameter of pocket, D_4 is the inner diameter of profiled bearing ring and I_c is the land width in circumferential direction across the mean thrust pad diameter (Mang, 2014; Liu, 2017). The bearing load capacity and lubricant flow rate can be obtained from:

$$W = \frac{Z_{\varphi p}}{16}\left(\frac{P_{en}}{1+\xi}\right)\left(\frac{D_1^2 - D_2^2}{\ln(D_1/D_2)} - \frac{D_3^2 - D_4^2}{\ln(D_3/D_4)}\right) \tag{5.33}$$

$$Q = \frac{Z_{\varphi p}}{12}\left(\frac{P_{en}}{1+\xi}\right)\left(\frac{h_{min}^3}{\eta}\right)\left(\frac{1}{\ln(D_1/D_2)} - \frac{1}{\ln(D_3/D_4)}\right) \tag{5.34}$$

where $\varphi_p = \left(\frac{2\pi}{Z}\right) - \left(\frac{2I_c}{D}\right)$, $D = \frac{D_1 + D_4}{2}$, ξ the resistance ratio, and η the dynamic viscosity.

The friction torque T is given by integration over the entire land outside the recess area,

$$T = \frac{\pi \eta \omega}{2h}\left(D_1^4 - D_4^4\right)$$ (5.35)

The total power loss H consists of viscous dissipation H_v and pumping loss H_p, therefore

$$
\begin{aligned}
H &= H_v + H_p \\
&= T\omega + P_p Q \\
&= \frac{\pi \eta \omega^2}{2h}\left(D_1^4 - D_4^4\right) + \frac{\pi h^3 p_p^2}{6\eta \ln(D_1/D_4)}
\end{aligned}
$$ (5.36)

The load-carrying capacity W, associated flow rate Q and pumping loss H are often expressed in nondimensional terms by defining a normalized or nondimensional load $\overline{W}$, nondimensional flow rate $\overline{Q}$ and nondimensional pumping loss $\overline{H}$, known as bearing pad coefficients (Rippel, 1960). Figure 5.16 shows the three bearing pad coefficients for various ratios of recess diameter D_4 to bearing diameter D_1.

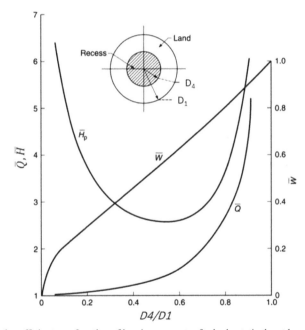

Figure 5.16: Bearing pad coefficient as a function of bearing geometry for hydrostatic thrust bearing. After Rippel (Rippel, 1960).

5.4 Gas-lubricated Bearings

5.4.1 Definition and Background

Gas-lubricated bearings are hydrodynamic bearings that use a thin film of pressurized gas, usually air, to provide a low friction load-bearing interface between friction surfaces. Although gases can perform as liquids in lubrication of bearing, in fact, the unit loads that a gas film can support are much lower. This is because the viscosity of a gas is several orders of magnitude lower than that of a liquid lubricant. As low viscosity means low friction, bearings running on gas can move at remarkable speeds without overheating.

In addition to the above, gas lubrication also has advantages in solving several contamination problems. First, the gas is clean and plentiful as well. Second, a gas bearing can use a fluid within a part of a closed system, thus, eliminating contamination of the system from the outside.

Third, the environment itself might degrade a liquid lubricant, which is the reason why gas bearings are being applied to compressors in nuclear systems. Finally, a gas bearing can easily be made quite stiff so that it adequately resists the bearing motion when the load changes (Gross, 1963).

Historically, the behavior of the shaft inside the sleeve bearings with an incompressible fluid was fully examined in the second half of the 19th century (Petrov, 1934). This was followed by the work introduced by Reynolds in 1886, "The hydrodynamic theory of lubrication and its application to the Tower's experiments" (Reynolds, 1886). The first patent on a gas bearing was granted in the U.S. in 1894. The hydrodynamic theory of lubrication for the smooth cylinder in a coaxial support was completed in the works of Petrov (Petrov, 1934). The possibility to use air as a lubricant was tested by Kingsbury (Kingsbury, 1897) and Harrison (Harrison, 1913). We should also mention that the rapid growth of gas bearing technology from research to application has been stimulated in large part by the Gas Bearing Technology Program administered by the Office of Naval Research and jointly sponsored by several agencies of the Department of Defense, AEC and NASA.

There are two broad categories of gas bearings: The self-acting (hydrodynamic) type and the externally pressurized (hydrostatic and hybrid) type. Further details about gas bearing classifications will be introduced below.

5.4.2 Fundamentals and Design Features

As mentioned before, gas bearings are hydrodynamic bearings that use the ambient gas as their working fluid. The gas used in lubrication can be air or any other gas. The hydrodynamic behavior of a gas bearing is significantly influenced by the viscosity and density of fluid and, thus, bearing performance is environment dependent, like all hydrodynamic bearings.

Gas bearings range in size from those supporting 3 mm diameter shafts in dental drills to an air-supported carriage on a 6 m long pitch-measuring machine. The gas-film thickness may vary from about 1 to 200 μm, and surface speeds range from 0 to more than 100 m/s. A representative gas bearing will have a surface area of about 6 cm^2, a gas-film thickness of about 2 μm, and will run at about 40 m/s surface speed while carrying a load of up to 5 kg (Gross, 1967; Gross, 1963).

The main assumption of the hydrodynamic lubricant theory is to neglect any change in the properties of the fluid and the pressure over the whole lubricant layer. A laminar flow occurs in bearing layer of lubrication. Equation (5.37) is the basic equation for determining the pressure of the lubricant in the sleeve bearing. The equation was obtained based on the assumptions of the hydrodynamic lubricant theory and the condition of the lubricant sticking to the surfaces of the bearing, taking into account the thinness of the lubricant layer in comparison with the dimensions of the bearing (Pavlovich Bulat and Bulat, 2013):

$$\frac{\partial}{\partial x}\left(h^3 \frac{\partial p}{\partial x}\right) + \frac{\partial}{\partial y}\left(h^3 \frac{\partial p}{\partial y}\right) = 6\eta\left\{ (U_0 + 3U_1)\frac{\partial h}{\partial x} + 2V \right\} \tag{5.37}$$

where h (x, y) is a function that determines the size of the gap at a given point, p is the average pressure in the gap; U_0 and U_1 represent the velocity of the surfaces that form the gap and the term (2V) is used to record movements of one of the gap walls, whereby the value of the function h (x, y) changes.

Equation (5.37) for a thrust bearing is written in non-dimensional form as:

$$\frac{\partial}{\partial X}\left(PH^3 \frac{\partial P}{\partial X}\right) + \lambda^3 \frac{\partial}{\partial Y}\left(PH^3 \frac{\partial P}{\partial Y}\right) \tag{5.38}$$

or
$$\Lambda \frac{\partial(PH)}{\partial X} + S\frac{\partial(PH)}{\partial T} \tag{5.39}$$

where the bearing number Λ is given by,

$$\Lambda = \frac{12\eta\bar{u}\ell}{P_a h_{min}^2} \tag{5.40}$$

and the squeeze number S,

$$S = \frac{12\eta\omega\ell^2}{p_a h_{min}^2} \tag{5.41}$$

where $X = x/\ell$, $Y = y/b$, $\lambda = \ell/b$, $T = \omega t$, $P = p/p_a$, and $H = h/h_{min}$. The ℓ and b are the bearing length (in the direction of motion) and width, respectively, ω is the frequency of vertical motion, p_a is the ambient pressure, and h_{min} is the minimum film thickness. The bearing number is also called the compressibility number (Bhushan, 2013).

Generally, design of gas bearing requirements arise as part of the design phase of a more inclusive device, such as a motor, pump or turbine. The overall requirements establish trade-offs between short and long range, and high performance and low cost (Gross, 1967).

From the bearing standpoint, the design requires an appropriate balance between load supported, stiffness, stability, gas consumption, space permitted and power that may be consumed in the bearing (Tang and Gross, 1962). It is necessary to consider the effect of external restrictions leading into the film as well as that of recesses, pools, grooves or slots about the source. On the other hand, design information must consider both steady-state and dynamic conditions. The needed design information and the parameters to be considered in each application are given in Table 5.1.

Additionally, it is important to consider the following information for the material to be employed in gas bearings (Sternlicht and Arwas, 1965).

1) Compatibility between journal and bearing
2) Metallurgical stability of journal and bearing
3) Coefficient of friction of journal and bearing material
4) Coefficient of thermal expansion

Table 5.1: Required design information.

Steady state performance information		
Type	**Design data required**	**Design specifications**
Self-Acting	Load carrying capacity Bearing or journal friction Attitude angle Minimum film thickness	Bearing geometry Speed Fluid characteristics Ambient pressure
Externally Pressurized	Load carrying capacity Bearing or journal friction Attitude angle Minimum film thickness Supply flow requirement	Bearing geometry Speed Fluid characteristics Ambient pressure Supply pressure External flow resistance
Dynamic performance information		
Self-Acting	Region of stable operation Stiffness characteristics Damping characteristics Other film force characteristics	Bearing geometry Speed Fluid characteristics Ambient pressure Rotor flexibility Time varying loads
Externally Pressurized	Region of stable operation Stiffness characteristics Damping characteristics Other film force characteristics	Bearing geometry Speed Fluid characteristics Ambient pressure Rotor flexibility Time varying loads Supply pressure External flow impedance

5) Radiation stability
6) Oxidation rate (in presence of oxygen)
7) Thermal conductivity
8) Reaction with environment (e.g., corrosion due to steam)

The characteristics of gas lubricated bearings can be seriously altered by environmental effects, including the presence of dust, condensable vapor and other types of impurities. Proper control of the environment may well determine the useful life of gas lubricated bearings.

5.4.3 Classification and Configurations

Gas bearings are generally divided into two main categories according to the manner in which the fluid pressure is developed as (Gross, 1963):

Self-acting bearings (or Gas-Dynamic Bearings GDB)

In which the fluid film pressures that serve to separate the journal from the bearing (or the thrust collar from the bearing) are generated by the viscous shear of the gas film.

Externally pressurized bearings (or Gas-Static Bearings GSB & Hybrid bearings)

Where the pressurized gas is introduced into the clearance space to achieve separation between the shaft and bearing. Load capacity is governed by the available gas supply pressure and the bearing design. However, squeeze-film bearings were introduced as a third category of gas bearings in which the load capacity is developed due to relative normal surface motion.

The three classes of bearings may each be subdivided into configurations for supporting thrust and radial loads. The principal configurations are the same as those used with liquid lubricating films. Selected examples of bearing configurations, associated with plain or curved surfaces, are introduced below.

Radial, radial-axial and thrust bearings

As the name implies, the radial bearing prevents displacement of the rotating shaft in the transverse direction. Respectively, thrust bearing prevents the displacement in the longitudinal direction and the radial-thrust bearing—in both directions simultaneously (Pavlovich Bulat and Bulat, 2013), see Fig. 5.17 (Zhang and Shan, 2007).

Tilting pad journal bearings (TPJB)

These bearings have excellent dynamic properties stemming from very small cross-coupling terms of stiffness matrix. Cross-coupling stiffness terms generate tangential forces drawing the journal around bearing center, thus promoting rotor instability. In contrast to bearings with fixed sliding surfaces, which have cross-coupling terms of the same order as the main one, in TPJB the cross-coupling terms are two or three orders lower than main terms. It means, that TPJB is inherently stable and possible rotor instability can be caused only by external destabilizing forces, generated, e.g., in labyrinth seals.

Foil gas bearings

Like conventional gas bearings, foil bearings are hydrodynamic bearings that use the ambient gas as their working fluid. These bearings are composed of compliant surfaces and generally employ a gaseous lubricant. In both geometries a rotating member drags the viscous process fluid into a converging gap, raising the fluid pressure and providing a load carrying capacity, as shown in Fig. 5.18 (Dykas et al., 2008). Gas bearings do not function in a vacuum. Further, they rely on surface speed of the moving shaft, or in the case of axial thrust foil bearings, the runner surface, to generate the fluid film pressure. Since gases are thermally stable to high temperatures, one distinct advantage of foil bearings is their ability to operate from cryogenic to very high temperatures without experiencing a major change in bearing properties (DellaCorte, 2013). Figure 5.18 illustrates the configuration of a typical journal foil bearing and Fig. 5.19 depicts a thrust foil bearing that is used to control axial motion.

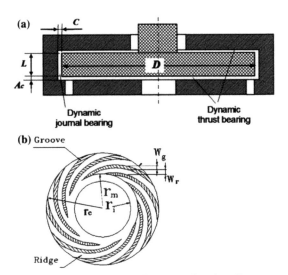

Figure 5.17: Schematic drawing of a dynamic thrust air bearing. (a) section view, (b) parameter and groove distribution of the bearing. After Zhang et al. (Zhang and Shan, 2007) with permission form Springer publishing.

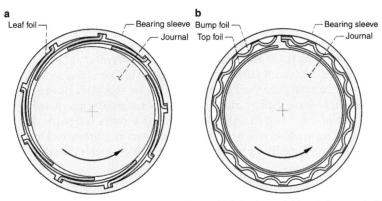

Figure 5.18: Typical foil journal bearings: (a) early overlapping leaf foil bearing and (b) early bump style foil bearing. After DellaCorte (DellaCorte, 2013) with permission from Springer publishing.

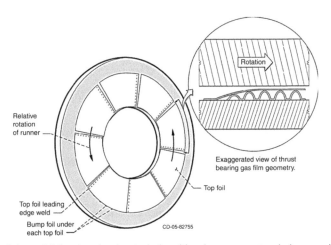

Figure 5.19: Sketch of thrust foil bearing showing typical multi-pad arrangement and close-up view of gas film. After DellaCorte (DellaCorte, 2013) with permission from Springer publishing.

5.5 Solid Film Lubrication

Over the last few decades, there was an evolution in many industrial and engineering applications. Many of these were new applications and were designed to operate in extreme environments such as high vacuum, microgravity, high/low temperatures, extreme pressure, space radiation and corrosive gas environments. In fact, these extreme environments are beyond the tolerable and usable domain of liquid and grease lubrications. In such cases, solid film lubrication may be the only choice for controlling friction and wear.

Solid lubrication is realized by imposing a solid material or self-lubricating material of low shear strength and high wear resistance between the frictional surfaces in relative motion. Reducing the coefficient of friction requires minimizing the shear strength of the interface, the surface energy, the real area of contact, and the plowing or cutting contribution. On the other hand, reducing wear generally requires one to minimize these factors while maximizing the hardness, strength and toughness of interacting materials (Miyoshi, 2001; Rabinowicz, 1995). However, a combination of solid and liquid lubricants also play an important role in controlling friction and wear performance in specialized sliding interfaces where the combined properties enhance the lubricant performance (Reeves et al., 2012).

5.5.1 Solid Lubrication Mechanisms

Typically, solid lubricants behave and function similar to liquid or grease type lubricants in sliding contact. The lubricity and durability of a solid lubricant film are controlled by a mechanism that involves imposing the solid lubricant between sliding surfaces. Solid lubricant films can be adsorbed, bonded or deposited on the contacting surfaces. Often solid lubricants are utilized as thin films on metallic substrates by burnishing, sputtering or plasma spraying (Kato et al., 1989; Gardos, 1988; Bryant et al., 1964).

It is clear that the solid lubricant should adhere strongly to the contacting surface, otherwise it would be easily rubbed away, resulting in a short service life. Also, we should maintain a continuous supply of solid lubricant films between the sliding surfaces. This can be achieved by incorporating the solid lubricant into the matrix of one of the sliding components by forming a matrix composite. Thus, self-lubricating matrix composites are materials in which solid lubricants, such as graphite or MoS_2, are introduced as reinforcing materials into the matrix during preparation in order to enable the self-lubricating properties (Lovell, 2013).

5.5.2 Advantages and Applications of Solid Lubrication

In special applications, solid lubricants may be preferred to other types of lubrication for several reasons. Table 5.2 represents the main advantages and the reason of utilizing solid lubrication in some special applications.

5.5.3 Disadvantages of Solid Lubrication

1. Poor thermal conductivity of most solid lubricants which results in poor dissipation of friction heating from sliding interfaces.
2. Friction coefficient of solid lubricants is significantly affected by the environment and contact conditions.
3. Solid lubricants have finite wear lives, and their replenishment is more difficult than that of liquid lubricants.
4. Oxidation and aging-related degradation may occur over time.
5. Some solid lubricants undergo irreversible structural-chemistry changes at high temperatures or oxidative environments which in turn lead to loss of its lubricity.

Table 5.2: Advantages and applications of utilizing solid lubrication.

Application environment and/or condition	Advantages and/or requirements	Applications
High dust areas Aqueous corrosive	Resist abrasion Resist aqueous and corrosive environments	Automobiles, off-road vehicles and equipment, construction equipment, textile equipment, agricultural and mining equipment, buildings, bridges, industrial facilities, dental implants, aircraft, space rovers, lunar base and equipment and Martian base and equipment
Clean environment	Avoid contaminating product or environment	Microscopes, cameras, spectroscopes, medical and dental equipment, artificial implants, food-processing machines, semiconductor manufacturing equipment, metalworking equipment, hard disks and tape recorders, textile equipment, paper-processing machines, equipment in lunar base and equipment in Martian base
Inaccessible or hard-to-access areas	Maintain servicing or lubrication in inaccessible or hard-to-access areas	Artificial implants, buildings, bridges, industrial facilities, nuclear reactors, consumer durables, semiconductor manufacturing equipment, aerospace mechanisms, aircraft, space vehicles and satellites
Vacuum	Used where fluid lubricants are ineffective	Vacuum products, analytical equipment, coating equipment, space mechanisms, satellites, space platforms and space antennas
High temperatures	It can extend the operating temperatures of systems beyond 523 K while liquid lubricants may decompose or oxidize	Furnaces, metalworking equipment, compressors, turbines, nuclear reactors and molten metal plating equipment
Cryogenic conditions	It can extend the operating temperatures of systems down to cryogenic temperatures while liquid lubricants may solidify or become highly viscous and not be effective	Liquid nitrogen pumps, butane pumps, liquid propane pumps, refrigeration plants, lunar and Martian bases, space mechanisms, satellites, space vehicles, space propulsion systems, space antennas, space-telescope mounts, space platforms
Radiation	Relatively insensitive to radiation	Nuclear reactors and plants, X-ray equipment, space mechanisms, satellites, space vehicles, space platforms, space antennas and lunar base
High pressures or loads	More effective than fluid lubricants at intermittent loading, high loads, and high speeds	Metalworking equipment, bridge supports, plant supports and building supports
Weight-limited applications	Weighing substantially less than liquid lubrication (lubrication distribution systems and seals are not required)	Nanotechnology-related or space-related equipment (satellites, spacecraft and rovers)
Unstable environments or service conditions	Stable performance under changes in critical service and environmental conditions	When equipment is stored or is idle for prolonged periods, under intermittent loading conditions or in corrosive environments

5.5.4 Friction and Wear Testing of Solid Lubricants

Solid lubricants according to desired specifications require a wide range of testing. However, friction and wear properties of solid film lubricants are the most important evaluation for engineering and design (Schneider, 2002). The standard test method ASTM D2625 covers the determination of the endurance (wear) life and load-carrying capacity of solid film lubricants. The test is performed under sliding motion by means of a Falex pin and V-block test machine, see Fig. 5.20. This test rig is designed to evaluate wear, friction and extreme pressure properties of lubricants, coatings and other materials (ASTM, 2015).

Likewise, the evaluation of the tribological film life, the coefficient of friction and the failure load can be evaluated according to DIN 65593. The tests are conducted under oscillating motion by means of a block-on-ring friction and wear test rig. In such a rig, the lower specimen is a fixed disk while the upper specimen can be either a ball, a roller or a ring disk. The upper specimen oscillates at a constant frequency and stroke amplitude under a constant load against the lower specimen. Wear life of solid film lubricants is also evaluated according to ASTM D2981 (ASTM, 2019). This test method is used for determining the wear life properties of bonded solid lubricants under an oscillating motion by means of a block-on-ring friction and wear test machine. A selected data of friction coefficients for steel lubricated by solid lubricants is provided in Table 5.3.

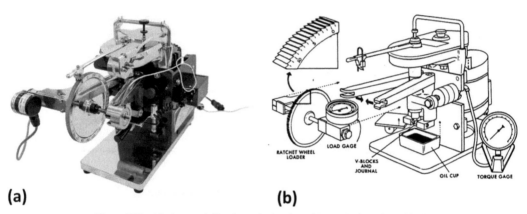

(a)   **(b)**

Figure 5.20: (a) photo and (b) schematic drawing of FALEX pin and Vee-block.

Table 5.3: A selected data of friction coefficients for steel lubricated by solid lubricants.

Solid lubricant	Coefficient of friction (static)	Coefficient of friction (kinematic)
Non (Steel-on-steel)	0.4–0.8	0.40
Molybdenum disulfide, MoS_2	0.05–0.11	0.05–0.093
Tungsten disulfide, WS_2	0.098	0.09
Selenium disulfide, SeS_2	–	0.25
Mica	–	0.25
Graphite	–	0.25
Boron nitride, BN	–	0.25
Calcium stearate, $C_{36}H_{70}CaO_4$	0.113	0.107
Sodium stearate, $C_{18}H_{35}NaO_2$	0.192	0.164

5.6 Grease Lubrication

5.6.1 Definition

Grease is defined by the ASTM international as a semi-fluid to solid product of a dispersion of a thickener in a liquid lubricant (International, 2009). The dispersion of the thickener forms a two-phase system and immobilizes the liquid lubricant by surface tension and other physical forces. Halling, J. also introduced a simple definition, that a grease is a stabilized mixture of a liquid lubricant and a thickening agent and may include additives to improve or impart particular properties (Halling, 1979). Similar definitions were also introduced by Boner (Boner, 1976), Zerbe (Zerbe, 1967), and Booser (Booser, 1984).

5.6.2 *Scientific Fundamentals*

In general, greases are chemicals that can either be mineral oils produced from refining crude oil, synthetic materials made from chemical synthesis, natural products like vegetable oils, silicones or fluorinated fluids. The common features of all these types are the low pouring point, high flash point, low volatility, good oxidative stability and compatibility with metals, plastics and elastomers.

Basically, greases are composed of mineral (petroleum) and synthetic base oils (85–90% of the volume) thickened with metal soaps and other additives. Base fluids are usually categorized into two major categories, mineral (petroleum) oil or synthetic, each of which can be broken down into the following subcategories:

Mineral oils base grease

Mineral oils, paraffinic and naphthenic, are derived through refining petroleum oil with different levels of hydrocarbon saturation. Paraffinic oils are more saturated than naphthenic base oils which results in more stable grease with less base fluid bleed. Thus, they have been found to provide superior oxidation resistance and seal compatibility than naphthenic oils and are often used in the formulation of greases, for which these properties are important. They can be used at temperatures up to 175°C, depending on the thickener, e.g., barium metal soap gives a maximum usable temperature of 175°C.

Synthetic oils base grease

The most common types of synthetic base fluids are polyalphaolefins (PAO), esters, polyalkylene glycols, dimethylsiloxane and perfluoroethers. PAO and esters are the most commonly used of these synthetic base fluids. Synthetic base fluids are generally used to provide increased performance in applications exposed to the extremes of temperature or oxidative environments. The best materials in these greases are silicones thickened with ammeline and silica, and they can be used at temperatures up to 300°C. All synthetic lubricants are eligible, but in practice, the cost of such materials restricts their use to applications having special requirements. Typically, for many applications, performance of mineral oils may be important than cost.

Thickeners

There are three general categories of thickeners used in grease manufacturing (Scott and Root, 1996):

i) Insoluble solids: Insoluble powders are introduced into a base fluid under high shear conditions until they are thoroughly dispersed and a thickened grease is formed without any real chemical reaction. These formulations could be related to the childhood favorite, the "mud pie". Common solids include organo-clay, fumed silica, carbon black and many different types of pigments.

ii) Polymers: Polymer-thickened greases are composed of a base fluid and a polymeric material with a gel-forming capability properly dispersed to produce the desired thickness of grease. The thickener is a low-molecular-weight organic polymer, which is usually formed *in situ* in the base fluid and yields no by-products that must be removed. Examples of polymer thickeners are polyurea and polytetrafluoroethylene (PTFE). PTFE is applied to grease formulation because it has shown to withstand high temperatures and oxidizing environments and because it is extremely unreactive once it is formed.

iii) Soaps: The term soap refers to the saponification reaction that results in the thickener. In this reaction, a metal hydroxide (base) is reacted with carboxylic acid to form a metal carboxylate salt (soap) and water. When this reaction is carried out in a lubricating fluid, it results in the formation of a crystalline soap dispersed in the fluid (Scott and Root, 1996). From this soap base, performance additives and the appropriate amount of additional oil are added to form a semi-solid or gel-like material. Most soap-type greases are sheared, or "milled", in order to form a consistent thickness throughout the material.

Additives

Additives can impart certain characteristics that may be desirable in some applications in which greases will be applied. Thus, in order to enhance the properties of the oil and thickener, additives are incorporated into the grease. There are general classes of additives that can be employed such as anti-oxidants, anti-wear, extreme pressure additives, corrosion inhibitors and tackifier. A detailed discussion of the lubricant additives will be introduced in upcoming sections.

5.6.3 Applications

Greases are used in a variety of applications where a fluid lubricant would run out and where a thick lubrication film is required. This includes lubrication of roller bearings in railway car wheels, rolling mill bearings, steam turbines, spindles, jet engine bearings and various other machinery bearings.

The lubricating process of a grease in a rolling-element bearing is such that the thickener phase acts essentially as a sponge or a reservoir to hold the lubricating fluid. In an operating bearing, the grease generally channels or is moved out of the path of the rolling balls or rollers, and a portion of the fluid phase bleeds into the raceways and provides the lubricating function.

The major advantages of a grease-lubricated rolling-element bearing are simplicity of design, ease of maintenance and minimal weight and space requirements. On the other hand, the limitation of using greases in lubrication is heat and speed. Grease is greatly limited in the amount of heat it will or can dissipate as compared with oil. Therefore, grease lubrication is typically restricted to low and medium capacity gears and bearings operating in short or intermittent intervals. Gear motors and all types of small and miniature gears, such as power tools, household and office equipment, and automotive servo drives are the primary applications for grease.

5.6.4 Testing of Grease

Grease evaluation is performed by many standard tests suggested by the Institute of Petroleum IP (IP, 2019), American Society of Testing and Materials ASTM, Defense Standard and other regulatory test methods and protocols. A wide suite of test rigs and test methods are available to assess grease performance, as tabulated in Table 5.4. Selected tests are described below.

Consistency

Consistency (or cone penetration) test is performed using a standard cone which is allowed to penetrate a sample of grease maintained at 25°C. The consistency is evaluated, over the full range of NLGI numbers, in terms of the depth of penetration after a period of five seconds. The NLGI consistency number (or NLGI grade) expresses a measure of the relative hardness of a grease, as specified by the standard classification of lubricating grease established by the National Lubricating Grease Institute NLGI, as shown in Table 5.5.

Drop point

In general, the dropping point is the temperature at which a drop of the grease passes from a semisolid to a liquid state under the conditions of the test. This change in state is typical of greases containing soaps of conventional types as thickeners. Greases containing materials other than conventional soaps as thickeners can, without change in state, separate oil. This test method is useful to assist in identifying the grease type and for establishing and maintaining bench marks for quality control. The results are considered to have only a limited significance with respect to the service performance as the dropping point is a static test. In practice, drop points are usually considerably higher than the normal maximum service temperatures (Halling, 1979).

Extreme pressure

For specification purposes (e.g., heavy-duty services), a grease must have acceptable extreme pressure EP properties in order to protect against frictional damage to the rubbing surfaces. For such purposes, EP test method ASTM D2596 is used to differentiate between lubricating greases having low, medium and high levels of extreme-pressure properties.

Oil separation

Oil separation (or bleeding) refers to the tendency of an oil to separate out from grease during storing. However, if the amount of free oil is small, the basic structure of the grease is not affected. In many cases,

Table 5.4: Standard test methods, IP and ASTM, for evaluating greases.

Test	IP	ASTM
Churning	IP 266	–
Consistency, Cone Penetration	IP 50	D217, D1403, D7342
Copper Corrosion	IP 112	D4048
Corrosion Preventative Properties	IP 220	D1743
Drop Point	IP 31, IP 131, IP 396	D566, D2265
Dynamic Rust (Emcor)	–	D6138
Evaporation Loss	IP 183	D972
Extreme Pressure	–	D2596
Fretting Wear Protection	–	D4170
High Temp Life Performance	–	D3527
Leakage Tendencies	–	D4290
Load Carry Capacity	–	D2509
Low Temperature Torque	IP 186	–
Oil Separations	IP 121	D1742, D6184
Oxidation Stability	IP 142	D942
Roll Stability	–	D1831, D8022
Rolling Bearing Performance	IP 168	–
Water Spray	–	D4049
Water Washout	IP 215	D1264
Wear, 4-Ball	IP 239	D2266, D4170

Table 5.5: NLGI consistency numbers.

NLGI number	ASTM D217 at 25°C tenths of a millimeter	Appearance	Consistency food analog
000	445–475	fluid	cooking oil
00	400–430	semi-fluid	apple sauce
0	355–385	very soft	brown mustard
1	310–340	soft	tomato paste
2	265–295	"normal" grease	peanut butter
3	220–250	firm	vegetable shortening
4	175–205	very firm	frozen yogurt
5	130–160	hard	smooth pate
6	85–115	very hard	cheddar cheese

when a lubricating grease separates oil, the remaining composition increases in consistency. This can affect the ability of the product to function as designed. The oil separation test method covers the determination of the tendency of a lubricating grease to separate oil during storage in both normally filled and partially filled containers. The test is performed by measuring the loss of oil from a sample of grease, supported on a filter paper, when acted on by gravitational or centrifugal forces.

Oxidation stability

This test method determines the net change in pressure resulting from consumption of oxygen by oxidation and gain in pressure due to formation of volatile oxidation by-products. It may be used for quality control to indicate batch-to-batch uniformity. It predicts neither the stability of greases under dynamic service conditions, nor the stability of greases stored in containers for long periods, nor the stability of films of greases on bearings and motor-parts. The test has some correlation with the storage life of greases but is not reliable for predicting the service life.

5.7 Additives

Lubricating oils have certain limitations by themselves. Therefore, lubricant additives are, in many cases, introduced to achieve an exceptional level of performance in a lubricant. Thus, lubricant additives are specific components which are added to a base oil for the purpose of protecting or enhancing the base oil, cleaning and protecting internal system components or neutralizing internal contaminants. They typically range between 0.1 to 30 percent of the oil volume, depending on the machine. Additives have three basic roles: Enhancing existing base oil properties, imparting new properties to base oils or suppressing undesirable base oil properties. The most common types are listed below:

Viscosity index improvers

They enhance the base oil to provide better stability with regard to changes in the fluid's viscosity through temperature changes. Viscosity index improvers can also be critical when operators face significant changes in operating temperatures, such as aviation applications and arctic operations. A higher viscosity index enables the lubricant to either get thicker or thinner at a slower rate as operating temperatures fall or rise.

Friction modifier

Used to achieve either reductions in friction in applications like engines or a specific level of friction in transmissions or fluid couplings.

Table 5.6 shows the effects of certain additives on the friction coefficient and wear of PTFE sliding on steel at 0.01 m/sec (Briscoe, 1992). Comparing the first and last rows of data shows how it is possible to increase wear resistance by more than three orders of magnitude while raising the sliding friction coefficient of the material by almost about 0.03.

Anti-wear

These additives are designed to form a relatively thick, tenacious coating which adheres to the contacting surfaces. The additives provide a sacrificial layer in the event of a slight metal-to-metal contact. They form organic, metallo-organic or metal salt films on the surface which are not easily removed by shear or cavitational forces which control wear. This additive is most important during mixed-film lubrication, where the fluid separates most of the surface, but some contact is likely to occur. The most common anti-wear additives used in practice are organochlorine, organosulphur, organophosphorus [tricresyl phosphate (TCP) and dibutyl phosphite (DBP)] and organometallic (ZDDP, MoDTP, MoDTC) and organic borate compounds (Xue and Liu, 1994).

Anti-oxidation

Anti-oxidation (or oxidation inhibitor) additives are necessary as equipment operation causes heat, moisture and other contaminants to degrade the base oil. They prevent or reduce the rate of formation of oxidation

Table 5.6: Effect of additives on the friction of blended PTFE (Briscoe, 1992).

Material composition	Wear rate improvement	Coefficient of friction
Unfilled PTFE	1	0.10
15 wt% graphite	588	0.12
15 wt% glass fibre	2857	0.09
12.5 wt% glass fibre and 12.5 wt% MoS_2	3333	0.09
55 wt% bronze and 5 wt% MoS_2	4000	0.13

Note: Wear rate improvement is calculated as the ratio of unfilled PTFE to that of the given material.

products such as acidic products and insoluble compounds. Different types of these additives are used, depending on the temperature of the required application.

Extreme pressure additives

Referred to as EP additives, these are required for applications, such as gearboxes, which are under heavy loads. An EP additive adheres to the gear surface and is activated by temperature to provide a cushion between contacts of the gear teeth. Solid lubricants, such as MOS_2, can be used in applications where there are very extreme loading conditions that create temperatures too high for traditional Sulphur phosphorus EP additives (Gao et al., 2004).

Tackifier

A tackifier is an additive that makes grease "stickier," thus providing increased adhesion to metal surfaces under conditions of high impact and water contact. They are used to assist the lubricant in adhering to a surface, so a fluid film can be maintained.

Detergent (or Dispersant)

Detergent (or Dispersant) additives are used mainly in engine oils to keep internal surfaces clean from contaminants. Detergents are designed to coat internal surfaces of components during normal expected operation to prevent deposits from forming. Detergents may have a limited ability to clean existing system deposits.

Pour-point depressants

The pour point of a lubricant is the temperature at which a lubricant becomes semi-solid and no longer maintains its expected flow characteristics. Pour-point depressants enable a lubricant to flow at very low temperatures in order to prevent lubricant starvation to components. Complex polymers and other types of products are used as pour-point depressants.

Emulsifier

It allows mineral oil to be mixable with water. It is frequently used in metal-cutting oils and in some lubricants for wet applications.

Foam inhibitor

They are additives that inhibit the formation of foam from the churning of air in the component by facilitating the release of air from the lubricant. The formation of foam can significantly reduce the fluid film strength by enabling air pockets to penetrate between internal surfaces. Foaming is normally caused by low oil levels or leaking fittings that enable air to enter the system. A significant foaming can lead to increased wear of the surfaces and lead to foam inhibitor additive depletion. Silicone-based compounds are usually used as foaming inhibitors.

Corrosion/rust inhibitors

Internal metallic components are subject to corrosion in the presence of moisture and heat. Lubricant corrosion/rust inhibitors serve to slow the corrosion process on internal surfaces.

References

ASTM D02.L - Industrial Lubricants, Minutes of the Meeting, San Diego, 19 June 2001.

ASTM D2625-94. 2015. Standard Test Method for Endurance (Wear) Life and Load-Carrying Capacity of Solid Film Lubricants (Falex Pin and Vee Method). ASTM International. West Conshohocken. PA. HYPERLINK "https://www.astm.org/" www.astm.org. dio: 10.1520/D2625-94R15.

ASTM D2981-94. 2019. Standard Test Method for Wear Life of Solid Film Lubricants in Oscillating Motion. ASTM International. West Conshohocken. PA. HYPERLINK "https://www.astm.org/" www.astm.org. dio: 10.1520/D2981-94R19.

Ahmed, N.S. and Amal, M.N. 2013. Lubrication and Lubricants.

Bayer, R.G. 2004. Mechanical Wear Fundamentals and Testing. New York: Marcel Dekker, Inc.

Bhushan, B., Gupta, B.K., van Cleef, G.Q., Capp, C. and Coe, J.V. 1993. Fullerene (C60) films for solid lubrication. Tribology Transactions 36(4): 573–580. Doi: http://dx.doi.org/10.1080/10402009308983197.

Bhushan, B. and Gupta, B.K. 1997. Handbook of Tribology: Materials, Coatings, and Surface Treatments: Krieger Publishing Company.

Bhushan, B. 2000. Modern Tribology Handbook. New York: CRC Press.

Bhushan, B. 2013. Introduction to Tribology. Second edition ed. New York: John Wiley & Sons Inc.

Bhushan, B. 2013. Principles and Applications of Tribology. 2nd edition ed. New York: John Wiley & Sons, Inc.

Boner, C.J. 1976. Modern Lubricating Greases. Broseley, Shropshire, UK.: Scientific Publications.

Booser, E.R. 1984. Theory and Design. In CRC Handbook of Lubrication. Theory and Practice of Tribology Boca Raton, Florida: CRC Press.

Briscoe, B.J. 1992. Friction of organic polymers. *In*: Pollock, H.M. (ed.). Fundamentals of Friction: Macroscopic and Microscopic Processes, Springer, Dordrecht.

Bryant, P.J., Gutshall, P.L. and Taylor, L.H. 1964. A study of mechanisms of graphite friction and wear. Wear 7(1): 118–126. Doi: https://doi.org/10.1016/0043-1648(64)90083-3.

Cameron, A. and Robertson, W.G. 1962. On the derivation of Reynolds equation. Industrial Lubrication and Tribology 14(6): 14–42. Doi: https://doi.org/10.1108/eb052697.

Campbell, M.E. 1972. Solid Lubricants: A Survey: Technology Utilization Office, National Aeronautics and Space Administration.

Conry, T.F., Wang, S. and Cusano, C. 1987. A Reynolds-eyring equation for elastohydrodynamic lubrication in line contacts. Journal of Tribology 109(4): 648–654. Doi: 10.1115/1.3261526.

Davim, J.P. 2011. Tribology for Engineers. Cambridge, UK: Woodhead Publishing Limited.

DellaCorte, C. 2013. Foil gas bearings. pp. 1240–1245. *In*: Wang, Q.J. and Chung, Y.-W. (eds.). Encyclopedia of Tribology. Boston, MA: Springer US.

Dowson, D. and Higginson, G.R. 1959. A numerical solution to the elasto-hydrodynamic problem. Journal of Mechanical Engineering Science 1(1): 6–15. Doi: https://doi.org/ 10.1243/JMES_JOUR_1959_001_004_02.

Dowson, D. and Higginson, G.R. 1961. New roller-bearing lubrication formula. Engineering (London) 192: 158–159.

Dowson, D. 1968. Elastohydrodynamics. Proc. Institution of Mechan. Eng.

Dykas, B., Bruckner, R., DellaCorte, C., Edmonds, B. and Prahl, J. 2008. Design, fabrication, and performance of foil gas thrust bearings for microturbomachinery applications. Journal of Engineering for Gas Turbines and Power 131(1): 012301–012301-8. Doi: 10.1115/1.2966418.

Echávarri Otero, J., Ochoa, E. de la G., Tanarro, E.C. and López, B. del R. 2017. Friction coefficient in mixed lubrication: A simplified analytical approach for highly loaded non-conformal contacts. Advances in Mechanical Engineering 9(7): 1687814017706266. Doi: https://doi.org/ 10.1177/1687814017706266.

Erdemir, A. 2001. Solid lubricants and self-lubricating films. In Modern Tribology Handbook, 40. New York: CRC press.

Errichello, R. 2015. Elastohydrodynamic Lubrication (EHL): A Review.

Fein, R.S., Rowe, C.N. and Kreuze, K.L. 1959. Transition temperatures in sliding systems. ASLE Transaction 2: 50–57.

Gao, F., Kotvis, P.V. and Tysoe, W.T. 2004. The surface and tribological chemistry of chlorine- and sulfur-containing lubricant additives. Tribology International 37(2): 87–92. Doi: https://doi.org/10.1016/S0301-679X(03)00040-9.

Gardos, M.N. 1988. The synergistic effects of graphite on the friction and wear of MoS_2 films in air. Tribology Transactions 31(2): 214–227. Doi: https://doi.org/ 10.1080/10402008808981817.

Gates, R.S. and Hsu, S.M. 1995. Silicon nitride boundary lubrication: Effect of oxygenates. Tribology Transactions 38(3): 607–617. Doi: https://doi.org/ 10.1080/10402009508983450.

Gilmore, R., Baker, M.A., Gibson, P.N., Gissler, W., Stoiber, M., Losbichler, P. and Mitterer, C. 1998. Low-friction TiN–MoS_2 coatings produced by dc magnetron co-deposition. Surface and Coatings Technology 108-109: 345–351. Doi: https://doi.org/10.1016/S0257-8972(98)00602-1.

Gross, W.A. 1963. Gas bearings: A survey. Wear 6(6): 423–443. Doi: https://doi.org/10.1016/0043-1648(63)90279-5.

Gross, W.A. 1967. Paper 9: Gas Bearings: Journal and Thrust. Proceedings of the Institution of Mechanical Engineers, Conference Proceedings 182(1): 116–150. Doi: https://doi.org/ 10.1243/PIME_CONF_1967_182_012_02.

Halling, J. 1979. Principles of Tribology. London: Macmillan Press Ltd.

Harrison, W.J. 1913. The hydrodynamic theory of lubrication with special reference to air as a lubricant. In Trans. Cambr. Phil. Soc.

Hirani, H. 2016. Mixed lubrication. *In*: Fundamentals of Engineering Tribology with Applications. Cambridge University Press.

Hsu, S.M. and Gates, R.S. 2005. Boundary lubricating films: Formation and lubrication mechanism. Tribology International 38(3): 305–312. Doi: https://doi.org/10.1016/j.triboint.2004.08.021.

International, ASTM. 2009. ASTM D4175 standard terminology relating to petroleum, petroleum products, and lubricants, in Annual Book of ASTM Standards. Philadelphia, 2009: ASTM International.

IP. 2019. Standard Test Methods (STM) for analysis and testing of petroleum and related products, and British Standard 2000 Parts. ISBN: 9781787250659.

Kanakia, M.D. and Peterson, M.B. 1987. Literature review of the solid lubrication mechanisms. In Interim Report, BFLRF #213. San Antonio: Southwest Research Institute.

Kato, K., Osaki, H. and Kayaba, T. 1989. The lubricating properties of tribo-coating films of Pb-Sn alloys in high vacuum. Tribology Transactions 32(1): 42–46. Doi: https://doi.org/ 10.1080/10402008908981860.

Kingsbury, A. 1897. Experiments with an air lubricated bearing. In J. Am. Soc. Nav. Engnrs.

Kopeliovich, D. 2012. Solid lubricants. Substance & Technology. http://www.substech.com/dokuwiki/doku.php?id=solid_lubricants.

Kumar, P., Khonsari, M.M. and Bair, S. 2008. Full EHL simulations using the actual ree–eyring model for shear-thinning lubricants. Journal of Tribology 131(1): 011802–011802-6. Doi: 10.1115/1.3002328.

Lansdown, A.R. 1996. Lubrication and lubricant selection—A practical guide. Neale, M.J., Polak, T.A. and Priest, M. (eds.). Tribology in Practice Series. London and Bury St Edmunds, UK: Mechanical Engineering Publications.

Liu, Z., Wang, Y., Cai, L., Zhao, Y., Cheng, Q. and Dong, X. 2017. A review of hydrostatic bearing system: Researches and applications. Advances in Mechanical Engineering 9(10): 1687814017730536. Doi: 10.1177/1687814017730536.

Lovell, M.R., Menezes, P., Ingole, S.P., Nosonovsky, M. and Kailas, S.V. 2013. Tribology for Scientists and Engineers. New York: Springer-Verlag.

Lubrecht, A.A., Venner, C.H. and Colin, F. 2009. Film thickness calculation in elasto-hydrodynamic lubricated line and elliptical contacts: The Dowson, Higginson, Hamrock contribution. Proceedings of the Institution of Mechanical Engineers, Part J: Journal of Engineering Tribology 223(3): 511–515. Doi: 10.1243/13506501JET508.

Mang, T. and Dresel, W. (eds.). 2007. Lubricants and Lubrication. Second ed: WILEY-VCH Verlag GmbH & Co. KGaA, Weinheim.

Mang, T. (ed.). 2014. Hydrostatic bearings. pp. 961–961. *In*: Encyclopedia of Lubricants and Lubrication. Berlin, Heidelberg: Springer Berlin Heidelberg. Doi.org/10.1007/978-3-642-22647-2.

Miyoshi, K. 2001. Solid Lubrication Fundamentals and Applications. New York, United States: Marcel Dekker, Inc.

Moonir-Vaghefi, S.M., Saatchi, A. and Hedjazi, J. 1997. Tribological behaviour of electroless Ni-P-MoS$_2$ composite coatings. Zeitschrift fuer Metallkunde 88(6): 498–501.

Morales-Espejel, G.E. and Wemekamp, A.W. 2007. Ertel–Grubin methods in elastohydrodynamic lubrication—a review. Proc. IMechE Part J: J. Engineering Tribology 222: 15–34. Doi: 10.1243/13506501JET325.

Parinam, A. and Kumar, P. 2012. New minimum film thickness formula for EHL rolling/sliding line contacts considering shear thinning behavior. Proceedings of the Institution of Mechanical Engineers, Part J: Journal of Engineering Tribology 227(3): 187–198. Doi: 10.1177/1350650112462173.

Pavlovich Bulat, M. and Bulat, P.V. 2013. Basic classification of the gas-lubricated bearings. World Applied Sciences Journal 28(10): 1444–1448. Doi: 10.5829/idosi.wasj.2013.28.10.13924.

Petroff, N.P. 1883. Friction in machines and the effects of the lubricant. Engng. J. (in Russian), St Petersburg: 71–140, 228–279, 377–436, 535–564.

Petrov, N.P. 1934. The hydrodynamic theory of lubrication. Saint-Petersburg: Saint-Petersburg: State Publishing of Technical and Theoretical Literature.

Rabinowicz, E. 1984. The least wear. Wear 100(1): 533–541. Doi: https://doi.org/10.1016/0043-1648(84)90031-0.

Rabinowicz, E. 1995. Friction and Wear of Materials. Second edition ed. New York: John Wiley and Sons.

Reeves, C.J., Menezes, P.L., Jen, T.C. and Lovell, M.R. 2012. Evaluating the tribological performance of green liquid lubricants and powder additive based green liquid lubricants. STLE Annual Meeting & Exhibition, STLE St. Louis, USA.

Ren, Z., Glodez, S., Fajdiga, G. and Ulbin, M. 2002. Surface initiated crack growth simulation in moving lubricated contact. Theoretical and Applied Fracture Mechanics 38(2): 141–149. Doi: https://doi.org/10.1016/S0167-8442(02)00091-5.

Reynolds, O. 1886. On the theory of lubrication and its application to Mr. Beauchamp Tower's experiments. Phil. Trans. R. Soc.

Reynolds, O. 1886. On the theory of lubrication and its application to Mr. Beauchamp Tower's experiments, including an experimental determination of the viscosity of olive oil. Royal Society, Phil. Trans.

Rippel, H.C. 1960. Cast bronze bearing design manual. Evanston, Ill.: Cast Bronze Bearing Institute.

Schneider, S. 2002. Solid Film Lubricants—Specifications, Properties and Testing. 13th International Colloquium Tribology, Technische Akademie Esslingen, Germany.

Scott, B. 2004. Actual eyring models for thixotropy and shear-thinning: Experimental validation and application to EHD. Journal of Tribology 126(4): 728–732. Doi: 10.1115/1.1792693.

Scott, P.W. and Root, J.C. 1996. Lubricating Grease Guide. 4 ed. Kansas City, USA: National Lubricating Grease Institute.

Stachowiak, G.W. and Batchelor, A.W. (eds.). 1993. 7—Elastohydrodynamic lubrication. pp. 335–424. In: Tribology Series. Vol. 24. Elsevier. Doi.org/10.1016/S0167-8922(08)70581-9.

Stachowiak, G.W. and Batchelor, A.W. 2006. 7—Elastohydrodynamic lubrication. pp. 287–362. *In*: Engineering Tribology (Third ed.). Burlington: Butterworth-Heinemann. Doi.org/10.1016/B978-075067836-0/50008-0.

Sternlicht, B. and Arwas, E.B. 1965. State-of-the-Art of Gas-Bearing Turbomachinery. Latham, New York.

Tang, I.C. and Gross, W.A. 1962. Analysis and design of externally pressurized gas bearings. ASLE Transactions 5(1): 261–284. Doi: 10.1080/05698196209347447.

Tower, B. 1884. Report on Friction Experiments. Proc. Instn. Mech. Engrs.

Tribonet. 2018. Reynolds Equation: Derivation and Solution. Available Online: http://www.tribonet.org/wiki/reynolds-equation/.

Václav, Š. and Václav, V. (eds.). 1992. Chapter Three: Types of lubricants and their compositions. pp. 125–254. *In*: Tribology Series. Vol. 23. Elsevier. Doi.org/10.1016/S0167-8922(08)70350-X.

Walck, S.D., Donley, M.S., Zabinski, J.S. and Dyhouse, V.J. 1994. Characterization of pulsed laser deposited PbO/MoS$_2$ by transmission electron microscopy. Journal of Materials Research 9(1): 236–245. Doi: 10.1557/JMR.1994.0236.

Wang, Y. 2013. Friction in conformal contact interface. pp. 1311–1315. *In*: Wang, Q.J. and Chung, Y.-W. (eds.). Encyclopedia of Tribology, Boston, MA: Springer US.

Xue, Q. and Liu, W. 1994. Tribochemistry and the development of AW and EP oil additives—a review. Lubrication Science 7(1): 81–92. Doi: 10.1002/ls.3010070107.

Zerbe, C. 1967. Lubrication and Lubricants. Von E.R. Braithwaite, Elsevier Verlag, Amsterdam 1967. Leinen, 568 S., 191 Abb., 46 Tab., Preis ca. DM 95. Materials and Corrosion 18(7): 659–659. Doi: 10.1002/maco.19670180716.

Zhang, Q.D. and Shan, X.C. 2007. Dynamic characteristics of micro air bearings for microsystems. Microsystem Technologies 14(2): 229. Doi: 10.1007/s00542-007-0414-1.

Chapter 6

Tribology of Polymer and their Composites

6.1 Introduction

Polymers are being used increasingly in tribological applications due to their elasticity, accommodation to a shock loading, low friction and wear resistance. The tribology of polymers is different from the tribology of metals for many reasons. In contrast to metals, polymers are viscoelastic and their properties depend on time. External liquid lubricants, which work well for other classes of materials, are easily absorbed by polymers. Furthermore, complexity arises as polymers are easily influenced by operating conditions and the prevailing environment. Nevertheless, it is a fascinating area because polymers can be modified, both on the surface and in bulk, by various chemical and physical means to suit a particular application. For this reason, they became attractive candidates and very promising materials for tribologists with an ability to control their friction and wear behaviors. This resulted in various tribosystems composed of polymers, metals and ceramic materials in sliding or rolling tribological contacts.

In fact, traditional tribology, as well as traditional experimental methods which were originally developed for metals, may not be well applicable for polymers. Interfacial and operational conditions, such as transfer film formation, thermal heat, and contact pressure have a different effect in the case of polymers. The earliest reference to the tribology studies on polymers that can be found is by Shooter and Thomas (Shooter and Thomas, 1949) in 1949. The success of introducing bulk polymers in bearing applications led to the work of employing polymers as composites, where their tribological and mechanical properties were modified by using filler additives. Till the day, thousands of studies introduced by hundreds of researchers on understanding the fundamental mechanisms of friction and wear of polymers and composites have been conducted. The present chapter aims to review three basic branches in polymer tribology: Wear, friction and lubrication, with a special concern for mechanisms and factors influencing wear behavior.

6.2 Wear of Polymers

During the past decade, the use of polymers in tribological applications, bearings, gears, biomaterials, etc., has increased markedly. Studying the wear mechanisms of polymers in contact with counterface surfaces became important from a practical standpoint.

A tribosystem usually comprises a polymer and a counterface that interact in operative environment under given conditions of applied load, speed, temperature, etc., resulting in the polymer wear process. A general classification of wear types in polymers is still an open matter. Earlier researches have established that the wear of polymers can be subdivided into three main groups: Adhesion, abrasion, and surface fatigue. Each wear mechanism is governed by its own laws and, on many occasions, it may act in such a way as to affect the others. It is important to emphasize that it is not always easy to differentiate between these types of wear; they are inter-related and rarely occur separately. However, other wear forms, such as corrosive, erosive, or fretting wear, are also included by other researchers.

6.2.1 Abrasive Wear

Abrasive wear in polymers is caused by hard asperities on the counterface which dig into the rubbing surface of soft polymers and remove material, resulting in micro-machining, wear grooves, tearing, ploughing, scratching and surface cracking, as shown in Fig. 6.1a. The wear debris produced usually take the shape of fine chips or flecks, similar to those produced during machining, Fig. 6.1b.

The abrasive wear of polymers is inversely proportional to the product of the nominal tensile breaking stress σ_u and the elongation-to-break ε_u, as shown in Fig. 6.2. The abrasion of polymers may also correlate with its cohesive energy, flexure modulus, yield strain or energy-to-rupture (Giltrow, 1970). A wide range of studies on the effect of counterpart surface on wear of polymers demonstrated that the abrasive wear process involves plastic deformation and shear, and it was found that, for abrasion, the dominate material property is the energy to fracture of the polymer (Lancaster, 1969).

Figure 6.1: (a) Rubbing surface of polyamide 66 sliding against dry steel showing grooves run across the surface of the wear pin parallel to the sliding direction, (b) Wear debris. After Abdelbary (Abdelbary, 2011).

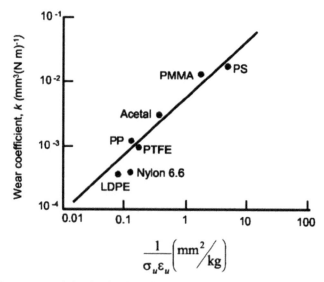

Figure 6.2: Ranter-Lancaster correlation for abrasion. After Lancaster (Lancaster, 1969) with permission from Elsevier.

Based on the above, a number of equations to express the abrasive wear of polymers have been proposed. There are three main stages involved in the production of wear debris (Ratner et al., 1964):

1) deformation of the surfaces to an area of contact is determined by the indentation hardness of the polymer, H.

2) relative motion opposed by the frictional force (f), $f = \mu L$, where L is the normal load and μ represents the coefficient of sliding friction.

3) disruption of material at the contact points involving an amount of work equal to the integral of the stress-strain relationship.

An approximate measure of the latter is the product of the breaking stress and the elongation to break. As these three processes occur sequentially, the total wear can be regarded as being proportional to the probability of completion of each stage. Thus, for the sliding distance X, the worn volume, V, is given by the following expression (Lancaster, 1969):

$$V = \frac{\mu L X}{H(\sigma_u \varepsilon_u)} \tag{6.1}$$

The product $(\sigma_u \varepsilon_u)$ is related to the area under the stress-strain curve, hence, the toughness or impact strength of the material. Therefore, correlations have been sought between abrasive wear and impact strength, or notched impact strength. The particular importance of the parameter $\frac{1}{\sigma_u \varepsilon_u}$ has been demonstrated by Lancaster, who obtained a linear relation between $\frac{1}{\sigma_u \varepsilon_u}$ and the resulting wear during single traversals of different polymers. It should be noted that both elongation and breaking strength are sensitive to strain rate and temperature variation. However, other attempts have been made to relate notched impact strength to the wear of a number of polymers, but the correlation obtained was not convincing.

As a result of these arguments, it is considered that the resistance of a polymer to abrasive wear can be increased by changing its mechanical properties. For example, both the breaking strength and elongation to break tend to increase with increasing molecular weight, up to a limiting value. In particular, with HDPE (high density polyethylene), the impact strength increased with molecular weight to a maximum at an average molecular weight of about 1.5×10^6. The abrasion test (Margolies, 1971) showed that the abrasion resistance improved greatly with increasing molecular weight, reaching a maximum and constant value at the molecular weight of 1.75×10^6 and greater. Thus, abrasion resistance and impact strength show similar correlation with molecular weight.

6.2.2 Adhesive Wear

This form of wear is likely to be the most significant in the wear of polymers when sliding repeatedly over the same wear track on a smooth metal counterface. During repeated sliding, high local pressure is experienced between the polymer surface and the counterface, causing plastic deformation leading to the formation of an adhesive junction. Further motion results in continuous formation and rupture of these junctions. A thin film of the soft polymer is transferred onto the hard mating surface. As the transfer film builds up, the surface topography of the counterpart changes and an equilibrium situation may be reached, in which the amount of material removed from the bulk polymer by adhesive wear is equal to the rate of subsequent detachment of wear particles from the transfer film. Another consequence of polymer transfer is a change in roughness of both surfaces in contact. The roughness of the polymer rubbing surface undergoes large variation during the primary stage, running-in, of wear until the steady state wear is reached, while metal surface roughness is modified due to transfer of polymer. In many cases, the adhesive wear is independent of surface roughness, it may often occur in very smooth surfaces as well as rougher ones.

Experiments were carried out to study the wear of polyamide sliding against steel and stainless steel counterfaces in dry conditions (Abdelbary, 2011a). During running-in, the thickness of the transfer film that adheres to the steel counterface increases with sliding time until a maximum value is reached. The final limiting thickness of the transfer film is probably a characteristic feature of each rubbing pair and the operating variables, as shown in Fig. 6.3.

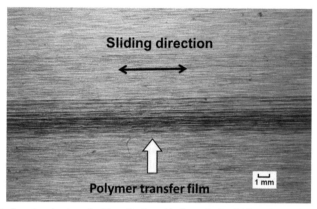

Figure 6.3: Steel counterface showing transfer film of polyamide formed after 20 km of dry sliding, under a load of 90 N. After Abdelbary (Abdelbary, 2011).

6.2.3 Surface Fatigue Wear

Surface fatigue wear has been recognized as an important process that takes place when the polymer undergoes repeated stressing during reciprocating sliding or rolling motion. Each asperity of friction surface experiences repeated loading and unloading from the asperities of the counterface. The repeated cycles lead to subsurface crack initiation in the polymer surface, which, on further deformation, leads to crack propagation parallel to the surface, and debris can be spalled off the surface by continued motion. The new surface of the material also experiences the same cyclic stressing, which leads to progressive process and flaking off of fragments becoming rapid. The precise type of stress cycle involved in a fatigue progress depends on the mechanical properties of the concerned polymer.

For many decades, researchers aimed their work at understanding the mechanisms of fatigue wear in polymers as well as finding out a clear relationship between the wear factor and the other fatigue parameters (such as number of loading cycles to failure, normal load, asperity radius of curvature, coefficient of friction, and moduli of elasticity of the contacting materials). Equation (6.2) was derived (Hollander and Lancaster, 1973) in order to predict wear factors of polymers from measuring the fracture mechanics parameters characterizing fatigue crack growth. It was found that the fracture toughness k_{IC} is the most important material property controlling wear in abrasive conditions, and that the wear factor should be proportional to $1/k_{IC}^2$. They developed an adhesive-fatigue wear equation by applying the concepts of linear elastic fracture mechanics to the crack propagation phase of the wear process. Assuming that the wear factor is inversely proportional to the number of cycles N_f required to propagate an incipient crack of length a_o to some critical length, at which the peak cycling load can detach a wear particle, they introduced the following relation:

$$WF \, \alpha b . a_o^{(n/2)-1}(\Delta\sigma_F)^n \pi^{n/2}[(n/2)-1] \qquad (6.2)$$

where WF represents the wear factor ($WF = V/X.L$), σ_F is the peak stress, b and n are empirical constants.

6.2.4 Other Forms of Wear

Delamination wear

The mechanisms of this form of wear are:

a) Plastic deformation of the surface.

b) Initiation of sub-surface cracks due to plastic deformation.

c) Crack propagation parallel to the surface.

d) Detachment of long thin wear sheets when the crack reaches a critical length and breaks through to the surface.

Figure 6.4: Delaminated polymer after 50 km of sliding against dry steel, under cyclic load. After Abdelbary (Abdelbary, 2011).

Microscopically, the loose wear sheets will appear flake-like in shape with striations over the surface due to the adhesive wear, as shown in Fig. 6.4. The presence of the aggressive environment may also act to accelerate the delamination process by boosting crack propagation.

Fretting Wear: Wear of polymers arising due to fretting action is called "fretting wear". Fretting wear occurs in mechanical systems where polymers undergo continuous vibration with load or no load on them. Accumulated wear debris, which is produced due to fretting, does not easily escape from the contact and play an important role in accelerating the wear process by three body abrasion. This eventually leads to the loss of clearance and significant localized damage in the mechanical applications. Researches (Tan et al., 2011; Guo and Luo, 2001) performed in order to investigate the fretting wear of unfilled polymers in oscillatory sliding against various metallic counterparts demonstrated that amplitude, frequency, temperature, contact pressure and the material structure have critical influence on the wear rate.

Chemical or Corrosive Wear: Corrosive wear occurs when both corrosion and wear mechanisms are involved during sliding of polymers in a chemical environment. This particular situation can result in a total worn volume that is much greater than the additive effects of each process taken alone. Generally, this wear process occurs in two stages. First, corrosion takes place on the polymer surfaces due to the attack of chemical environment. Then, the sliding process removes the corroded surface layer by abrasion. The fresh surface will, again, suffer from corrosive attack, and the wear rate is accelerated by the abrasive sliding action. A few examples of corrosive wear are as follows:

a) Water absorption: Can lead to deterioration in the mechanical properties and swelling of many thermoplastics.

b) Oxidation of polymers: Results in reduction in the average molecular weight and chain disengagement. Consequently, causes increases in the modulus of elasticity, density and percent crystallinity. Oxidation of polymeric materials can be determined by detecting the carbonyl groups formed.

6.3 Friction of Polymers

The discrepancy between metallic and polymeric friction is due to the differences in the elastic-plastic behavior of metals and the viscoelastic behavior of polymers. For all materials, including polymers, "Adhesion" and "Deformation" are the two main components that should be considered when studying friction. During sliding, the creation and termination of interfacial junctions are influenced by the nature of contact areas, surface chemistry, and stress in the surface layers at the given load conditions. Assuming that the real contact area of the junction is A_{r1}. According to law of friction (Tabor, 1982), the shear force F_a that resists mutual sliding is given by:

$$F_a = \tau_s \cdot A_{r1} \qquad (6.4)$$

where τ_s is the shear stress required to produce sliding between the rubbing surfaces, which coincides with the strength of adhesion at the interface between the asperities and the bulk shear strength of the polymer. These strength properties are of the same order of magnitude for nearly all of the thermoplastic polymers and appear to vary with temperature, especially near the glass transition temperature.

Deformation is caused by asperities on the hard counterface penetrating into the softer polymer surface, and ploughing out a groove by plastic flow in the polymer. The main factor that involves the deformation of polymers is the dissipation of mechanical energy that depends on the mechanical properties of the polymer, sliding conditions, environmental conditions and others. The energy of deformation represented by the ploughing process is attributed to the second component of friction, F_d, and given by:

$$F_d = \sigma_y \cdot A_{r2} \qquad (6.5)$$

where σ_y is the polymer yield pressure and A_{r2} is the area of the grooved track.

In the case of a polymer sliding on a metallic counterface, the material is transferred from the polymer to the metal and, after sufficient running, the friction approaches that of polymer sliding on polymer. In this case, the adhesion component for the polymer greatly exceeds the deformation one due to the transfer film generated on the metallic counterface (Briscoe and Sinha, 2013). However, it is not an easy task to investigate the effects of both components separately, and many studies have shown that it is often sufficient to consider that the friction is just the simple sum of the predictions of both adhesion, F_a, and asperity deformation, F_d, terms; hence, the total friction force F is given by:

$$F = F_a + F_d \qquad (6.6)$$

In Equation (6.6), the first term is usually the more important with metals, but with polymers sliding over rough surface or in the presence of a lubricant, the deformation component can become quite significant. The relative importance of the contribution of each of the two parameters to the total frictional force depends on the type of motion involved (rolling or sliding), the surface topography and the mechanical properties of the polymeric materials.

Relating the coefficient of friction, during sliding of polymers over a metallic counterface, to the mechanical properties of the polymer is not an easy task. An assumption that both surfaces involved are relatively smooth and that all the deformation terms are negligible should be considered. The friction force is given by a simple model:

$$F = \tau_s \cdot A_r \qquad (6.7)$$

where A_r is the real area of contact. If the stress at the asperities in contact is sufficient to cause plastic deformation, then $A_r = \dfrac{L}{P}$ where L is the applied normal load and p is the flow pressure of the polymer. Thus, for plastic deformation

$$\mu = \frac{\tau_s}{p} \qquad (6.8)$$

If the deformation is elastic, the estimation of real area of contact becomes more difficult and some assumptions about the size, shape and distribution of the asperities should be done. Thus, for elastic deformation

$$\mu = K \frac{\tau_s}{E^x} \cdot L^{x-1} \qquad (6.9)$$

where E is the Young's modulus, K and x are surface parameters.

Consequently, it is important to define the type of deformation involved with polymers. Equation (6.10) (Halliday, 1955) was derived in order to find an expression relating the mechanical properties of a material to the maximum angle of slope of an asperity, θ, which can just be flattened into the general plane of the surface by contact with a smooth, and rigid counterface.

$$\tan\theta_{\text{lim}} = K\frac{H}{E}(1-v^2) \tag{6.10}$$

where

H Indentation hardness
v Poisson's ratio
K Constant ($K = 0.8$ for the onset of plastic flow, $K = 2$ for full plastic flow)

The calculated values of the limiting angle θ show that the angles for polymers are appreciably greater than those for metals. Therefore, polymers are much more likely to encounter elastic deformation in contact compared to metals.

6.4 Severity Parameters for Friction and Wear of Polymers

Polymers are sensitive to the effect of numerous factors caused by the effects of sliding interaction and external medium. A reliable relationship between material properties and operating conditions is essential in order to acquire a better understanding of the friction and wear behavior of polymers under different circumstances. In the former sections, we discussed the fundamentals of wear and friction of polymers, and it is important to debate the main factors which affect the sliding of polymeric materials. Thus, most of these factors are considered in the present section.

6.4.1 Sliding Speed

Generally, in polymer-metal contact, the effect of speed is to increase the wear of polymers due to an increase in the contact surface temperature generated at the points of rubbing contact. However, for thermoplastics, there is a critical sliding speed over which the localized flash temperature, generated at the interface, can be reached and the wear rate reduces slightly due to surface melting and thermal softening. This, flash, temperature is caused by adhesive friction at the instantaneous contact interface. Generally, it is not possible to measure the flash temperature during running with any accuracy. This may explain the complex dependence of the friction coefficient on the sliding speed in polymer tribosystems compared to that in metals. A strong dependence of the friction coefficient μ on the sliding speed is observed when the temperature of the polymer approaches that of glass-transition (Ainbinder and Tyunina, 1974). Nevertheless, there is no significant dependence on the sliding speed at lower temperatures. Thickness and stability of polymer transfer film is also affected by the sliding speed. Figures 6.5 and 6.6 represent a number of researches that investigate the relationship between sliding speed and polymer tribology (Byett and Allen, 1992; Unal et al., 2004).

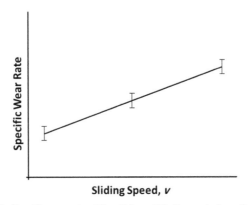

Figure 6.5: Specific wear rate of dry sliding of Derlin on steel, $v = 0.1–0.7$ m/s.

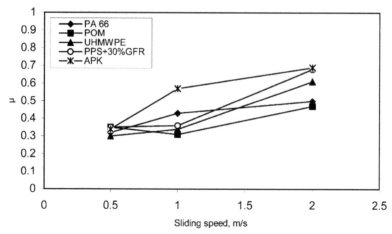

Figure 6.6: Friction coefficient vs. sliding speed for some industrial polymers. After Unal et al. (Unal et al., 2004) with permission from Elsevier.

6.4.2 Sliding Temperature

Low thermal conductivity of viscoelastic materials, including thermoplastic polymers, is an important limitation for sliding applications, which alters the important result of frictional heating parameter on polymer tribology. Heat is usually generated by friction from the deformation of the polymer during its sliding on real contact spots. Sequential formation and shearing of adhesive bonds is also another source of heat generation.

The mechanical properties of a polymer show a transition from the glassy state, with high strength and high stiffness, into the rubbery state, with lower strength and lower stiffness, upon heating. The transformation of frictional energy to heat is the reason of the increase in temperature of polymer rubbing surface especially at spot-to-spot contact interfacial temperature (T^*). High contact temperatures are not limited to high relative sliding speed of rubbing surfaces. The interesting example is that of sliding at relatively low sliding speeds in fretting applications. The high fretting frequencies of short duration ($< 10^{-3}$ s) that occur over small (spot contact) dimensions ($< 10^{-4}$ m) have a similar heating effect (Kalin and Vižintin, 2001).

6.4.3 Counterface Roughness

A number of wear screening studies have provided evidence of a strong correlation between the tribology of polymeric materials and the roughness of the metallic counterface. It is widely accepted that the wear mechanism and the friction resistance of polymers are governed by the counterface roughness. In polymer-metal sliding, as the counterface roughness decreases the friction coefficients of polymers decrease. However, after reaching a minimum value of R_z (the mean peak-to-valley), further decrease in the roughness causes a high friction. The reason is that adhesion forces become the dominant factor, whereas for higher surface roughness, abrasive wear prevails.

It was generally thought that the smoother the counterface, the lower the wear rate. However, earlier studies (Dowson et al., 1985) indicated that there is an optimum scale of surface finish which gives a minimum dry wear factor for polyethylene sliding on dry stainless steel, Fig. 6.7. The roughness affected the type of transfer film laid down and it was likely that high wear rates found for very smooth surfaces are due to highly adhered polymer "lumps". This was potentially very important in the field of biomedical applications, as the typical surface finish for the femoral head of hip joint is 0.02–0.05 μm Ra, which naturally increases the cost of these joints as they have to be subjected to lengthy polishing operations.

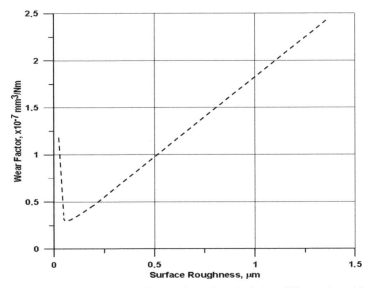

Figure 6.7: Schematic representation to the wear factor for Polyethylene sliding on dry stainless steel.

The wear characteristics of twelve different polymers were studied using a pin-on-disc testing machine (Bellow and Viswanath, 1993). The polymers were rubbed against an AISI 1018 steel disc with a surface roughness, R_a, ranging from 0.05 to 0.4 μm. For high density polyethylene (HDPE) and Teflon PTFE, the wear rate increased with the increase of counterface roughness. However, the test results did not indicate any direct relationship between the wear factor and counterface roughness. On the other hand, as the counterface roughness increased, the coefficient of friction was found to decrease but with an increase in wear volume. In some cases, it was found that the transfer of the polymer film to the metallic counterface contributes to changes in the surface profiles. Consequently, a definite effect of the counterface surface roughness on the degree of wear and the coefficient of friction was reported.

In addition, it was demonstrated that a single imposed imperfection in the stainless steel counterface has a pronounced effect on the wear of ultra-high molecular weight polyethylene UHMWPE (Dowson et al., 1987). The counterface imperfections were in the form of either transverse or longitudinal scratches or single indentation generated by diamond markers. It was realized that a single transverse scratch can increase the wear rate of the polymer to at least one order of magnitude. The increase in wear was due to the piled-up metallic material flanking each transverse scratch after the ploughing action by a diamond marker. In a total joint replacement, such scratch damage can be caused by bone fragments which act as "third body" abraded particles. This is one reason why femoral heads with a high hardness, such as alumina, have been preferred over less scratch resistant materials, such as stainless steel and titanium.

6.4.4 Applied Load and Contact Pressure

It is broadly known that the mechanism of friction and wear of polymers varies depending on the applied normal load. At high loads, thermal softening of the polymer and plastic deformation at the asperity interactions have a dominant role in determining the real area of contact. For many thermoplastic polymers, earlier experiments have shown that the coefficient of friction is constant at high loads, from 10 to 100 N (Tabor and Shooter, 1952; Bowers, 1953). In the range of relatively lower loads, from 0.02 to 1 N, as the load decreases, elastic deformation of surface asperities governs the sliding process with a consequent increase in the coefficient of friction (Rees, 1957). That is, the influence of applied load on the friction coefficient in polymer sliding corresponds to transition from the elastic contact to plastic one. We should draw attention to the fact that the increase in sliding temperature due to applied load may also affect the viscoelastic transitions in polymers and, thereby, the mechanism of friction.

The relation between friction force F and applied normal load L can be given by:

$$F = \mu L^n \tag{6.11}$$

where μ is the coefficient of friction and n is an exponential constant that has a value of 0.8 for Polyamide and 0.96 for Cellulose acetate. Many attempts were made in order to formulate the wear-load relationship.

Equation (4.1), as one of the earlier wear equations, was introduced by Archard (Archard, 1959) in order to relate the worn volume to the applied load, sliding distance and hardness of rubbing material. The expression has been widely accepted for metals and, in some cases, also applied to polymers. However, in some cases, the wear factor WF increases exponentially with the normal load L or contact stress. Thus, the relationship between wear factor WF and load may depends on exponent parameter n and that the linear dependence occurs when n $\cong$ 3 as well as the effect of load on the wear (Bijwe et al., 2005).

$$WF \, \alpha \, L^{\frac{n}{3}} \tag{6.12}$$

The linear wear relationship in metals, which implies that the wear factor is independent of contact pressure, may only be an approximate relationship in polymers. This is due to the viscoelastic properties of polymers. When polymers are brought into contact with a metallic counterface, real contact is only localized near some rough asperities. While the number of localized spots increases due to high contact pressure, the size of these spots also grows at low contact pressure. We should bear in mind that at extremely high contact pressures the real contact area approximates the apparent contact area, thus, the micro-asperity effects become less important.

Studying the effect of contact pressure on the wear behavior of thermoplastic polymers has been an interesting research topic for many authors. Kar et al. (Kar and Bahadur, 1978) demonstrated that the contact pressure is the reason for temperature rise in various viscoelastic transitions in polymers. At low pressures, where the heating due to friction is small, the wear rate of polymer is proportional to the applied pressure, while at higher pressure the wear rate increases abruptly. He attributed that to the possibility of thermal effects at the interface, which is high enough to cause melting or softening of the sliding polymer surface. In studying the effects of contact stress on the wear factor of isotropic UHMWPE, results obtained by Barbour et al. (Barbour et al., 1995) showed a decrease in the wear factor as the contact stress increases. They concluded that at high contact stresses a wear particle will not be able to escape easily from the interface, and may transfer back to the polymer pin or act as three-body wear, and vice versa.

6.4.5 Material Properties

Friction and wear behavior of polymers is strongly influenced by their material properties: Mechanical, physical and thermal. Yield strength, hardness and elastic modulus as mechanical properties show a significant effect at the glass transition temperature. The mechanical properties of polymers, such as elastic modulus, hardness, tensile strength, elongation to break, etc., at various temperatures influenced the wear rate of polymers against metal counterpart. The mechanical properties decrease with increasing temperature, resulting in minimum wear resistance. Both mechanical stresses and thermal effects play a role in the mechanism of surface failure of polymers.

Physical properties of thermoplastic polymers also have a marked effect on their wear and friction behavior. The work of adhesion is governed by the chemical composition and interfacial energy of the polymer tribosystem. Moreover, surface energy and fracture energy have a role in determining the adhesion of polymer to the counterface. On the other side, solids, including polymers, having high surface energy exhibiting a higher coefficient of friction. A polymer of low surface energy tends to transfer to that of high surface energy. Raman spectra of worn polymer surfaces (Xue, 1997; Hendra et al., 1991) indicated that a more viscous and continuous film of polymer has formed on the polymer surface, related to the lower surface energy of stainless steel compared to alloyed carbon steel. The thickness of the transferred polymer layer increased during sliding, which indicates that polymer wear can occur even on the transferred layer of a similar polymer. This confirmed that the bonding between the worn material and the counterface is due to mechanical interlocking.

In dry sliding, thermoplastic polymers can easily lose their strength and deform at high temperature due to poor thermal conductivity and low specific heat, thus limiting their applications, especially in extreme working conditions. In other words, thermal properties of polymers make them preferable in tribo-systems of low operating temperature and moderate (or light) loads.

6.4.6 Humidity and Surface Wettability

Usually, wear is considered as a mechanical process. However, it is mostly a study of rubbing surfaces in contact; thus, its characteristics may be influenced by the operating environment. In particular, the amount of atmospheric moisture has a pronounced effect on the wear and friction of polymers. Changes in environmental conditions play a significant role by affecting the polymer film transfer to the counterface.

As a matter of fact, there are numerous sliding applications in which water is either deliberately introduced to the tribology system as a coolant or lubricant, or present as a working fluid. A general agreement was confirmed that the presence of water results in lubricating the rubbing surfaces and reducing the coefficient of friction (Tanaka, 1984; Yamamoto and Takashima, 2002). Moreover, the wear rate of many polymers and their composites under water lubricated sliding was always higher than that under dry sliding, as shown in Fig. 6.8.

The effect of water on the wear behavior of polymers can be generalized mainly in the following points:

i. Water molecules diffuse readily into the free volume of the amorphous phase of the polymer, leading to plasticization, swelling and softening, which reduce the hardness and strength of polymer. Water diffusion also reduces the attractive forces between polymer chains, thus leading to easy material removal during sliding and an increase in the wear rate (Yamamoto and Takashima, 2002).

ii. Water has the effect of washing action for the counterface surface; therefore, a transfer film similar to that under dry sliding condition fails to form and the polymer slides directly against the metallic counterface (Yu et al., 2008).

iii. Water might induce an increase in the chemical corrosion wear of the metallic counterface, which would lead to a modification of the surface profile, and finally a greater wear rate of polymer is expected (Meng et al., 2009).

Morphological analysis (Abdelbary et al., 2013) of the rubbing surface of polyamide 66 confirmed that, under dry sliding conditions, some apparently plastic flow traces and plowed grooves on the polymer worn surface were detected. Uniform and continuous transfer film of the polymer was formed on the metallic counterface. This suggests that the wear process was governed by the plastic deformation and mechanical microploughing. On the other side, under water lubricating conditions, the worn surfaces were almost smooth and slightly ploughed grooves parallel to the sliding direction were observed. This suggested that the wear process was governed by the mechanical microploughing and abrasive wear mechanisms. Thermal analysis of wear tests suggest that the temperature rise of the rubbing surfaces under dry sliding conditions were higher than those under water lubricated conditions. Clearly, this is the result of the cooling action of water. Due to the high contact temperature during sliding, the polymer was softened, enhancing the ability of transfer film formation and resulting in relatively higher wear resistance than in water lubricated condition where the transfer film is absent.

Another important factor contributing to the lubricated wear of polymers is the wettability of the rubbing surfaces, which is measured by the contact angle θ_C. Contact angle is defined as the angle that the tangent to the drop plane makes with the surface plane, in stable equilibrium conditions. The contact angle characterizes the surfaces of different materials and whether they are defined as wettable (high hydrophilic, $0^o \le \theta_C \le 45^o$) or not wettable (high hydrophobic, $90^o \le \theta_C \le 180^o$) (Steijn, 1967). The increase in the polymer wear rates in water lubricated condition can be related to good wettability of the metallic surface. Moreover, chemical corrosion of the counterface may reduce the surface roughness within the wear track and increase the surface wettability and polymer wear rates.

An earlier study showed that when nylon slides against a hydrophilic surface, such as glass, water can be an effective lubricant, but when nylon slides on a hydrophobic counterface, like PTFE or polyethylene,

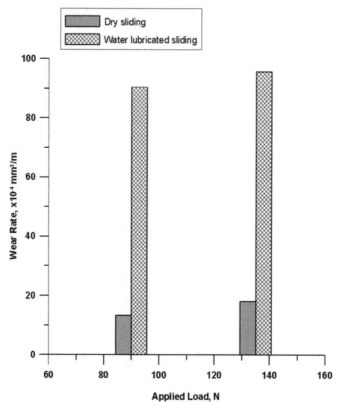

Figure 6.8: Effect of water lubrication on the wear of polyamide sliding against steel counterface. After Abdelbary (Abdelbary, 2014a) with permission from Elsevier.

water becomes completely ineffective (Cohen and Tabor, 1966). The strongest evidence in support of the effect of wettability on friction is obtained from a series of tests in which polyethylene sliding on itself was lubricated by a different alcohol-water solution. The results indicated that the lubrication was ineffective until the surface tension dropped below a certain critical value.

6.5 Lubrication

Tribological behavior of polymers in lubricated conditions may significantly differ from dry contacts due to the effects arising from the presence of a lubricant. In general, lubricants may interact with the mating surfaces in many ways and alter the wear behavior of polymers.

The primary wear mechanism of polymeric materials in dry sliding under moderate loads is adhesive wear. This type of wear is affected by two factors, namely, the amount of work being done on the polymeric surface and the intrinsic durability of the sliding interface. The wear rate of decay of the polymeric material surface, is determined by the rate of attrition and subsequent replacement of the transfer film as new layers of the surface of the polymeric material are abraded by sliding contact with the metal component. In order to enhance the tribological performance of polymer-metal tribo-systems, externally or internally, lubricants are often used.

Conventionally, external lubricants, such as lubricating fluids or grease, are utilized in tribo-systems to increase the wear resistance and reduce frictional losses between moving contacting surfaces. However, such external lubricants must often be replaced periodically and may be unevenly distributed over the wear surface, resulting in increased cost and inefficiency of the tribo-system. Hence, it may be desirable to replace or eliminate external lubricants by using internal lubricants in polymeric components in order to improve wear resistance and friction resistance. Internally lubricated polymers can replace polymers that wear too quickly and can even replace lubricated metals in some applications, providing advantages such as weight reduction, noise reduction, low production cost due to moulding compared to machining costs, part consolidation and eliminated corrosion issues.

6.5.1 External Lubrication

In general, lubrication plays a dominate role in the tribological behavior of polymers. Friction coefficient and wear rate can vary significantly as an effect of introducing an external lubricant to the polymer tribosystem. The main way by which a lubricant influences friction and wear of thermoplastics is by reducing adhesion of the polymer on the counterface. There are three general mechanisms for this; (a) absorption on the contact surfaces, (b) modifying the surface chemically, (c) physical separation of the surfaces. The first two mechanisms tend to reduce the strength of the bonds at the junctions and refer to boundary lubrication, while the third tends to reduce the number of junctions and refer to hydrodynamic and elasto-hydrodynamic lubrication.

Conventional boundary lubrications tend to be relatively ineffective in reducing the friction of polymers. It was found that the reduction in friction of polymer-polymer tribosystem, lubricated by various long chain compounds, was very limited and that the best results were obtained for those molecules which gave the most adherent and coherent absorbed films (Bowers et al., 1954). For example, the difficulty to lubricate nylon was attributed to the reason that the absorption sites are so far apart that adsorption of a sufficient condensed film was prevented. The nylon-steel tribosystem was easier to lubricate because a close-packed film could be formed on the steel by any long-chain aliphatic polar material could be absorbed. Fort (Fort, 1962) demonstrated that the friction decreased linearly with increasing lubrication chain length for homologous series of liquid acids, alcohols and alkanes.

The ineffectiveness of boundary lubricants was due to one of two possible reasons: First, the shear strength of long chain organic compounds is of the same order of magnitude as those of polymers, particularly for linear, crystalline thermoplastics. Another factor that may be involved is that, if the lubricant is absorbed by the polymer, the resulting plasticization might change the mechanical properties of the surface layer, particularly its shear strength. Cohen et al. (Cohen and Tabor, 1966) have examined this factor for Nylon 66 (PA66) and water, where the long exposure of nylon to water may result in absorption to the extent of 10%.

Abouelwafa (Abouelwafa, 1979) noted that, in wear tests of polymers, hydrodynamic lubrication can take place under some conditions. The basic assumption for this situation is that the wear specimens are not completely rigid. However, even though the apparent area of contact can be arranged to be very small, and the contact stresses correspondingly high, complications may still arise as a result of the formation of hydrodynamic fluid films. The governing equation for this form of hydrodynamic lubrication to take place was introduced by Lancaster (Lancaster, 1972a):

$$d_{crit} = 2.73 h_o^{\frac{2}{3}} \left(\frac{F}{\eta v} \right)^{\frac{1}{3}} \tag{6.13}$$

where

d_{crit} Critical wear scar diameter (m).

h_o Fluid film thickness at the trailing edge of the pin (m).

η Viscosity of the lubricant fluid (Ns/m^2).

v Sliding speed (m/s).

F Applied force (N).

In order to minimize hydrodynamic effects, it is preferable to choose the operating parameters so that "wear scare diameter" is always less than the critical wear scare diameter (d_{crit}) above which wear is effectively ceased. The likely magnitude of film thickness h_o was found to be in the order of 46.4×10^{-6} meters, which is much higher than the height of the asperities for the rough surface used during his investigations. This was an evidence that contact between those surfaces was largely relieved by hydrodynamic action during the tests.

Elastro-hydrodynamic lubrication often plays more important role in the lubrication of polymers than boundary lubrication. A comparison between the coefficients of friction in one particular set of conditions using a mineral oil lubricant were: steel – 0.07, glass – 0.02, graphite – 0.007, polymethylmethacrylate (PMMA) – 0.005. Thus, as the modulus of elasticity of the material decreased, the friction also decreased. This effect is likely to be an important factor in many applications, such as dynamic seals. It is also applicable to the lubrication of prostheses for which ultra high molecular weight polyethylene (UHMWPE) is most commonly used.

While it is generally found that both wear and friction are simultaneously reduced by the use of a lubricant, it is possible that the lubricant may decrease friction while increasing wear in polymers. An example of this is one in which the wear of the system is controlled by transfer or tribofilm formation; the addition of a lubricant can increase wear by inhibiting the formation of the film. Furthermore, in some cases, lubricant may be absorbed by polymer, affecting the mechanical properties these polymers. A secondary effect of a lubricant is cooling of the interface, modification of the stresses associated with the contact, and flushing of wear debris or contamination from the contact region.

A secondary effect of external lubrication is that the lubricant will also act as a coolant and will reduce sliding temperatures. The lubricant type will determine its influence on thermal failure. A partial external lubricant, such as grease, will reduce the level of friction in the mesh but will have no effect on operating temperatures.

A number of researchers investigated the use of water as an external lubricant in polymer tribosystems. For example, Tsukamoto et al. (Tsukamoto et al., 1993) employed water as a lubricant for polymer gears. Significant increases in durability and wear performance were observed. It was demonstrated that water has several advantages, such as lower cost, lower energy consumption, smaller viscosity-temperature change rate, safety for use, and clear environmental benefit compared to oil. In contrast to Tsukamoto's findings, Abdelbary et al. (Abdelbary et al., 2013) found a dramatic increase in the wear rate of Polyamide sliding against stainless steel in water lubrication conditions up to about 3 to 6 times those found in dry sliding conditions, as shown in Fig. 6.8. It was suggested that the high wear rates of polyamides are attributed to the plasticization caused by water absorption. The interaction of polymer with water suggests that water molecules diffuse into the polymer surface and loosen the hydrogen bonds between polymer chains, forming hydrogen bonds with amide groups. Consequently, water weakens the intermolecular forces and deteriorates the polymer's mechanical properties. The reduction of attractive forces between polymer chains allows for easy material removal and high wear rate in water lubricated sliding contacts. The worn surfaces of test specimens are smooth, as shown in Fig. 6.9, this could imply that the wear mechanism of Polyamide is dominated by abrasive wear.

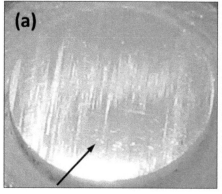 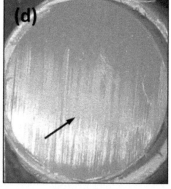

Burnished area   **Scratches**

Figure 6.9: Burnished area (a), and scratches (b) of wear surfaces of polyamide sliding against stainless steel. After Abdelbary et al. (Abdelbary et al., 2013) with permission from Elsevier.

6.5.2 Internal Lubrication

There could be different effects of the internal lubrication of polymers on the friction and wear due to modified mechanical properties and surface energy. Unlike in pure polymers, the action of internal lubricants in lowering friction is not unique and their efficiency strongly depends on the operating conditions.

Internally lubricated polymers offer obvious advantages: The lubricant is released slowly but steadily, exactly where it is required. Internally lubricated thermoplastics are used in classical bearing applications as well as in non-typical applications. The most common lubricating solids are Graphite, Molybdenum disulphide MoS_2 and PTFE. Both graphite and MoS_2 are natural materials, whereas PTFE is a synthetic organic material. Briscoe et al. (Briscoe et al., 1974) have added lead oxides and copper oxides to low density and high density polyethylene to improve wear and friction characteristics. They reported a 20-fold reduction in the wear rate of high-density polyethylene for adding 5% by weight of copper oxides and 35% by weight of lead oxides when compared to the wear rate of unfiled polyethylene. The general conclusion from other literatures (Cohen and Tabor, 1966; Reinicke et al., 1998; Samyn et al., 2007) was that the lubricating action depended on achieving complete coverage of the polymer surface with lubricant. At room temperature, the friction measurements for the impregnated polymers showed that the longer the time of heating the lower the friction becomes. Direction of sliding and counterface surface roughness markedly influence the lubricating action.

Abouelwafa (Abouelwafa, 1979) investigated the effect of silicone additives on wear, friction characteristics and related mechanical properties of low and ultra high density molecular weight polyethylene (LDPE and UHMWPE) sliding against mild and stainless steel, whether in wet or dry environment. It was found that the dominant mechanism in the early stages of the wear process was abrasive wear. Sequentially, adhesion wear gradually developed and the wear tests followed the steady state wear process, during which adhesion is the dominate mechanism. The rate of transfer film formation of the internally lubricated polyethylenes was less than that of the pure material due, of course, to the lower rate of impregnated polyethylenes but also perhaps due to the presence of the silicone fluid between the polymer and counterface, as shown in Fig. 6.10. The wear rate of the silicone impregnated polymers was found to be related to the silicone concentration, but the improvement resulting from the addition of small amounts of silicone fluid was very small. Figure 6.11 suggested that a significant improvement in wear rate was obtained after adding just 5% of silicone fluid.

It was only when at least 5% of silicone fluid was added that a significant improvement in wear rate was noticed, see Fig. 6.11. in addition, the effect of adding 10% silicone to LDPE was much higher than its effect for UHMWPE. The wear rate of the internally lubricated LDPE resulting from the addition of ten percent of silicone fluid is only about one hundredth of that for pure material. Also, the corresponding factors for UHMWPE ranged from one quarter to one twentieth for the same percentage of silicone fluid, as tabulated in Table 6.1.

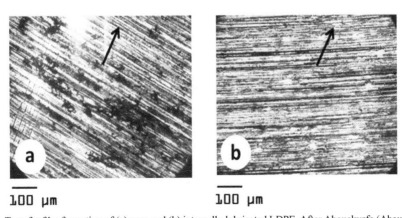

Figure 6.10:   Transfer film formation of (a) pure and (b) internally lubricated LDPE. After Abouelwafa (Abouelwafa, 1979) with permission.

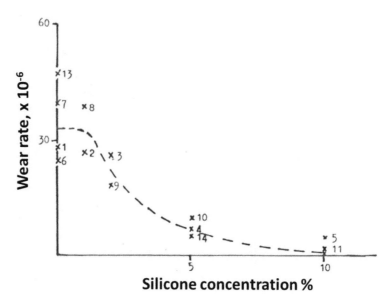

Figure 6.11: Effect of the concentration of internal lubricating fluid (silicone) on the LDPE wear rate. After Abouelwafa (Abouelwafa, 1979) with permission.

Table 6.1: Wear rates of pure and internally lubricated polymers (Abouelwafa, 1979).

Polymer	Counterface roughness, μm	Total sliding distance, km	Wear rate x10⁻⁶ mm³/Nm
LDPE	0.30	18.2	38.9
LDPE + 10% Silicone	0.31	724.7	0.304
UHMWPE	0.33	123.1	0.202
UHMWPE + 10% Silicone	0.30	648.2	0.041

Applied load = 52 N, Sliding speed = 0.24 m/s, Dry sliding.

Table 6.2: Comparison between the wear rates of internally and externally lubrication.

Polymer	Counterface	Counterface roughness μm	Load N	Speed m/s	Sliding distance km	Wear rate x10⁻⁷ mm³/Nm
UHMWPE	Mild steel	0.33	52	0.24	123	2.02
UHMWPE + 10% Silicone	Stainless steel	0.34	52	0.24	1128	0.11
UHMWPE lubricated with Bovine Synovial fluid	Stainless steel	–	68	0.018	37	0.9
UHMWPE lubricated with water	Stainless steel	0.05	88.8	0.24	274	0.17

Using silicone fluid as internal lubricant in polymers also has the effect of reducing the mechanical properties of polymers. The addition of 10% silicone resulted in about 30% reduction for most of the mechanical properties of LDPE and UHMWPE. For cases in which the polymeric bearing is required to operate for long period of time, the effect of silicone fluid on the creep characteristics of the polymer is potentially very important.

The comparison between the effect of internally lubricated polymers and externally lubricated polymers on wear characteristics shows that internal lubrication has the most pronounced effect, see Table 6.2.

6.5.3 Factors Contributing to Water Lubricated Polymers

The application of polymers, with water lubrication or contamination with water is a very important and practical issue. In lubricated contacts, water can influence friction and wear processes in several ways. The interactions between the lubricant and the lubricating surfaces should be fully understood in order to provide the machine elements with satisfactory life. In the presence of water, friction and wear of polymers are influenced by many factors which can be summarized as:

i) Water Absorption: The sensitivity of some polymers towards moisture or water has always been the matter of concern for the polymer scientists. Not only changes in mechanical properties but dimensional changes as a result of swelling also emerge because of water absorption. Thus, water absorption characteristics of the polymeric materials are very important to consider in the design of water lubricated bearings. In bearing applications, swelling of a polymer due to water up-take could potentially act as a break, resulting in increased frictional heating and wear.

Although assessment of water absorption of polymers should be based on long term equilibrium, most specification values are only based on a few hours of exposure. The results show that most of the polymers exhibited water absorption levels of less than 0.2%, while POM, PMMA, PET, PA 66 and PA6 showed larger values, with PA 66 and PA 6 exhibiting the largest water absorption respectively. Among polymers, Polyamides have the highest water absorption, which is attributed to the presence of amide groups in their molecular chains, favoring water absorption by forming hydrogen bonds with water molecules owing to their high polarity (Prakash and Hogmark, 2011).

ii) Wettability: Surface wettability is an important factor in determining the adhesion strength of polymer film to a substrate. The wettability is usually considered by measurement of the contact angle (θ_C). Contact angle is the angle that a tangent to the drop plane makes with the surface plane, in stable equilibrium conditions. Water, as a polar solvent, is the main liquid that is commonly used to measure contact angle, however, other liquids are also used to measure their respective contact angles.

A surface is called *hydrophobic* when the water contact angle is more than or equal to 90° ($90° \leq \theta_C \leq 180°$), whereas it is called *hydrophilic* when water contact angle is less than 45° ($0° \leq \theta_C \leq 45°$). The surface having a water contact angle between 45° to 90° is defined as an *amphoteric* hydrophobic/hydrophilic surface. Apart from chemical functionality, contact angle also depends on surface roughness.

Due to the surface tension of water, lubricating water adsorbed on hydrophilic surfaces reduces the coefficient of friction when sliding against hydrophobic surfaces. In this case, water droplets carry a load leading to low coefficient of friction between the surfaces. The performance of this lubrication is based on reducing the coefficient of friction by utilizing the repulsive force between water and the hydrophobic surface.

The increase in the polymer wear rates in water lubricated conditions can be related to the good wettability of the metallic surface. In aqueous media, chemical corrosion of metallic counterface may reduce the surface roughness within the wear track and increase a surface wettability and polymer wear rate (Srinath and Gnanamoorthy, 2007). Polymers with hydrophobic groups (CH_2- and CH_3-) can be modified by oxygen-plasma-treatment to those carrying a hydrophilic group ($-OH$). This modification makes the surface hydrophilic with a small contact angle, which is effective in full aqueous lubrication (Pawlak et al., 2011).

For polymer–water mixture systems, there are two situations of lubrication depending on the normal load (Kremnitzer, 1983). At low loads, the lubricating efficiency of the fluid depends on surface tension and wettability of the fluid. Low fluid surface tensions generally lower the coefficient of friction due to the formation of a uniform lubricating layer at the interface. On the contrary, at high loads elevated contact stresses tend to extrude the interposed fluid out of the interface zone leading to a direct solid–solid contact and, as a result, a high friction situation is generated.

iii) Solubility Parameter: The solubility parameter, δ, has been used for many years to select solvents for coating materials. Solubility of a polymer in a medium is characterized by its Relative Energy Difference (RED) in regards to that medium. A widely-used approach to determine the solubility of polymers in solvents is proposed by Hansen (1967). Hansen Solubility Parameter (HSP) has proven to be a powerful and practical way to understand issues of solubility, dispersion, diffusion, chromatography and more. HSP consists of several individual parts based on total energy of vaporization of a liquid. Materials which have similar solubility parameters have high affinity for each other and the extent of their similarity influences the extent of interaction between them; a principle often quoted as "like seeks like". The investigation of the HSP of polymers and their RED in comparison with water can provide information about the extent of interaction of the polymer chains at the interface with water and, therefore, can be of high importance for studying the tribological behavior of polymers in water lubricated contacts (Prakash and Hogmark, 2011).

The usefulness of polymers in many tribological applications is critically dependent on the solubility parameter. It has been shown that the solubility parameter is connected to other physical properties such as the surface tension and wettability, the ratio of the coefficient of thermal expansion to compressibility, the ultimate strength, and the glass transition temperature of polymers. Therefore, the ability to estimate the solubility parameters can often be a useful tool to predict the systems' physical properties and performance. In lubricated conditions, solubility of polymers can potentially provide information about the extent of the interaction of the polymer chains at the interface with the lubricant. The latter can significantly influence the tribological behavior of polymers in water, as penetration of water molecules at the surface alters the polymers' affinity to water and influences the wettability characteristics and possibly surface free energy of the polymers.

6.6 Sliding Mechanics of Polymers

6.6.1 Transfer Film Formation

In the sliding of a polymer over a dry metallic counterface, if the adhesive bonds formed between the rubbing pairs become stronger than the cohesive strength of the bulk polymer, some parts of the polymer will transfer onto the counterface, forming a transfer film. Another part of the worn polymer is removed as wear debris. Sequential polymer transfer occurs and bands build up by the agglomeration of wear particles. The bands grow in length and widen as sliding progresses. The transfer film may be in the form of irregular lumps adhered to the smooth metallic surface or a continuous uniform layer which may deform during repeated contacts, leading to a smoother counterface or it may take any other form.

There are three main factors determining the transfer of a polymer: The deformation of surface asperities under load, the fracture of the material in the substrate and the adherence of the material to the other surface. These effects depend on the type of wear mechanisms, abrasion for rough and adhesion for smooth counterfaces.

One of the interesting consequences of polymer transfer is a change in the topography of both rubbing surfaces in contact. The roughness of a polymer surface undergoes large variation during the running-in wear until the steady state wear is reached, while the roughness of the counterface surface is modified due to the transfer of the polymer. Counterface roughness measurements have shown that the transfer film reaches a limiting value after a certain period of sliding time. As the transfer film fills the valleys of the metallic counterface, thus. effectively smoothing the surface, a point will be reached at which equilibrium between the polymer transferred to the counterface and that removed from the wear track is achieved.

In water lubricated sliding, it has been stated that the polymer transfer film, which is usually found in dry sliding, fails to form and the slide occurs directly between the polymer surface and the steel counterpart. The effect of water lubrication on the sliding interaction is due to inhibition of the transfer film formation on the counterface. The washing action of water makes the virgin surface of the polymer that is continuously exposed to the counterface. In such sliding conditions, all the wear rates of tested polymers were much higher (up to 6 times) than those tested in dry sliding (Abdelbary, 2014b).

6.6.2 Wear Regimes

In polymer tribosystems, when the volume loss V of a polymer is plotted continuously against the sliding distance X, a typical characteristic curve can be divided into two distinct zones. The first regime of sliding, called "Unsteady State Wear" or "Running-in", is related to the removal of the artificial (manufacturing) surface of the polymer specimen, which is characterized by a highly residual strain layer. The second phase, Steady State Wear, has a lower and linear wear rate and is characterized by the formation of a stable transfer film layer of the polymer which is worn in a steady and uniform way. However, in long-term wear tests on a polymer, using a pin-on-disc configuration, Dowson et al. (Brown et al., 1974b; Atkinson et al., 1985) introduced a third wear regime referred to as "Section B" wear. After prolonged, 120 to 150 km, sliding distance on a steady-state wear curve, they observed a remarkable increase 10–30% in the wear rate. It was suggested that Section B is a surface fatigue wear, taking place after a number of cycles to failure proportional to the sliding distance. Evidence of transverse cracks in the polymer surface suggested the possibility of a fatigue process. Similar discontinuities of wear rate have been observed in thrust bearing tests at higher bearing loads. These indicate that the sliding distance to the point of transition is strongly dependent on the applied load. Since, as has been suggested, a transition marks the onset of a fatigue mechanism of the wear, it is to be expected that this would be strongly dependent upon the load.

Abdelbary et al. (Abdelbary et al., 2011) found that performing a single transverse crack upon the rubbing surface of the polymer during a steady state wear phase resulted in the generation of Section B wear, which is characterized by a relatively higher wear rate. The wear test, at first, starts with an un-cracked pin, and after 80 km of sliding distance, a surface crack was formed. The test was run again for another 40 km sliding distance, as shown in Fig. 6.12. It was observed that Section B starts after 80 km of sliding, i.e., just after the formation of a surface crack. Although the increase in wear rates was about 15%, it is still a considerable value since there were no changes in any of the other test conditions during these tests in particular. These results demonstrate how important it is to detect surface fatigue cracks in a polymer-on-metal bearing configuration which play a role in increasing the wear of sliding components.

According to aforementioned discussion, three different regimes are now accepted for describing a typical wear process in polymers: Running-in, steady state and a third accelerated stage where the wear rate increases in an exponential way leading to a fatigue failure, as illustrated in Fig. 6.13.

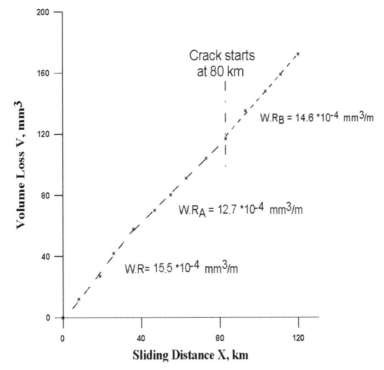

Figure 6.12: Variation of Polyamide 66 volume loss (V) with sliding distance (X). After Abdelbary (Abdelbary, 2011).

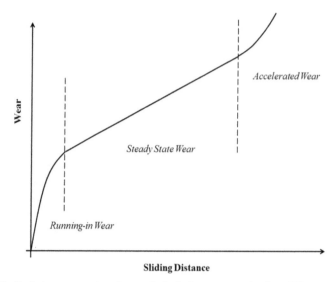

Figure 6.13: Typical wear curve in polymer tribological system contains three different wear regimes.

6.6.3 pv-limit

Many parameters influence the sliding mechanisms in polymer tribosystems. The upper limit of operation is one of the important descriptors of the rubbing severity, above which a severe or catastrophic wear may occur. This limit is usually based on the maximum allowable value of pv, where p is the average contact pressure and v is the sliding speed. This factor is widely used as a performance criterion for bearing materials and represents a limiting value for the maximum allowable pressure of contact versus the sliding speed under normal running condition. Lancaster (Lancaster, 1972) defined the pv limit as the value above which the wear rate increases rapidly and the maximum pv as the maximum value for a continuous operation at a specified wear rate.

Each polymer has a unique pv limit, as measured by some test, usually a "washer" test, as shown in Fig. 6.14.

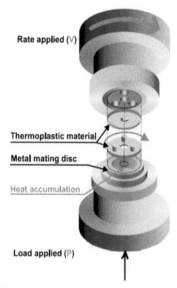

Figure 6.14: pv limit measurement equipment, washer test.

In some cases, manufacturers perform a series of tests in which, at a timed sliding speed, the bearing pressure is increased in steps until a limiting *pv* is reached. The limit is determined when the polymer sample cannot dissipate the energy accumulated and its internal temperature increases to the point of failure. This is then repeated at other sliding speeds and the limiting *pv* curve for the material can be drawn, as shown in Fig. 6.15. The *pv* limit curve separates pressure and velocity combinations into two regions of wear behavior, one that is generally considered to be acceptable for applications and the other that is not.

Evans and Lancaster (Evans and Lancaster, 1979) carried out a series of pin on ring experiments with polyacetal pins sliding against different counterface materials. They demonstrated that, in each case, there was a critical sliding speed above which the polyacetal wear rate increased rapidly. This sliding speed corresponded to a point where the calculated maximum polymer surface temperature *T** reached the melting point of the polyacetal (175°C). In these tests, the duration of sliding was kept deliberately short in order to limit the mean surface temperature rise. The wear rate at given *pv* values as well as the maximum recommended *pv* values of selected polymers are presented in Tables 6.3 and 6.4, respectively.

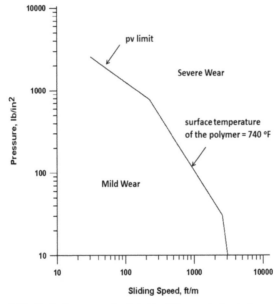

Figure 6.15: Typical *pv* curve of polyimide sliding against a steel counterface.

Table 6.3: The wear rate at given *pv* values for selected polymers. After Lancaster (Lancaster, 1972) with permission from Elsevier.

Sliding conditions			Wear rate of polymer x10⁻⁶ mm³/N.m			
Speed m.s⁻¹	Pressure MPa	pv value MPa.m.s⁻¹	PTFE	Acetal	UHMWPE	Nylon 66
0.25	0.28	0.07	20			4
0.10	1	0.10	800			
0.01	1.4	0.014	107	2.8		5.8
0.10	5	0.50		1.5	0.50	20
0.24	2.5	0.606			0.25	
0.25	5	1.25		2.1		16
0.24	10	2.4			0.20	
0.24	10	2.4			0.27	
0.25	10	2.5		12	0.51	
12.4	0.15	1.86			2.9	

Table 6.4: The recommended *pv* limits for selected polymers. Adapted from Franklin (Franklin, 2001) with permission from Elsevier.

Name	Polymer	Maximum *pv* (N/m.s)
POM	Polyoxymethylene	130000
POM + 30% GF	POM + 30% glass–fibres	350000
POM + 20% PTFE	POM + 20% polytetrafluorethylene	400000
PA6.6	Polyamide 6.6	90000
PA6.6 + 30% GF	PA 6.6 + 30% glass–fibres	400000
UHMWPE	Ultra high molecular weight polyethylene	70000

6.6.4 Thermal and Chemical Degradation

According to the American Society for Testing and Materials' (ASTM) definitions, thermal degradation is "a process whereby the action of heat or elevated temperature on a material, product, or assembly causes a loss of physical, mechanical, or electrical properties". In polymer tribosystems, polymers are often exposed to high friction heating during sliding. They undergo both physical and chemical changes resulting in unwanted changes to their properties. Thus, thermal and chemical stability are among the most important limiting factors in the tribological application of plastics at high temperatures. Degradation can present an upper limit to a polymer's service temperature as much as the possibility of a mechanical property loss, softening and melting, because it can occur at temperatures much lower than those at which a mechanical failure is likely to occur. As polymers degrade by wear, their performance is generally observed with the surface temperatures approaching and exceeding the glass transition temperature. On the other hand, an elevated temperature can increase reaction rates, influence phase changes, increase diffusion, and enhance flow characteristics of materials (Villetti et al., 2002).

Thermal degradation of polymers is a molecular deterioration occurring at high temperatures. The components of the long chain of an overheated polymer can begin to separate and react with one another to change the properties of the polymer. This generally involves changes in the molecular weight of the polymer. Typical property changes include ductility reduction, chalking, color changes, cracking and general reduction in most other desirable physical properties (Madorsicy and Straus, 1959).

In general, all polymers will undergo some form of degradation during usage, and this will result in a deterioration of their mechanical and physical properties. The ability of a polymer to resist this decline is called the 'stability' or 'stabilization' of the polymer. Therefore, the radiation-processing may be applied to inhibit unwanted material property changes. The irradiation of polymeric materials with ionizing radiation (gamma rays, X rays, accelerated electrons, ion beams) leads to the formation of very reactive intermediates. These intermediates can follow several reaction paths, which results in rearrangements and/or formation of new bonds. During irradiation, carbon fluorine bonds are primarily broken, creating C free radicals along the carbon backbones of the polymer chains. These unstable free radicals may eventually either react with each other to form crosslinks between chains, or be subject to scission reactions rupturing the backbones of these chains (Menzel and Blanchet, 2005). The phenomena involved in radiation-degradation are exceedingly complex. In some instances, a suitable stabilization technology is yet to be developed. In other cases, a functional stabilization scheme may exist, but its fundamental basis and its details are solely in the hands of one or more private firms, who keep the information proprietary.

Chemical degradation, or thermal decomposition, is a process of extensive chemical change caused by heat. In many polymers, the chemical degradation processes are accelerated by oxidants. In polymer-metal tribosystems in an air environment, the most common reaction mechanism involving chemical degradation is the bond breaking in the polymer chain. The chain scission results in the production of a monomer, and the process is often known as depolymerization. These physico-chemical changes that take place during sliding influence the tribological properties of the polymer. Such changes include the break-up of the polymer structure, and the formation of new chemical compositions (Pogosian et al., 2006).

With the rapid development of nanotechnologies, it is well known that nanoparticles can enhance thermal degradation of polymers and composite materials. Polymers can interact with a surface having inorganic nanoparticles forming Hydrogen or covalent bonds, which can increase the adhesion of nanoparticles with the polymer, resulting in higher dispersion degrees. In most cases, this leads to a radical enhancement of the chemical degradation properties (Bikiaris, 2011).

6.6.5 Generation of Wear Debris

The generation of wear debris is one of the most significant evidences of the wear in polymer tribosystems. The wear mechanism is defined by contact conditions, mechanical properties of the bulk polymer and how these parameters lead to subsequent events of the wear debris generation or transfer film formation (Briscoe and Sinha, 2013). Furthermore, the entrapment of wear debris particles in the interface or the formation of transfer film is determined by the test configuration.

In open tribosystems, debris particles can be immediately removed from the sliding contact as loosen particles. Detached debris particles may be involved in a three-body wear, thus, they can accelerate the deterioration of the polymer surface due to the wear. On the other hand, in closed tribosystems, wear debris remains circulating into the sliding contact and, consequently, interacts to form a polymer transfer film. This process is particularly important in fretting and reciprocating wear, where debris particles are not allowed to escape from the contact. Wear debris in closed tribosystems act as a "lubricant" and the wear rate is reduced under certain circumstances. However, under lubricated sliding conditions, a lubricant fluid tends to flush the wear debris from the interface and prevent their buildup, which would result in more contact with the abrasives or asperities.

Under repeated loading, the shape of polymer wear debris is attributed to the fatigue of micromechanical abrasion, tribo-chemical reactions or adhesion. In fatigue wear of polymers, where the subsurface cracks propagated due to the repeated deformation of the material at friction. Wear debris are formed as a result of the growth and intersection of the small cracks. Nucleate in the subsurface at a depth of several micrometers, which are perpendicular to the sliding direction. This wear mechanism producing such plate-like debris appeared to be a fracture process. Figure 6.16 shows wear debris particles of polyamide 66 sliding against dry steel counterface (Abdelbary, 2011b).

The debris formed by abrasion are frequently shaped as fine cutting chips, similar to those produced during machining, although at a much finer scale. The scratches, grooves, and dents on a metallic counterface may penetrate into a softer polymer. Deep scratching will produce debris of the polymer. It is obvious that the rate of debris production by abrasion depends strongly on the shape, orientation, and manner of the constraint of the abrasive. However, for relatively ductile polymers, wear debris can be spiral in shape, as shown in Fig. 6.17. For harder polymers, the wear debris tend to be in the form of chips, generated by local brittle fracture of the material.

Brown et al. (Brown et al., 1974a) studied the wear of UHMWPE against stainless steel in a dry sliding condition. The most common debris produced were in the form of thin streamers (up to 20 mm in length) of material attached to the trailing edge of the wear pin. Fine powdery white debris were also generally seen at the wear track during section B wear.

Nonetheless, the running-in wear phase of Nylon 66 sliding against stainless steel was investigated by Zalisz et al. (Zalisz et al., 1988). Microscopical examination of the loose wear particles suggested an important role of the running-in regime of a reciprocating sliding. Various types of wear particles were found during the investigations and four basic groups were distinguished:

a) "Needles"—highly slender, with sharp ends could smear into a very thin transfer layer with a characteristic crescent shape. They were found only on initially smooth surfaces.

b) "Cigars"—rather short, blunt-ended; elastic rather than plastic. Created by rolling between the rubbing surfaces of previously irregular wear particles which were products of abrasion by a rough metallic surface on a polymer above its glass transition temperature.

Figure 6.16: Wear debris of polyamide 66, after 80 km sliding distance against dry steel, cyclic load F_{mean} = 90 N. After Abdelbary (Abdelbary, 2011).

Figure 6.17: Wear debris of polyamide 66, after 60 km sliding distance against dry steel, F = 90 N. After Abdelbary (Abdelbary, 2011).

c)   "Compacts"—compact, random in shape, and brittle, the result of abrasion by a rough metallic surface onto a polymer only below its glass transition temperature. They were characterized by very low adhesiveness to the metallic surface and so could not create a transfer film.

d)   "Flats"—irregular, flat and resembling floating ice, usually with marks of oxidation. They only appeared when a transfer film fully covered the sliding track. They are a characteristic of a steady state phase of reciprocating sliding, independent of the initial roughness of the counterface.

6.6.6 Surface and Subsurface Cracking

In rolling and sliding of polymers, each asperity of a rubbing surface experiences a cyclic stress from asperities of the counterface. As a result, stress cycling and plastic strain accumulate and many surface and subsurface cracks are ultimately initiated. With further cycling, these cracks propagate deeply into the substrate or to join their neighbour cracks to create one crack large enough to break from the bulk causing pitting and spalling. The process continues, resulting in a progressive loss of material from the polymer worn surface. The nature and number of crack initiation sites in the surface depends on the type

Figure 6.18: Scanning electron micrographs of UHMWPE rubbing surface shows surface cracks running perpendicular to sliding direction (Load = 52 N, Speed = 0.24 m/s). After Abouelwafa (Abouelwafa, 1979) with permission.

of loading exerted on the surface and the sliding conditions (temperature, pressure, humidity, etc.) that exist in the tribosystem during this exertion. In addition, the cracks under rolling conditions tend to be more macroscopic or coarser than those under sliding and frequently result in larger wear debris (Terheci, 2000). These aspects suggested that the only way to determine the existence of surface cracks under sliding conditions is by means of a microscopic examination of the worn polymer surface, such as those shown in Fig. 6.18.

Keer et al. (Keer et al., 1982) studied the fatigue effect of vertical surface cracks in order to correlate the crack geometry to a surface fracture and to introduce possible mechanisms for a crack propagation. It was concluded that if a small surface breaking crack is formed above a horizontal subsurface crack, the vertical crack gains a strong tendency to propagate downwards towards the horizontal crack. It may either break through or the horizontal crack may extend further; the tendency of the crack to propagate is strong, and a possible result will be for the material to flake off. Microscopic investigation of polyamide 66 worn surface showed many transverse vertical cracks (Chen et al., 2000). The depth at which the main cracks propagated varied from 150–450 μm. Moreover, between these main cracks there were some sub-cracks which propagated to about a quarter of the depth of the main cracks.

6.7 Wear and Friction of Polymer Composites

The successful applications of unfilled polymers in engineering purposes led to the work of using polymers as composites where modifications to their properties could be achieved by adding one or more reinforcing fillers. The fillers may have the form of particles and provide high hardness, strength or elasticity; the matrix, which is a polymeric material, adds self-lubrication. This unique combination combines the best and suppresses the worst properties of the individuals.

Composites attracted an increased attention for tribological applications, owing to their excellent structural performance, and excelled over conventional materials in many engineering applications, such as gears, bearing cage, shoe soles, automotive brakes and so on. Yet, to date, there have been many extensive researches investigating the role of fillers and reinforcements in the wear behavior of polymeric composites, and several wear reduction mechanisms of fillers have been proposed. A common conclusion of the majority of friction and wear investigations is that the tribology of polymer composites cannot be an intrinsic property. In fact, it depends strongly upon the constituent variables such as the material of fiber, fiber orientation, matrix polymer, etc. Also, it depends on process variables, such as counterface material, normal load, sliding velocity, etc. (Vishwanath et al., 1992). Although reinforcing a polymer with fibres enhances the tribological properties of the bulk polymer, sometimes it can worsen them. It has been indicated that in most polymeric composites, a high stiffness and a low thermal conductivity result in a high temperature at the sliding contact during friction. Beyond a certain critical temperature, wear rates were found to be increased very sharply (Pihtili, 2009).

6.7.1 Mechanisms of Failure

a) Fiber Fracture: It has several forms, as illustrated in Fig. 6.19, such as:

(i)   Fiber debonding: It takes place when the stress exceeds the strength of the weakest fibers in the composite. An isolated fiber break causes shear stress concentration at the interface close to the tip of the broken fiber. The interface may then fail, leading to fiber debonding from the surrounding matrix. Consequently, the debonded area magnifies the tensile stress of the matrix leading to a transverse crack.

(ii) Fiber Pull-out: In some cases, cracking of fibers and resin can occur before complete separation of the fracture surface. The two surfaces are held together by fiber which bridges the fracture plane and the work done in separating the surfaces makes a major contribution to the total energy of fracture.

(iii) Fiber Bridging: It occurs after a sever matrix progressive crack in a specified part of the matrix like air bubbles or an inclusion. This mode happens only if the fiber strength is very high and the matrix is defected in a certain part and subjected to a stress concentration in the defected zone.

b) Delamination: It is the separation occurring between composite layers (laminates). One of the main sources of delamination is the introduction of a shearing stress between layers. The shearing stress arises due to the tendency of each layer to deform independent of its neighbors because of its orthotropic (or generally anisotropic) properties. Such shearing stresses are largest at the edges of a laminate and may cause delamination there.

c) Matrix Cracking: It occurs when the stress exceeds failure levels or under a cyclic loading, if the matrix is subjected to a strained controlled fatigue and the applied cyclic strain exceeds its strain limit. Also, at low strain, matrix crack would stop at the interface.

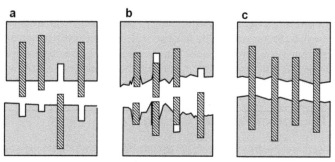

Figure 6.19: Systematic illustration of possible forms of fibre fracture: (a) debonding; (b) fibre pull-out; (c) fibre bridging.

6.7.2 Continuous Unidirectional Fiber Composites

The addition of continuous fibers is a commonly-used technique for enhancing the tribological properties of many polymer materials. In particular, continuous unidirectional fiber-reinforced polymers are being introduced for a variety of mechanical and biomedical applications due to their great wear resistance either normal or parallel to the sliding direction. When unidirectional polymer composites slide against a rough counterface, the normal and tangential forces from the rough surface induce tensile or combined tensile and shear stresses in the surface fibers, depending on the orientation of the fiber, which are functions of the counterface asperity geometry, the composite properties, and the load acting on the asperity (Ovaert, 1997). However, the influence of the fiber orientation on the wear rate also depends on the type of the composite material under consideration, as well as the type of the tribological system under which it operates (Cirino et al., 1988).

Carbon-fiber-reinforced polyetherketone (CF-PK) and armid-fiber-reinforced epoxy (AF-EP) were tested in order to clarify the effect of a fiber orientation on the dry abrasive wear behavior of continuous unidirectional fiber-reinforced polymers (Ovaert, 1997). The fiber orientation with respect to the sliding direction was described by two rotational parameters α and β. The α rotations are defined as counterclockwise rotations about 2-axes, and the β rotations are defined as counterclockwise rotations about 3-axes. The three proposed orientations were: (a) Normal N, $(\alpha, \beta) = (0, 0)$; (b) Parallel P, $(\alpha, \beta) = (0, 90)$; Antiparallel AP, $(\alpha, \beta) = (0, 90)$, as illustrated in Fig. 6.20. It was observed that the direction of sliding relative to the fiber as well as the fiber orientation relative to the sliding surface have a strong effect on the wear behavior of both composites. Other factors, such as the material type and the direction of sliding, were also proved to influence the wear resistance. The microscopy of wear mechanisms indicated that the basic micro-wear occurring in the N orientation was in forms of fiber slicing, fiber-matrix debonding, fiber cracking and fiber bending. On the other hand, internal crack propagation, fiber cracking, fiber-matrix debonding and fiber fracturing were observed in the P orientation. The important influential factor was whether an abrasive particle pulls the fiber out as it passes or tends to compresses the fiber. Both mechanisms generate different levels of wear, where the second one will tend to generate a higher material removal.

Examination of the friction and wear properties of various carbon fiber reinforced polymers sliding against metals in water, aqueous solutions and organic fluids indicated that the wear of carbon fiber reinforced polymers as well as unfilled polymers under water lubrication was generally greater than that under dry conditions (Lancaster, 1972b). The higher wear rate in water is attributed to the fact that the

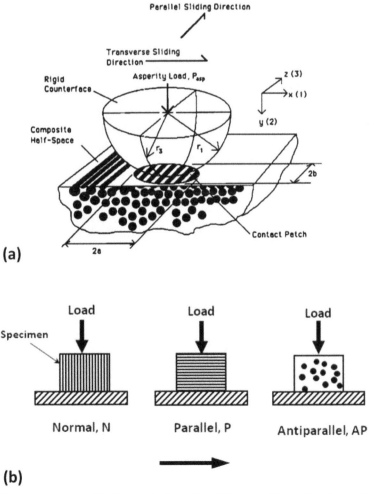

Figure 6.20: (a) Contact geometry and (b) Systematic illustration of fiber orientations. After Ovaert (Ovaert, 1997) with permission from Taylor & Francis.

counterface could not be modified by transfer, because the transferred film could not observed on the counterface rubbed under water lubrication.

6.7.3 Short Fiber Reinforced Composites

The application of different fillers gives an opportunity to improve the tribological behavior of polymers. For example, to reduce the adhesion, internal lubricants such as polytetrafluoroethylene (PTFE) and graphite flakes are frequently incorporated. Short aramide fibers (AF), glass fibers (GF) or carbon fibers (CF) are used to increase the creep resistance and the compressive strength of the polymer matrix system used (Zhang et al., 2002; Abdelbary and Mohamed, 2020). The selection of suitable fibers is often a compromise between the properties of the polymer and its friction and wear behavior.

It is reported that polymer composites, reinforced with carbon, glass and aramide fibers, are often one to four times stiffer than pure polymers (Suresha et al., 2010). Furthermore, it is found that the wear rate increases with load and the wear rates are increased by about a factor of 10 over those for unreinforced material. Carbon fibers and glass fibers significantly reduce the coefficient of friction of nylon composites up to 50% compared to unreinforced nylon. However, it should be noted that mechanical and chemical interactions of the aforementioned issues are very complicated, accordingly, further studies to understand these relationships in more detail are a subject of current and future investigations.

In order to enhance tribological properties of polymers, one typical concept is to reduce their adhesion to the counterface and to improve their hardness, stiffness and compressive strength. This can be achieved effectively by using various fibers. Table 6.5 summarizes the specific wear rate of various fillers modified and/or short fiber reinforced thermoplastic composite systems.

The length of fibers has been shown to be one of the principal factors affecting the tribological behavior of short fibers. Relatively longer fibers seemed to be more effective in improving the abrasive wear resistance of thermosetting composites (Zhang et al., 2002). The longer fibers were damaged locally under wear against the abrasive particles, and the rest of the fibers contributed to the wear resistance of those composites. Under dry sliding conditions, the friction and wear response of short carbon fiber reinforced epoxy (SCF/EP) composites was correlated to the length of the fibers (Zhang, 2007) . The tribological tests were performed using nominal fiber lengths of 90 µm and 400 µm while the matrix was a bisphenol-A type epoxy resin. Graphite flakes with a size of 20 µm and TiO_2 particles with an average diameter of 300 nm were applied as additional fillers. The fiber length shows distinct effect on the wear rate. When the average fiber length was increased from 67 to 236 µm, the wear rate was significantly reduced nearly by three times, as illustrated in Fig. 6.21.

Table 6.5: Examples of various newly developed, filled polymer systems with excellent tribological properties under various loading conditions. After Friedrich et al. (Friedrich et al., 1995) with permission from Elsevier.

Material compositions (vol%)							Specific wear rate
PTFE	**PPS**	**PEEK**	**Graphite**	**CF**	**Bronz**	**Al_2O_3**	
51.6	31.8	–	4.8	11.8	–	–	1.53
51	31.6	–	3.9	13.5	–	–	1.40
51.9	32.4	–	2.6	13.1	–	–	1.20
12.4	–	61.8	11.7	14.1	–	–	4.31
9.7	–	49.7	12.5	28.1	–	–	6.33
76.8	–	–	19.8	–	–	3.4	22.4
84.1	–	–	–	12.6	–	3.3	1.25
52.5	28	–	–	19.5	–	–	1.69
78.6	–	–	–	21.4	–	–	1.75
80	–	–	–	10	10	–	0.565

Block-on-ring tests, p = 2 MPa; v = 1 m/s; t = 8 h

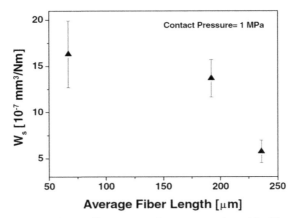

Figure 6.21: Effect of fibre length on the specific wear rate of epoxy composites. After Zhang et al. (Zhang et al., 2007) with permission from Elsevier.

The longer fiber composites always exhibited a better wear resistance than the shorter ones. This reflected that the longer fiber composites possessed a higher load-carrying capacity, which can probably be attributed to their greater strength and stiffness. Correspondingly, the fiber length was found to have a distinguished effect on the frictional coefficients of the wear samples after running-in periods. The longer fibers could slightly reduce the frictional coefficient in comparison to the shorter ones. SEM observation of worn surfaces showed that the shorter fiber composites had a much rougher worn surface in comparison to the longer ones. It was accepted that the wear behavior of short fiber composites was due to the process of fiber peeling-off. It seems that the strong frictional force may "dig out" individual fibers directly without obvious breakage. In contrast, the peeling-off of longer fibers from a matrix would not occur so easily. Parts of them were finally separated from the wear surface, whereas the rest remained embedded in the wear surface. It is likely that the rubbing surface became relatively smooth, and, consequently, the frictional coefficient and the wear rate dropped to some extent.

6.7.4 Particulate-filled Composites

The shape, size, volume fraction, and specific surface area of added particles have been found to affect mechanical properties of the composites greatly.

Micro- and nanotribology are new areas of tribology where one tries to improve tribological properties by using fillers with sizes in the micro- or nano-scale range, respectively. Regarding polymeric materials, it is well known that all macroscopic properties of polymers depend on the micro-nano structure and interactions at the molecular level. Thus, many kinds of inorganic materials, such as metal powders, minerals, oxides, carbon black, and solid lubricants, have been used recently in micro and nanometer sizes as fillers. Each filler type has different properties and these in turn are influenced by the particle size, shape and surface chemistry. Generally, fillers have shown effectiveness in reducing the coefficient of friction and also increasing the wear resistance in many cases. However, in some cases, the addition of particulate fillers to polymers has resulted in degradation in the wear resistance.

The effects of inorganic particles, such as SiC, Si_3N_4, SiO_2, ZrO_2, ZnO, $CaCO_3$, Al_2O_3, TiO_2, and CuO, on the tribological properties of polymers such as PEEK; PMMA; PTFE and epoxy have been widely investigated in several publications. Some fillers have shown promising results in terms of the tribological performance. For example, using TiO and ZrO_2 particles, as filler materials, markedly reduced the wear of PTFE polymer matrix (Tanaka and Kawakami, 1982). The filler content of ZrO_2 particles was 40 wt%, while the size ranged from several microns to about 50 μm. The TiO_2 was an agglomeration of fine particles of less than 0.3 μm in size, and the filler content was 20 wt%. Both particles could significantly decrease the wear rate of PTFE, although the frictional coefficient was increased. Examination of worn surfaces indicated that, ZrO_2 resulted in a greater abrasiveness against a metallic counterface than TiO_2, due to the bigger particle size and probably the higher hardness of ZrO_2. It was hypothesized that smaller filler particles were transported within the wearing PTFE in its process of transferring to the counterface. These particles are incapable of preventing large-scale reorganization of the PTFE structure at its frictional surface. Finally, the limited transfer wear process would provide only a weaker wear-reducing action. Al_2O_3 fillers also exhibited similar results when incorporated into PTFE (Sawyer et al., 2003). The addition of 38-nm Al_2O_3 nanoparticles at 20% by weight reduced the wear rate of PTFE further down to $1.2 * 10^{-6}$ mm^3/Nm, and, upon further modification, attained $1.3 * 10^{-7}$ mm^3/Nm with an 80-nm Al_2O_3 filler particle at an optimum 5 wt% concentration.

The size of the filler particles plays an important role in simultaneously enhancing friction and wear in particular. Many studies have focused on how a single-particle size affects the mechanical properties of composites. Smaller particles seem to contribute better to the improvement of tribological properties under sliding wear conditions than larger particles. Reportedly, a few inorganic micro fillers, that have length dimensions greater than 1 μm, have shown benefits in nearly all of their uses in tribological applications. Also, submicron particles showed considerable effects on both the mechanical properties and the wear resistance of composites (Chauhan and Thakur, 2013). By reducing the particle size to a nano-scale level, the wear performance of composites may be significantly different from that of micro-particle filled systems. Nanoparticle-filled polymers, so-called polymer nano-composites, are very promising materials for various sliding applications. The size effect has been found to be important for a tribological behavior because nanoparticles are effective in increasing wear resistance in much smaller volume percentages than the microparticle fillers. This behavior makes the nanoparticles more attractive than the microparticles as fillers in polymers for tribological applications. It is of high importance that the nanoparticles are uniformly dispersed rather than being agglomerated in order to yield a good property profile in general.

6.7.5 Laminated Fiber Reinforced Composite

Laminated fabric reinforced polymer composites are gaining popularity in many mechanical industrial applications because they exhibit very good mechanical strength properties in both longitudinal transverse directions. Also, fabrics are easy to handle and have relatively low fabrication costs. Plain wave, woven roving, satin weave and many other fabric geometries are commonly-used fabrics. Various factors such as type and amount of the matrix resin and fabric, its orientation and weave, fiber–matrix adhesion along with the fabric geometry control the tribo-performance of these composites.

The effect of a laminate orientation with respect to the sliding direction is one of the important parameters to be considered when assessing the tribological properties of laminated composites. Glass laminate reinforced polyester (GLRP) composite was tested on three principal orientations with respect to the sliding direction (El-Tayeb and Mostafa, 1996), as illustrated in Fig. 6.22.

a)   normal-laminate   NL

b)   parallel-laminate   PL

c)   cross-laminate   CL

The friction coefficient and wear resistance of composite were found to be highly affected by the orientation of the laminates with respect to the sliding direction. With varying normal load, PL orientation gives the highest responses, followed by CL and NL, respectively. The location of an individual laminate with respect to a counterface surface controls the friction and wear processes. In case of the PL orientation, the counterface meets every hard and stiff single glass fibre normally or tangentially to its axis. The main wear process takes place at the projection of the fibre cross section, which reduces the contribution of the softer matrix in the wear process. The relative motion of counterface with respect to the laminates does not permit accumulation of wear debris; it sweeps them away and consequently allows exposure of the fresh fibre to the counterface. These processes lead to the highest wear resistance and friction coefficient of PL composite. On the other hand, NL composite shows relatively lower values of friction coefficient and wear resistance. In this orientation, the fibres are backed by the soft matrix and, therefore, suffer a little bending at their ends, followed by an easy shear effect. Finally, the lowest values of friction coefficients

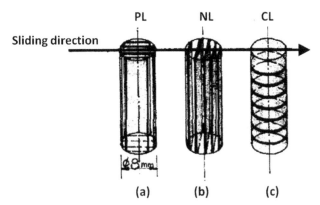

Figure 6.22: Position of glass laminate orientations with respect to the sliding direction: (a) normal-laminate (NL); (b) parallel-laminate (PL); (c) cross-laminate (CL) sliding direction. After El-Tayeb et al. (El-Tayeb et al., 1996) with permission from Elsevier.

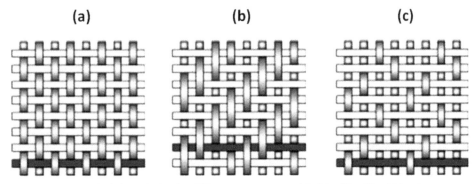

Figure 6.23: Schematic showing different weave patterns: (a) plain, (b) twill and (c) stain.

were found in the case of CL orientation. This is due to the fact that matrix components are entrapped between fibres within the laminates. Also, the configuration of weaving prevents an easy detachment of the fibre during sliding, leading to an intermediate wear resistance.

The fabric geometry also has a significant role in controlling the tribo-performance of laminated composites. Of the many fabric geometries available, woven roving, plain weave, twill wave and satin weave woven fabrics are being widely used in many of the composites, as shown in Fig. 6.23. As mentioned before, wear of fabric reinforced composites depends on two main factors. Firstly, the fibre damage mechanism by various processes, such as micro-cracking, micro-cutting and pulverization, and secondly, the efficiency of removing wear debris from the surface, resulting in "positive" wear.

The type of weave contributes to these mechanisms in various ways (Bijwe and Rattan, 2007):

a) Rigidity of weave: As the waves gets more rigid, the fibre damage in the composite increases.

b) Number of crimp points: A crimp (or fibre overlap area) indicates a point of weakness because of a poor fibre–matrix adhesion, thus, as the points of weakness increase, the fibre damage should increase.

c) Number of pockets: Indicates the efficiency to retain the debris in the pockets and beneath the crimp.

d) Amount of resin transfer on the disc: If a resin transfer is high, fibres will be less supported and this would lead to more fibre damage and less retention of debris.

According to the above, the wear resistance would be maximum if the weave is least rigid with least number of crimp points, maximum number of tight pockets to trap debris and least tendency for resin transfer on the disc. Yet there is still a need for a fundamentally comprehensive understanding of many other factors affecting the friction and wear behavior of laminated composites.

References

Abdelbary, A. 2011. Effect of vertical cracks at the surface of polyamide 66 on the wear characteristics during sliding under variable loading conditions. Ph.D. Mechanical Engineering Department, Alexandria University.

Abdelbary, A., Abouelwafa, M.N., El Fahham, I.M. and Hamdy, A.H. 2011. The influence of surface crack on the wear behavior of polyamide 66 under dry sliding condition. Wear 271(9): 2234–2241. Doi: https://doi.org/10.1016/j.wear.2010.11.042.

Abdelbary, A., Abouelwafa, M.N., El Fahham, I.M. and Hamdy, A.H. 2013. The effect of surface defects on the wear of Nylon 66 under dry and water lubricated sliding. Tribology International 59: 163–169. Doi: https://doi.org/10.1016/j.triboint.2012.06.004.

Abdelbary, A. 2014a. 1—Polymer tribology. In Wear of Polymers and Composites, edited by Ahmed Abdelbary, 1–36. Oxford: Woodhead Publishing.

Abdelbary, A. 2014b. 2—Sliding mechanics of polymers. In Wear of Polymers and Composites, edited by Ahmed Abdelbary, 37–66. Oxford: Woodhead Publishing.

Abdelbary, Ahmed and Yasser S. Mohamed. 2020. 4—Tribological behavior of fiber reinforced polymer composites. *In*: Sanjay, M.R., Suchart Siengchin, Jyotishkumar Parameswaranpillai and Klaus Friedrich (eds.). Tribology of Polymer Composites: Characterisation, Properties, and Applications. Elsevier. In press.

Abouelwafa, M.N. 1979. A study of the wear and related mechanical properties of silicone impregnated polyethylenes. Ph.D. University of Leeds.

Ainbinder, S.B. and Tyunina, É.L. 1974. Friction of polymeric materials. Polymer Mechanics 10(5): 731–737. Doi: https://doi.org/10.1007/BF00857958.

Archard, J.F. 1959. The temperature of rubbing surfaces. Wear 2(6): 438–455. Doi: https://doi.org/10.1016/0043-1648(59)90159-0.

Atkinson, J.R., Dowson, D., Isaac, J.H. and Wroblewski, B.M. 1985. Laboratory wear tests and clinical observations of the penetration of femoral heads into acetabular cups in total replacement hip joints: III: The measurement of internal volume changes in explanted Charnley sockets after 2–16 years *in vivo* and the determination of wear factors. Wear 104(3): 225–244. Doi: https://doi.org/10.1016/0043-1648(85)90050-X.

Barbour, P.S.M., Barton, D.C. and Fisher, J. 1995. The influence of contact stress on the wear of UHMWPE for total replacement hip prostheses. Wear 181-183: 250–257. Doi: https://doi.org/10.1016/0043-1648(95)90031-4.

Bellow, D.G. and Viswanath, N.S. 1993. An analysis of the wear of polymers. Wear 162-164: 1048–1053. Doi: https://doi.org/10.1016/0043-1648(93)90121-2.

Bijwe, J., Sen, S. and Ghosh, A. 2005. Influence of PTFE content in PEEK–PTFE blends on mechanical properties and tribo-performance in various wear modes. Wear 258(10): 1536–1542. Doi: https://doi.org/10.1016/j.wear.2004.10.008.

Bijwe, J. and Rattan, R. 2007. Influence of weave of carbon fabric in polyetherimide composites in various wear situations. Wear 263(7): 984–991. Doi: https://doi.org/10.1016/j.wear.2006.12.030.

Bikiaris, D. 2011. Can nanoparticles really enhance thermal stability of polymers? Part II: An overview on thermal decomposition of polycondensation polymers. Vol. 523.

Bowers, R.C., Clinton, W.C. and Zisman, W.A. 1953. Laboratory Naval Research, and School Naval Postgraduate. Frictional behavior of polyethylene polytetrafluoroethylene and halogenated derivatives. Washington, D.C.: Naval Research Laboratory: For sale by the Superintendent of Documents, U.S. G.P.O.

Bowers, R.C., Clinton, W.C. and Zisman, W.A. 1954. Friction and lubrication of Nylon. Industrial & Engineering Chemistry 46(11): 2416–2419. Doi: 10.1021/ie50539a055.

Braham, P. and Hogmark, S. 2011. Selected papers from those presented at the 14th Nordic Symposium on Tribology (Nordtrib 2010) Storforsen, June 8–11, 2010. Wear 273(1): 1. Doi: https://doi.org/10.1016/j.wear.2011.06.010.

Briscoe, B.J., Pogosian, A.K. and Tabor, D. 1974. The friction and wear of high density polythene: The action of lead oxide and copper oxide fillers. Wear 27(1): 19–34. Doi: https://doi.org/10.1016/0043-1648(74)90081-7.

Briscoe, B.J. and Sinha, S.K. 2013. Chapter 1—Tribological applications of polymers and their composites—past, present and future prospects. pp. 1–22. *In*: Friedrich, K. and Schlarb, A.K. (eds.). Tribology of Polymeric Nanocomposites (Second Edition). Oxford: Butterworth-Heinemann.

Brown, K.J., Atkinson, J.R. and Dowson, D. 1974a. The wear of HMWPE with particular reference to its use in artificial human joints. Advances in Polymer Friction and Wear 5B: 533–551.

Brown, K.J., Atkinson, J.R. and Dowson, D. 1974b. The wear of HMWPE with particular reference to its use in artificial human joints. Polymer Science and Technology 5.

Byett, J.H. and Allen, C. 1992. Dry sliding wear behaviour of polyamide 66 and polycarbonate composites. Tribology International 25(4): 237–246. Doi: https://doi.org/10.1016/0301-679X(92)90061-Q.

Chauhan, S.R. and Thakur, S. 2013. Effects of particle size, particle loading and sliding distance on the friction and wear properties of cenosphere particulate filled vinylester composites. Materials & Design 51: 398–408. Doi: https://doi.org/10.1016/j.matdes.2013.03.071.

Chen, Y.K., Kukureka, S.N., Hooke, C.J. and Rao, M. 2000. Surface topography and wear mechanisms in polyamide 66 and its composites. Journal of Materials Science 35(5): 1269–1281. Doi: 10.1023/A:1004709125092.

Cirino, M., Friedrich, K. and Pipes, R.B. 1988. The effect of fiber orientation on the abrasive wear behavior of polymer composite materials. Wear 121(2): 127–141. Doi: https://doi.org/10.1016/0043-1648(88)90038-5.

Cohen, S.C. and Tabor, D. 1966. The friction and lubrication of polymers. Proceedings of the Royal Society of London. Series A. Mathematical and Physical Sciences 291(1425): 186. Dio: https://doi.org/10.1098/rspa.1966.0088.

Dowson, D., El-Hady Diab, M.M., Gillis, B.J. and Atkinson, J.R. 1985. Influence of counterface topography on the wear of ultra high molecular weight polyethylene under wet or dry conditions. In Polymer Wear and its Control, 171–187. American Chemical Society.

Dowson, D., Taheri, S. and Wallbridge, N.C. 1987. The role of counterface imperfections in the wear of polyethylene. Wear 119(3): 277–293. Doi: https://doi.org/10.1016/0043-1648(87)90036-6.

El-Tayeb, N.S.M. and Mostafa, I.M. 1996. The effect of laminate orientations on friction and wear mechanisms of glass reinforced polyester composite. Wear 195(1): 186–191. Doi: https://doi.org/10.1016/0043-1648(95)06849-X.

Evans, D.C. and Lancaster, J.K. 1979. The wear of polymers. In Treatise on Materials Science & Technology, edited by Douglas Scott, 85–139. Elsevier.

Fort, T. 1962. Adsorption and boundary friction on polymer surfaces. The Journal of Physical Chemistry 66(6): 1136–1143. Doi: 10.1021/j100812a040.

Franklin, S.E. 2001. Wear experiments with selected engineering polymers and polymer composites under dry reciprocating sliding conditions. Wear 251(1): 1591–1598. Doi: https://doi.org/10.1016/S0043-1648(01)00795-5.

Friedrich, K., Lu, Z. and Hager, A.M. 1995. Recent advances in polymer composites' tribology. Wear 190(2): 139–144. Doi: https://doi.org/10.1016/0043-1648(96)80012-3.

Giltrow, J.P. 1970. A relationship between abrasive wear and the cohesive energy of materials. Wear 15(1): 71–78. Doi: https://doi.org/10.1016/0043-1648(70)90187-0.

Guo, Q. and Luo, W. 2001. Mechanisms of fretting wear resistance in terms of material structures for unfilled engineering polymers. Wear 249(10): 924–931. Doi: https://doi.org/10.1016/S0043-1648(01)00827-4.

Halliday, J.S. 1955. Surface examination by reflection electron microscopy. Proceedings of the Institution of Mechanical Engineers London.

Hendra, P.J., Jones, C. and Warnes, G. 1991. Fourier Transform Raman Spectroscopy Instrumentation and Chemical Applications. New York; London; Toronto: Ellis Horwood.

Hollander, A.E. and Lancaster, J.K. 1973. An application of topographical analysis to the wear of polymers. Wear 25(2): 155–170. Doi: https://doi.org/10.1016/0043-1648(73)90068-9.

Kar, M.K. and Bahadur, S. 1978. Micromechanism of wear at polymer-metal sliding interface. Wear 46(1): 189–202. Doi: https://doi.org/10.1016/0043-1648(78)90120-5.

Keer, L.M., Bryant, M.D. and Haritos, G.K. 1982. Subsurface and surface cracking due to hertzian contact. Journal of Lubrication Technology 104(3): 347–351. Doi: 10.1115/1.3253217.

Kremnitzer, S.L., Adams, M.J., O'Keefe, E. and Briscoe, B.J. 1983. A study of the friction and adhesion of polyethylene-terephthalate monofilaments in liquid media. Journal of Physics D: Applied Physics 16(1): L9.

Kyuichiro, T. and Kawakami, S. 1982. Effect of various fillers on the friction and wear of polytetrafluoroethylene-based composites. Wear 79(2): 221–234. Doi: https://doi.org/10.1016/0043-1648(82)90170-3.

Kyuichiro, T. 1984. Kinetic friction and dynamic elastic contact behaviour of polymers. Wear 100(1): 243–262. Doi: https://doi.org/10.1016/0043-1648(84)90015-2.

Lancaster, J.K. 1969. Abrasive wear of polymers. Wear 14(4): 223–239. Doi: https://doi.org/10.1016/0043-1648(69)90047-7.

Lancaster, J.K. 1971. Estimation of the limiting PV relationships for thermoplastic bearing materials. Tribology 4(2): 82–86. Doi: https://doi.org/10.1016/0041-2678(71)90136-9.

Lancaster, J.K. 1972a. Lubrication of carbon fibre-reinforced polymers: Part II—Organic fluids. Wear 20(3): 335–351. Doi: https://doi.org/10.1016/0043-1648(72)90414-0.

Lancaster, J.K. 1972b. Lubrication of carbon fibre-reinforced polymers: Part I—Water and aqueous solutions. Wear 20(3): 315–333. Doi: https://doi.org/10.1016/0043-1648(72)90413-9.

Madorsicy, S.L. and Straus, S. 1959. Thermal degradation of polyethylene oxide and polypropylene oxide. Journal of Polymer Science 36(130): 183–194. Doi: 10.1002/pol.1959.1203613015.

Margolies, A.F. 1971. Effect of molecular weight on properties of HDPE. SPE Journal 27.

Meng, H., Sui, G.X., Xie, G.Y. and Yang, R. 2009. Friction and wear behavior of carbon nanotubes reinforced polyamide 6 composites under dry sliding and water lubricated condition. Composites Science and Technology 69(5): 606–611. Doi: https://doi.org/10.1016/j.compscitech.2008.12.004.

Menzel, B. and Blanchet, T.A. 2005. Enhanced wear resistance of gamma-irradiated PTFE and FEP polymers and the effect of post-irradiation environmental handling. Wear 258(5): 935–941. Doi: https://doi.org/10.1016/j.wear.2004.09.051.

Mitjan, K. and Vižintin, J. 2001. Comparison of different theoretical models for flash temperature calculation under fretting conditions. Tribology International 34(12): 831–839. Doi: https://doi.org/10.1016/S0301-679X(01)00083-4.

Ovaert, T.C. 1997. Wear of unidirectional polymer matrix composites with fiber orientation in the plane of contact. Tribology Transactions 40(2): 227–234. Doi: 10.1080/10402009708983649.

Pawlak, Z., Urbaniak, W. and Oloyede, A. 2011. The relationship between friction and wettability in aqueous environment. Wear 271(9): 1745–1749. Doi: https://doi.org/10.1016/j.wear.2010.12.084.

Pihtili, H. 2009. An experimental investigation of wear of glass fibre–epoxy resin and glass fibre–polyester resin composite materials. European Polymer Journal 45(1): 149–154. Doi: https://doi.org/10.1016/j.eurpolymj.2008.10.006.

Pogosian, A.K., Bahadur, S., Hovhannisyan, K.V. and Karapetyan, A.N. 2006. Investigation of the tribochemical and physico-mechanical processes in sliding of mineral-filled formaldehyde copolymer composites against steel. Wear 260(6): 662–668. Doi: https://doi.org/10.1016/j.wear.2005.03.023.

Ratner, S.B., Farberova, I., Radyukevich, V.O. and Lure, E.G. 1964. Connection between the wear resistance of plastics and other mechanical properties. Soviet Plastics 7.

Rees, B.L. 1957. Static friction of bulk polymers over a temperature range. Research 10.

Reinicke, R., Haupert, F. and Friedrich, K. 1998. On the tribological behaviour of selected, injection moulded thermoplastic composites. Composites Part A: Applied Science and Manufacturing 29(7): 763–771. Doi: https://doi.org/10.1016/S1359-835X(98)00052-9.

Samyn, P., De Baets, P., Schoukens, G. and Van Driessche, I. 2007. Friction, wear and transfer of pure and internally lubricated cast polyamides at various testing scales. Wear 262(11): 1433–1449. Doi: https://doi.org/10.1016/j.wear.2007.01.013.

Sawyer, W.G., Freudenberg, K.D., Bhimaraj, P. and Schadler, L.S. 2003. A study on the friction and wear behavior of PTFE filled with alumina nanoparticles. Wear 254(5): 573–580. Doi: https://doi.org/10.1016/S0043-1648(03)00252-7.

Shooter, K.V. and Thomas, O.H. 1949. The frictional properties of plastics. In: Research. London.

Srinath, G. and Gnanamoorthy, R. 2007. Sliding wear performance of polyamide 6–clay nanocomposites in water. Composites Science and Technology 67(3): 399–405. Doi: https://doi.org/10.1016/j.compscitech.2006.09.004.

Steijn, R.P. 1967. Friction and wear of plastics. Metals Engineering Quarterly. American Society of Metals: 9–21.

Suresha, B., Kumar, K.S., Seetharamu, S. and Kumaran, P.S. 2010. Friction and dry sliding wear behavior of carbon and glass fabric reinforced vinyl ester composites. Tribology International 43(3): 602–609. Doi: https://doi.org/10.1016/j.triboint.2009.09.009.

Tabor, D. and Shooter, K.V. 1952. The frictional properties of plastics. Proceedings of the Physical Society. Section B 65(9): 661.

Tabor, D. 1982. Friction and wear: Calculation methods. I.V. Kragelsky, M.N. Dobychin and V.S. Kombalov. Tribology International 15(5): 283–284. Doi: https://doi.org/10.1016/0301-679X(82)90085-8.

Tan, Z.H., Guo, Q., Zhao, Z.P., Liu, H.B. and Wang, L.X. 2011. Characteristics of fretting wear resistance for unfilled engineering thermoplastics. Wear 271(9): 2269–2273. Doi: https://doi.org/10.1016/j.wear.2011.02.029.

Terheci, M. 2000. Microscopic investigation on the origin of wear by surface fatigue in dry sliding. Materials Characterization 45(1): 1–15. Doi: https://doi.org/10.1016/S1044-5803(00)00045-0.

Tsukamoto, N., Maruyama, H., Mimura, H. and Ebata, Y. 1993. Basic characteristics of plastic gears lubricated with water. Japan Society of Mechanical Engineers International Journal C 36.

Unal, H., Sen, U. and Mimaroglu, A. 2004. Dry sliding wear characteristics of some industrial polymers against steel counterface. Tribology International 37(9): 727–732. Doi: https://doi.org/10.1016/j.triboint.2004.03.002.

Villetti, M.A., Crespo, J.S., Soldi, M.S., Pires, A.T.N., Borsali, R. and Soldi, V. 2002. Thermal degradation of natural polymers. Journal of Thermal Analysis and Calorimetry 67(2): 295–303. Doi: 10.1023/A:1013902510952.

Vishwanath, B., Verma, A.P. and Rao, C.V.S.K. 1992. Effect of matrix content on strength and wear of woven roving glass polymeric composites. Composites Science and Technology 44(1): 77–86. Doi: https://doi.org/10.1016/0266-3538(92)90027-Z.

Xue, G. 1997. Fourier transform Raman spectroscopy and its application for the analysis of polymeric materials. Progress in Polymer Science 22(2): 313–406. Doi: https://doi.org/10.1016/S0079-6700(96)00006-8.

Yamamoto, Y. and Takashima, T. 2002. Friction and wear of water lubricated PEEK and PPS sliding contacts. Wear 253(7): 820–826. Doi: https://doi.org/10.1016/S0043-1648(02)00059-5.

Yu, S., Hu, H. and Yin, J. 2008. Effect of rubber on tribological behaviors of polyamide 66 under dry and water lubricated sliding. Wear 265(3): 361–366. Doi: https://doi.org/10.1016/j.wear.2007.11.006.

Zalisz, Z., Vroegop, P.H. and Bosma, R. 1988. A running-in model for the reciprocating sliding of Nylon 6.6 against stainless steel. Wear 121(1): 71–93. Doi: https://doi.org/10.1016/0043-1648(88)90032-4.

Zhang, H., Zhang, Z. and Friedrich, K. 2007. Effect of fiber length on the wear resistance of short carbon fiber reinforced epoxy composites. Composites Science and Technology 67(2): 222–230. Doi: https://doi.org/10.1016/j.compscitech.2006.08.001.

Zhang, Z., Friedrich, K. and Velten, K. 2002. Prediction on tribological properties of short fibre composites using artificial neural networks. Wear 252(7): 668–675. Doi: https://doi.org/10.1016/S0043-1648(02)00023-6.

Chapter 7

Tribology of Automotive Components

7.1 Introduction

Tribology of automotive components is the aspect of tribology focused on automobile performance and reliability. In addition, automotive components are undisputedly one of the classic tribological challenges. There are many hundreds of tribological components, from bearings, pistons, transmissions and clutches, to gears and drivetrain components, as illustrated in Fig. 7.1. All these components are interacting on their surfaces to perform a desired job. High temperature, high speed and the oxidative environment are the most severe working conditions for automotive components (Kumar and Agarwal, 2019).

The tribological efficiency of internal combustion engines ICE is largely dictated by the piston assembly, bearings and the valvetrain. Table 7.1 represents the many frictional interfaces in a typical ICE (Blau, 2013).

The piston assembly can be considered as providing a means of transforming the chemical energy of fuel into beneficial kinetic energy in an engine. The ring pack installed in the piston assembly seals the gases in and prevents them from escaping towards the crankcase when produced after the burning of fuel. The piston rings are often subjected to harsh operating conditions, like variation of load, temperature, speed and different lubrication modes. Considering the lubrication of piston rings, the relative sliding velocity between the compression rings and the liner wall influence the lubrication condition, substantially. The lubrication condition seems to be healthy as a hydrodynamic mode in the mid stroke due to increased velocity, entraining oil in the contact patch. At top and bottom dead centers, where the relative velocity is the least, satisfactory lubrication conditions would largely depend upon the squeeze film action. At this time, EHL to boundary lubrication regime may be anticipated in the vicinity. During a single stroke of

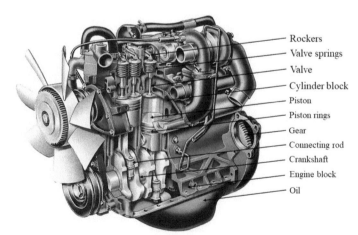

— Rockers
— Valve springs
— Valve
— Cylinder block
— Piston
— Piston rings
— Gear
— Connecting rod
— Crankshaft
— Engine block
— Oil

Figure 7.1: Main engine components.

Table 7.1: Tribological interfaces in a typical internal combustion engine (Blau, 2013).

Subsystem	Interface	Type of motion
Combustion chamber	Piston ring-cylinder bore	Reciprocating sliding
	Connecting rod small end bearing	Reciprocating sliding
	Piston skirt	Reciprocating sliding
	Piston ring groove	Fretting motion
Valve train	Valve stem in the valve guide	Reciprocating sliding
	Tappet shim/bucket lifter	Unidirectional sliding/rolling
	Roller-follower	Rolling and unidirectional slip
	Cam shaft bearing	Rolling and unidirectional slip
Fuel system	Fuel injector plunger (needle)	Reciprocating sliding
Crankshaft/connecting rods	Crankshaft main bearing	Rolling and slip
	Connecting rod large-end bearing	Unidirectional sliding
Exhaust gas recirculation system	Bushings on actuators for exhaust gas valves	Reciprocating sliding
Turbochargers	Nozzle vanes	Fretting motion
	Turbocharger axial bearings	Rolling or unidirectional sliding

a piston in an engine, the rings may experience boundary, mixed and hydrodynamic lubrication regimes (Khurram, 2016).

Regarding the future development of automotive tribology, there are many recommendations and research goals related to fuel efficiency, emissions, durability and profitability of powertrain systems (Tung and McMillan, 2004). These include: (a) development of a quantitative understanding of failure mechanisms, such as wear, scuffing and fatigue, (b) development of a variety of affordable surface modification technologies which are suitable for various vehicle components that are used under a variety of operating conditions, (c) development of a better understanding of the chemistry of lubricants and how additives affect the interactions between lubricants and rubbing surfaces.

The application of tribological principles is essential for the reliability and consistency of the automotive industry and has led to enormous advances in the field of tribology (Tung and McMillan, 2004). In the following section, the tribological behavior of selected automotive components will be discussed.

7.2 Piston Ring

The piston is at the heart of the reciprocating internal combustion engine, forming a vital link in transforming the energy generated by combustion of the fuel and air mixture into a useful kinetic energy (Kapoor et al., 2001). The essential roles of a piston ring are to seal off the combustion pressure, to distribute and control the lubricating oil, to transfer heat, and to stabilize the piston.

It is the most complicated tribological component in the internal combustion engine to analyze because of the large variations of load, speed, temperature and lubricant availability. Therefore, the tribological considerations in the piston rings contact have attracted much attention over several decades. Undoubtedly, this is due to the fact that the sliding of these components against the cylinder wall is the largest contributor to friction in the engine. For all categories of internal combustion engines, piston rings have to meet all the requirements of a dynamic seal for linear motion that operates under demanding thermal and chemical conditions.

7.2.1 Piston Ring Assembly

Piston rings form a ring pack, which usually consists of 2–5 rings, including at least one compression ring. The number of rings in the ring pack depends on the engine type, but there are usually 2–4 compression rings and 0–3 oil control rings (Andersson et al., 2002), as illustrated in Fig. 7.2.

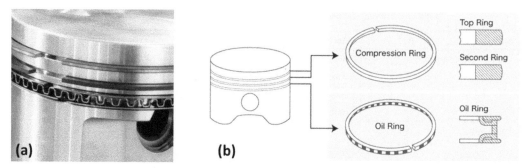

Figure 7.2: (a) A photo, and (b) A schematic of piston and piston ring.

A piston ring material is chosen in order to meet the demands set by the running conditions. Elasticity and corrosion resistance of the ring material are required. From a tribological point of view, gray cast iron, malleable/nodular iron and carbides/malleable iron are the main materials used for manufacturing piston rings. The graphite phase in the gray iron can be beneficial as a dry lubrication at dry starts or similar conditions of oil starvation. However, steel is now used for top compression rings due to its high strength and fatigue properties.

7.2.2 Piston Ring Groove

In a large number of piston designs, the piston ring belt consists of three ring grooves. Damaged sealing surfaces lead to increased blow-by and reduced effective combustion pressure. At the same time the increased flow of the hot blow-by gas interferes with the oil film between the sliding surfaces and may cause hot gas damage to the piston rings. To increase the wear resistance of the ring grooves in the pistons of heavy fuel oil engines, the grooves are typically either induction hardened or chromium plated.

The wear of the ring groove flanks can affect the effective geometry of the ring face against the cylinder liner. In order to improve the ring-groove wear resistance in steel composite piston crowns, a hard chromium layer can be applied. To protect the first ring groove, sometimes also the second, in high-performance diesel engines against wear, a so-called ring carriers made of high-alloyed cast iron are casted-in. Ring carriers are preferably made of Niresist, an austenitic cast iron with a thermal expansion coefficient almost equal to that of aluminium (Röhrle, 1995).

7.2.3 Tribology of Piston Ring

Perhaps, the tribology of piston rings is the most complicated issue to be analyzed in the internal combustion engine. This complexity is due to the fact that it is subjected to large and rapid variations of load, speed, temperature, and lubricant availability. In fact, lubrication in the interface between the piston ring and the cylinder wall may experience more than one type of lubrication regime within a single stroke of the piston. Boundary, mixed, full fluid film and elastohydrodynamic lubrication are possible regimes in internal combustion engines, as shown in Fig. 7.3 (Ruddy et al., 1982; Rycroft et al., 1996; Chong and De la Cruz, 2014). The Stribeck curve shown in Fig. 7.3 clearly shows the difference between lubrication mechanisms as the differentiation between full fluid-film lubrication and solid asperity interactions (Delprete and Razavykia, 2017). Another complexity of this tribosystem is the environmental effect inside the combustion chamber. Near to the walls, the results of the fuel burn like CO, HC, and CO_2 are formed as well as some sub-products like acids, H_2CO_3, and H_2SO_4 (Vatavuk and Ferrarese, 2013).

The wear of piston rings and cylinder walls has a significant effect on the performance of the piston assembly (Ma et al., 2018; Xiuxu et al., 2020). For that reason, experimental studies are aimed at investigating early life wear of the piston rings and cylinder wall to modify the profile and roughness of the interacting surfaces to achieve acceptable performance as part of the running-in process (Kapoor et al., 2001). There are a variety of surface treatments and coatings for the piston ring running-in processes. As

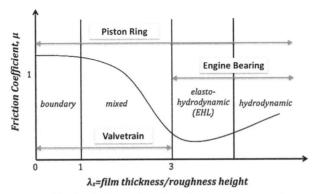

Figure 7.3: Stribeck curve distinguishes lubrication regimes for engine components, based upon the ratio of lubricant film thickness to the asperity heights on the counterface surfaces.

mentioned above, chromium plating and flame sprayed molybdenum are the most common wear resistant coatings, although plasma sprayed molybdenum, metal matrix composites, cermets and ceramics are getting their popularity as their technology progresses (Tung and McMillan, 2004).

However, the wear characteristics may play a more important role than the coating itself in some applications. For example, geometric profile of the surface in contact can define the lubrication regime and, consequently, the level of wear (Tomanik and Ferrarese, 2006; Martínez et al., 2010; Vatavuk and Ferrarese, 2013). Figure 7.4 shows the influence of the piston ring geometry on the interaction with the piston groove.

Coatings for rings are widely used to enhance their wear resistance, especially in abrasive and corrosive conditions where the running conditions are severe. One example of such a coating is chromium, which is particularly relevant for the compression ring. Piston ring surfaces are, in addition to chromium plating, thermally (plasma) sprayed with molybdenum, metal composites, metal-ceramic composites or ceramic composites, as a uniform coating or an inlay coating material.

On the other hand, surface finish of piston rings and cylinders can have a major impact on the wear performance of the engine. The main objectives of surface finishing techniques to improve oil retention at or within the surface is to minimize scuffing and to promote ring profile formation during the running-in period. Piston rings made from cast iron are often used in a fine turned condition and electroplated; flame sprayed and plasma sprayed rings are generally ground to the desired finish. The poor wettability by oil of chromium plated piston rings can be overcome by etching a network of cracks and pores into the surface by reversing the plating current, chemical etching, grit blasting, lapping or depositing the plate over a surface with fine turned circumferential grooves (Tung and McMillan, 2004).

Various experimental and theoretical studies have been conducted in order to develop consistent models to understand the lubrication mechanisms. Delprete et al. (Delprete et al., 2017) summarized the synthesis of the main technical aspects considered during modeling of piston ring-liner lubrication and friction losses. The literature review highlights of the effects of piston ring dynamics, components geometry, lubricant rheology, surface topography and adopted approaches, on frictional losses contributed by the piston ring-pack. It was postulated that a piston ring-liner interface lubrication is a complex phenomenon, which has a significant impact on the oil consumption and frictional losses. Analytical and numerical methods for examining piston ring–liner trajectory losses and lubrication have received widespread attention from researchers due to the fact that a piston ring-pack shows different behaviors in different situations. It was concluded that analytical modeling of a piston ring-liner can be considered as an applicable tool to assess the tribological performance of an engine.

Degradation of the engine lubricating oil with time has a considerable effect on the tribological behavior of the piston ring and, thereby, impacts fuel economy and durability. High pressure and high temperatures in the regions of interaction with the combustion gases, especially around the top compression ring, are the main causes for such a degradation. As the lubricant degrades, a change in the chemistry, in terms of additive depletion, oxidation and nitration and the viscosity, occurs and these parameters can alter the lubricant film thickness experienced by the piston ring and, thereby, the piston ring friction (Notay et al., 2019). The lubricant films were found to increase in thickness across the entire piston ring pack. This would ultimately affect piston assembly friction and have implications for overall engine performance, fuel economy and exhaust gas emissions.

Therefore, measurements of the lubricating oil film thickness between the piston ring and cylinder wall have been of great importance to help validate the mathematical models and provide information on the ring pack tribological performance. Laser induced fluorescence (LIF) measurement of the lubricant film thicknesses between the cylinder wall and the piston ring was first introduced by Ting (Ting, 1980). The principle uses laser light which absorbs in oil that has been doped with a fluorescent dye, as illustrated in Fig. 7.5. The oil flames at a different wavelength to the carrier signal and this return beam is converted into a voltage signal by using photomultiplier tubes. The intensity of the signal is related to the film thickness

Abnormal Piston Groove Wear

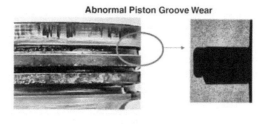

Influenced by geometric characteristic of piston ring inner profile.

Figure 7.4: Piston groove wear influenced by ring geometry (tumbling operation reduces the piston ring groove wear). After Tomanik et al. (2006). With permission from Springer Nature.

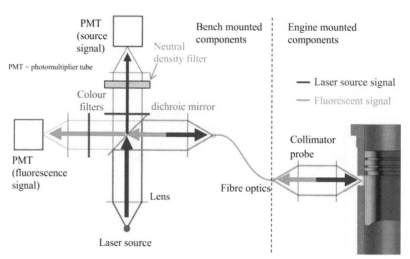

Figure 7.5: LIF optical set up. After Notay et al. (2019), with permission from Elsevier publishing.

for a given lubricant and dye. Recently, novel methods of using ultrasound to measure the lubricant film thicknesses has been developed. The ultrasonic methods involve transmitting and measuring reflecting ultrasonic pulses between different densities of materials.

7.3 Valvetrain Mechanism

The valvetrain (or valve train) is regarded as one of the most challenging systems in an internal combustion engine due to the severe operating conditions. The primary function of this mechanism is to transform the rotary camshaft motion into a linear valve motion in order to control the fluid flow into and out of the combustion chamber. However, there is another minor function, which is to control the auxiliary devices, such as distributors, fuel pumps, water pumps and power steering pumps.

Various different styles of valvetrain configurations are usually employed in automotive engines, depending on the nature of operation and type of an engine. In high speed cars, the direct acting valvetrain with overhead camshaft arrangement is often used, whereas the push rod valve train configuration is usually employed in heavy duty diesel engines. In case of valves, poppet valves are mostly employed in the automotive industry due to their enhanced efficiency. On the other hand, sleeve and rotary valves presented some difficulties in effective lubrication, increased friction, sealing problems and high oil consumption (Khurram, 2016).

Valvetrains can be broadly classified into either sliding contact or rolling contact. Over the last decade, the most frequently-used mechanism is the roller follower, which utilizes a rolling motion to transmit the required motion, allowing for a lower friction. This is because the coefficient of friction at the camshaft/follower interface is less for rolling than sliding friction by an order of magnitude. Other less frequent mechanisms include the finger follower and direct acting bucket follower. Figure 7.6 illustrates two types of valvetrain mechanisms showing sub-components.

In a valvetrain mechanism, the poppet valve configuration is the most popular and is used by virtually all the major automobile manufacturers for the inlet and exhaust valves of reciprocating internal combustion engines. The opening and closing of the valves is invariably controlled by a cam driven from the crankshaft in order to ensure synchronization of the valve motion with the combustion cycle and piston movement (Kapoor et al., 2001).

The tribological performance of a valvetrain plays an important role in the overall engine efficiency. Valvetrain systems due to their wide variations in the configurations and severe operating conditions usually pose serious challenges for the tribologists. The fundamental aims in the automobile industry to ensure the durability and reliability of engine components are improving fuel economy, reducing friction and wear, and controlling emissions.

The friction associated with a valvetrain is generally considered to be a small component of the mechanical losses. Typically, the valvetrain mechanism contributes about 0.06 to 0.1 of the total frictional losses in an engine (Gangopadhyay et al., 2004). However, friction is seen to rise significantly, up to 25%, at low engine speeds and for larger, slower running engines associated with the prestigious cars market (Tung and McMillan, 2004). Efforts are being made to explore every opportunity to reduce frictional losses.

In an actual engine operation, the valvetrain mechanism has to operate in severe operating conditions and mostly in elastohydrodynamic and mixed lubrication regimes where the asperity interactions can play an important role. This may cause excessive friction, leading to wear problems of mating surfaces. Therefore, there are opportunities to reduce friction by using novel materials, surfaces and lubricant formulations. Friction could also be reduced by lowering the spring load and the reciprocating mass.

As illustrated in Fig. 7.7, it is possible to identify two tribosystems in valvetrain mechanism micro and macro-tribosystems. A micro-tribosystem represents specific wear points or wearing contacts. These are the contacts between the various components of the mechanism, such as those between the valve and seat insert, rocker arm and valve tip and the cam and rocker arm. Micro-tribosystem parameters are related to the macro-tribosystem, for example, a change in the valve profile would alter the closing velocity of the valve and would affect the wear occurring in the valve and seat insert micro-tribosystem (Pashneh-Tala et al., 2014).

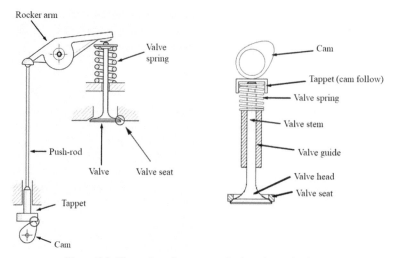

Figure 7.6: Illustration of two types of valvetrain mechanisms.

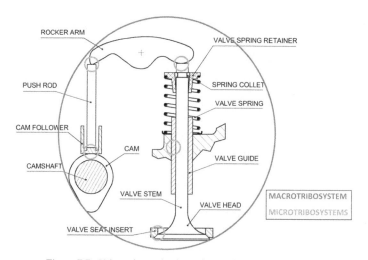

Figure 7.7: Valvetrain mechanism micro and macrotribosystems.

7.3.1 Valve and Valve Seat

The valve and valve seat in an internal combustion engine play a central role in the engine breathing, compression, performance and endurance. They are critical components of the valvetrain mechanism, because they control the entry of air, the output of exhaust gas, and provide sealing during the compression and combustion processes.

Wear is a problem at the valve/seat interface; the valve "recesses" into the seat and results in loss of engine timing. Loading is from a combination of the dynamic closing of the valve under cam action and the application of the firing pressure on the closed valve face, therefore, the seat is subjected to both high static and dynamic stresses. Furthermore, the high temperatures in the region of the valve (typically 300°C to 500°C) and lubrication tends to be from small quantities of oil that flow past the valve guide (Tung and McMillan, 2004). However, lowering valve temperature by only a slight amount may take it out of the critical range for failure for the component.

Engine valve and valve seat materials should have wear resistance, corrosion resistance, high strength, and high temperature stability. In general, engine valves are usually manufactured from iron-, nickel-, and cobalt-based metallic alloys. In particular, inlet valves are made from hardened low alloy martensitic steel for good wear resistance and strength. Likewise, the exhaust valves are often made from precipitation-hardened, austenitic stainless steel for corrosion resistance and hot hardness. On the other hand, the valve seats are made from cast or sintered high carbon steel. In some high performance engines, the seat is formed by induction hardening of the cylinder head material (Kapoor et al., 2001).

In order to overcome the tribological challenges at a valve/seat interface, a recent trend in valve design is reducing the mass of reciprocating components. It is assumed that lower masses reduce inertia and, therefore, friction losses will be reduced (Muzakkir et al., 2015). Fukuoka et al. (Fukuoka et al., 1997) developed a lighter valve DOHC (direct overhead cam) train using an aluminium tappet, an aluminium spring retainer and a thin sintered shim to reduce inertia loading on the cam. Cost effective Fe-spray coating was selected by authors in order to ensure the wear resistance of valve components. By utilizing these parts, they reduced inertia mass by 28% and achieved a 40% reduction in friction. Von Kaenel et al. (von Kaenel et al., 2000) preferred weight reduction of the valve using 0.5 mm to 1.0 mm stem wall thickness. They suggested filling hollow valve areas with liquids to improve the heat distribution in order to reduce temperature peaks in critical areas. Development of lightweight hollow valves was a major target of a research conducted by Gebauer et al. (2004). They optimized welding and manufacturing operations in order to develop a deep drawn hollow valve for automotive engine operation.

Sodium filled valves, an emerging manufacturing technology, have been utilized in engines significantly in recent years. Two main types of sodium filled engine valves have been developed to date: The hollow stem sodium filled valve (HSV) and the hollow head and sodium filled valve (HHSV) (Baek et al., 2014). Compared to conventional solid engine valves, the internally filled sodium valve significantly decreased the valve temperature. Also, the hollow head valve had an apparently better performance than the hollow stem valve in its cooling effect.

Based on drilling and friction welding technology, Lai et al. (Lai et al., 2018) presented a novel method for manufacturing hollow head and sodium filled engine valves. The manufacturing method is based on drilling and friction welding. Compared to the solid valve, the mass of an HHSV produced by this method was reduced by about 16%. The durability bench test results of four tested valves in the aspect of fatigue behavior are presented in Table 7.2.

The temperature distributions of exhaust valves for a solid valve and HSV (gasoline engine) in the same cylinder are presented in Fig. 7.8. It was found that the maximum temperature of the solid valve was 745°C while the temperature of valve head center reached 665°C. In contrast, for the hollow stem and sodium filled valve, the maximum temperature of the whole valve was 590°C, and the hottest point shifted toward the head area. Compared to the solid valve, the maximum temperature reduction was about 155°C. It is, therefore, suggested that the heat in the valve head was conducted through the hollow stem to the cylinder head cooling system via the valve guide by the filled sodium.

Wear scar width measurement versus the impact cycles is presented in Fig. 7.9. It is noticed that the wear scar width of valve 3 (solid valve) was the highest, showing the lowest wear resistance. Valve 4 wear scar width had a minimum value. Compared to the wear scar of valve 2 (nitrided 23-8N steel), the Stellite F hardfacing alloy on valve 4 performed better in terms of wear resistance ability. On the other hand, investigation of wear scar of valve seat insert demonstrated that the solid valve contact pair had a maximum total wear loss. The following was HHSV with 23-8N steel nitriding. The HHSV with Stellite F hardfacing alloy held a minimum total wear loss.

Microscopic examination of worn surfaces of valve and valve seat postulated that the wear mechanisms acting on the interface between the valve seating surface and the seat insert were found to be a combination

Table 7.2: Test results of valve durability bench (Lai et al., 2018).

Specimen name	Valve type	Valve head material	Cycles (10³)	Results
Valve 1	HHSV	23-8N	460	Broken
Valve 2	HHSV	23-8N	10,000	Run-out
Valve 3	Solid	X60	10,000	Run-out
Valve 4	HHSV	23-8N	10,000	Run-out

of oxidative wear and adhesive wear. The wear loss was produced by impact (impact wear) and sliding (adhesion and abrasion). Figure 7.10 shows worn surfaces of valve 2 (HHSV) seating face and energy-dispersive spectroscopy EDS test area.

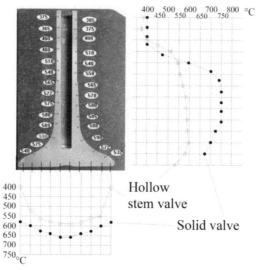

Figure 7.8: The temperature distributions of exhaust valves for solid valve and HSV. After Lai et al. (2018) with permission from Elsevier.

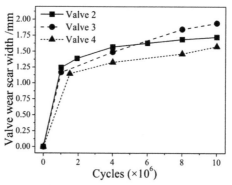

Figure 7.9: Wear scar width measurement versus the impact cycles. After Lai et al. (2018) with permission from Elsevier.

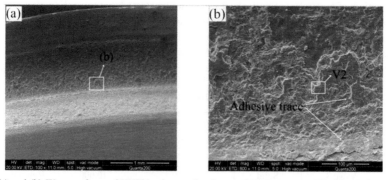

Figure 7.10: (a) and (b) Worn surfaces of HHSV valve seating face and EDS test area V2. After Lai et al. (2018) with permission from Elsevier.

7.3.2 Cam and Tappet

In an engine, the opening and closing of valves is invariably controlled by cam/tappet mechanism. This mechanism usually consists of two moving elements: The rotating cam and the tappet (follower) mounted on a fixed frame. The cam is driven from the crankshaft in order to ensure synchronization of the valve motion with the combustion cycle and piston movement. In general, there are different types of surface contact between a cam and a follower and, according to surface contact, they are classified into roller, flat, knife edge and spherical follower, as shown in Fig. 7.11.

Both flat and roller types are the most popular ones used in valvetrain systems, as shown in Fig. 7.12. The major concern with the flat tappet type is its considerable power loss. One of the recommended ways to reduce the power loss is to use a roller tappet instead of the conventional flat one.

Irons and steels are the most popular materials for cams and tappets, with a variety of metallurgies and surface treatments to assist running-in and prevent early life failure. However, there are opportunities to enhance tribological properties of cam/tappet interaction by using novel materials, surfaces and lubricant formulations. Therefore, ceramic tappets are becoming more common, especially in direct acting valvetrains, in an attempt to reduce frictional losses (Gangopadhyay et al., 1999; Gangopadhyay et al., 2004).

Friction and wear have been problems with cam/tappet mechanisms for many years, especially with finger tappet configurations. The famous failure modes of cams and tappets are pitting, polishing and scuffing, all of which are influenced by materials, lubrication, design and operating conditions. The mode of failure can vary significantly depending on the combination of materials chosen, their surface treatment and the lubricant and its additive package.

Fukuoka et al. (Fukuoka et al., 1997) observed a 40 per cent reduction in friction torque by reducing the reciprocating mass, i.e., replacing steel tappets with aluminium tappets with thinner walls, and an aluminium spring retainer. However, aluminium tappets require a coating on the side wall to prevent galling with the aluminium bore. Also, it was suggested that friction could be reduced by lowering the spring load and the reciprocating mass.

Attempts were performed to reduce friction and wear losses in direct-acting mechanical tappet through surface finish, surface texture and coatings (Gangopadhyay et al., 2004; Turturro et al., 2012). In this study, two sets of experiments were conducted. The first set of experiments was to evaluate the effect of surface finishing techniques on friction torque. On the other hand, the second set of experiments was to evaluate the effect of superfinishing and coating on the production tappet insert.

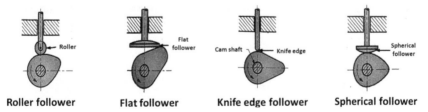

| Roller follower | Flat follower | Knife edge follower | Spherical follower |

Figure 7.11: Types of cam followers.

Figure 7.12: Examples of roller and flat tappet used in valvetrain system.

It is demonstrated that surface finish, surface texture and engine oil formulation play significant roles in reducing frictional losses. The tappet insert surface with a texture showed lower valvetrain friction than a non-textured surface having similar centre-line average roughness. Furthermore, the tappet insert wear rate was independent of the camshaft speed, although the friction torque continued to drop.

The contact between the cam and tappet operates under severe tribological condition for almost all of its cycle, due to the continuous variations of load, velocity and geometry. Indeed, this interface is the most critical situation in the valvetrain system due to the difficulties in obtaining effective lubrication for such contacts. Traditional design philosophy assumed that the cam and tappet operated entirely in the boundary lubrication regime. However, researchers over the last two decays have shown that mixed and elastohydrodynamic lubrication have a significant role to play in the tribological performance of cams and followers (Taylor, 1994). In fact, the nonconformal surfaces deflect under high pressure, which is induced in the separating lubricant film. As a result, the cam and tappet may experience boundary, mixed and elastohydrodynamic lubrication in a single cycle.

Based on experimental data obtained by Lee (Lee, 1993), a numerical model was developed by Torabi et al. (Torabi et al., 2017; Torabi et al., 2018) to show the performance improvement of a cam/tappet mechanism when using a roller type tappet compared to the flat-faced tappet. The model is established such that the speed, load, geometry, lubricant properties, and the surface roughness profile are taken as inputs and the lubricant film thickness and friction coefficient as a function of cam angle are predicted. Furthermore, the asperities are assumed to have elastic, elasto-plastic and plastic deformation. The model consists of solving mixed thermo-elastohydrodynamic lubrication at every cam angle of rotation, which itself includes changing film thickness, equivalent radius and entraining velocity in each step. Therefore, a four cyclic procedure of solution was suggested for a full rotation of cam cycle. The first cycle was assigned to predict the pressure film, while the temperature was determined across the lubricant film in the second cycle. In the third cycle, the asperity interaction was calculated. Finally, the last cycle is for cam rotation and changing the geometry, kinematic and dynamic parameters.

Figure 7.13(a) and (b) shows the variation of a dimensionless entraining velocity as a function of the cam angle for roller and flat tappets, respectively. Entraining velocity in roller tappet showed a relatively smooth hill shape curve without any sign change, as shown in Fig. 7.13(a). In such mechanisms, rolling the tappet will increase the lubricant entraining velocity. In contrast to the conventional flat tappet, shown in Fig. 7.13(b), the direction of lubricant motion does not change in the roller tappet case. In the flat-faced tappet, the zero entraining velocity causes a halt of lubrication due to the lubricant movement. Consequently, the lubricant film squeezing action prevents the surfaces for contacting each other. Increasing the cam speed increases the entraining velocity and contact load and, thus, the minimum film thickness slightly increases. This increase was in the order of 30% to 60% in the roller tappet.

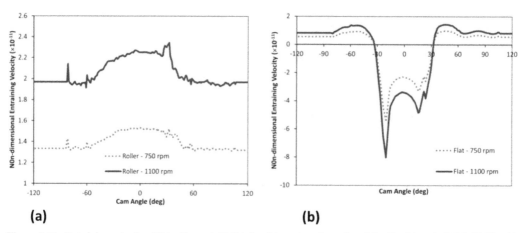

(a)   **(b)**

Figure 7.13: Entraining velocity of (a) roller and (b) flat-faced tappet configuration. After Torabi et al. (2017; 2018) with permission from Elsevier.

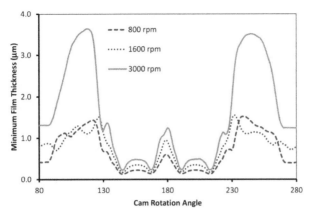

Figure 7.14: Minimum film thickness variation with cam degree in different speeds, flat-faced tappet. After Torabi et al. (2017; 2018) with permission from Elsevier.

The minimum film thickness specification for both types of tappets is one of the most important simulation's outputs that may be used for judgment on the friction and wear behavior. Measurements of minimum film thickness for roller and flat-faced tappet in different cam rotation speeds postulated that the film thickness in roller follower smoothly varies with cam angle. Also, it has two small peaks with heights about twice the flat part of the curve. On the other hand, for the flat-faced tappet the minimum film thickness variation has the same trend, while the peak height is about ten times the average value. Increasing the cam speed increases the entraining velocity and contact load and the minimum film thickness slightly increases, as shown in Fig. 7.14. This increase is more pronounced for the roller follower. These findings suggested that the friction coefficient substantially reduces by using roller tappet.

7.4 CrankShaft Bearing

Crankshaft (or crank) mechanism is a part of internal combustion engines, as shown in Fig. 7.15. It is utilized, along with the connecting rods, to translate the reciprocating motion of the pistons into rotational motion. The journal bearings that accomplish the transmission of load and keep the crankshaft in place are designed to run under hydrodynamic conditions (Becker, 2004).

The crankshaft rotates on protective journal bearing sleeves that provide a frictionless seal between the contacting metal surfaces. These sleeves are usually in the form of half shells with oil holes. The lubricating oil is supplied at pressure from the oil pump to a main oil gallery that runs along the side of the engine and then to the main bearing. The oil supply hole in the upper main bearing normally feeds into a groove so that a drilling in the crankshaft journal can transport the oil to the big-end bearing in the connecting rod. Oil supply to the pin joint small-end and piston pin boss is performed by "splash" from the sides or via internal drillings in the connecting rod or piston.

The minimum oil film thickness between the shaft and the journal bearing, h_o, is an important parameter in determining the phase of lubrication. Hydrodynamic lubrication exists when h_o is sufficient to completely separate the running surfaces. Under these conditions, both the shaft and bearing should be made of strong and stiff materials to resist deformation. Unfortunately, vehicle engines stop frequently, and the shaft will settle and be in contact with the bearing at the next start. Also it is possible for some solid particles to become entrained in the oil, and if these particles are larger than h_o they could damage the surfaces if allowed to circulate. It has been demonstrated that soft and compliant materials minimize the sticking of the shaft and bearing during idle periods, and soft materials can capture some small particles and remove them from circulation, a property called embeddability (Becker, 2004). Currently, minimum film thickness predictions in the range 0.5–1.0 μm have been made for engine bearings in passenger cars, suggesting that asperity interaction may occur between the journal and bearing for part of the engine cycle. A significant approach to improve engine performance is to design engine bearings which can operate in mixed lubrication regime (Priest and Taylor, 2000).

Figure 7.15: Crankshaft construction.

A key strategy in increasing engine efficiency is to reduce engine friction, including bearing friction. For instance, Schommers et al. (Schommers et al., 2013) point out that, considering a typical gasoline engine, up to 25% of the total energy losses can be influenced with engine friction. Therefore, one of the tribological challenges is to find a way to make bearings hard and stiff, and concurrently soft and compliant. The bearing material must be extremely strong because of the stresses caused by the explosions inside the internal combustion engine. Reducing friction is accomplished in part by the fact that dissimilar metals slide against each other with less friction and wear than similar materials do. The usual conciliation is to make the crankshaft out of steel or nodular cast iron. The bearing is then formed from a low carbon steel backing (for strength) and coated with a soft alloy (for embeddability). Lead-based coatings were used successfully for many years, but are no longer preferred for crankshaft applications due to their low strength and hazardous metal content. Modern engine bearings are made of tin–aluminum or lead–bronze alloys, which are stronger but have poorer embeddability, so oil cleanliness is critical to their long-term durability (Becker, 2004; Tung and McMillan, 2004).

The most common types of materials used in manufacturing crankshafts are heat-treated steels or spheroidal graphite irons, with hardness approximately three times that of the principal bearing material (Becker, 2004). Compared to forged shafts, shafts made from graphite cast iron usually offer superior manufacturing properties. Also, casted shafts are able to provide up to 10% weight benefits compared to forged shafts with the same dimensions, because of their lower density (Becker, 2004; Summer et al., 2015). However, technical engineers mostly associate high performance crank shaft bearing systems only with forged steel shafts sliding against the bearing material. This is in regard to their superior fatigue properties being a main driver in high loaded engine applications.

Many researches have been aimed at enhancing the friction performance of ICE (Ali et al., 2018; Grützmacher et al., 2018; Jang and Khonsari, 2019; Jog et al., 2020). Recently, Summer et al. (Summer et al., 2019) introduced and verified novel bearing coatings. In this study, a bimetal bearing without additional overlay has been tested, along with bearings consisting of a lead bronze lining and various coatings. Two different test methodologies have been used to visualize the bearing material performances: A ring on disc set up (RoD) and a bearing segment (BS) test system. Four bearing materials were employed and designated as:

- Reference sample S0, represents an advanced aluminum based bearing alloy with elevated tin content and further alloy elements to form intermetallic phases.
- Novel sample S1, uses a CuPb22Sn lining which is directly spray coated with a polymer-based overlay with sufficient temperature stability using MoS_2 and graphite fillers.
- Novel sample S2, stands for a sputtered AlSn20 overlay coated on the same CuPb22Sn lining material with a thin Ni barrier in between.
- Novel sample S3, composed of the same CuPb22Sn lining of S1 and S2, a Ni barrier and an electroplated PbSn18Cu overlay.

For all bearing materials, friction coefficient μ follows a Stribeck characteristic, where high friction is measured at low velocities and decreasing with increasing speed. Also, it has been postulated that the friction performance of the tested bearing polymer overlay depends on the proportion of fillers and polymer matrix in contact with the counterpart. High amounts of MoS_2 and graphite on the top surface result in low friction and vice versa. The systems using the novel polymer bearing material show low friction characteristic over the whole speed range, as shown in Fig. 7.16. Even at low speed, the increase of the coefficient of friction curve is not pronounced.

Accompanying analysis is used to explain the performance of the novel bearing material. Figure 7.17b and c, show surface conditions of the bearing surface after sliding in a Stribeck test program, a temperature test program and a start stop wear test program. The surface is composed of densely arranged particles on top, as seen in Fig. 7.17a. Elemental compositions of these particles introduced in Table 7.3 prove that MoS_2 (2 and 3), graphite (1) and TiO_2 (4 and 5) are present at the interface. This would be the reason for low friction losses during the corresponding test of the surface presented in Fig. 7.17a. Figure 7.17b shows a similar picture for surfaces derived from the temperature test program and Fig. 7.17c for the start stop test program after 1800 starts. Hereafter, during a very short time of running in, the particle enriched surface structure is formed quickly. These particles, then, remain stable during various sliding conditions resulting in low friction losses and simultaneously a high wear resistance.

It is concluded that the outstanding tribological enhancement in the novel polymer bearing could be linked to a dense filler structure of finely distributed MoS_2, graphite and Ti particles. This particle layer

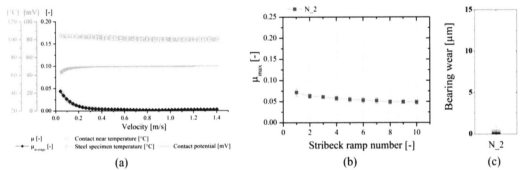

Figure 7.16: Tribometric results of Stribeck tests: (a) Tribodata plotted against speed, (b) evolution of µmax, (c) gravimetric bearing wear (error bars are derived based on mean deviation). After Summer et al. (2014; 2019) with permission from Elsevier.

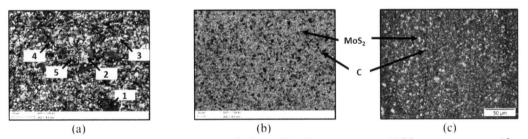

Figure 7.17: Post-test surface analysis of (a) RoD Stribeck test, (b) RoD temperature test, (c) BS start stop wear test. After Summer et al. (2014; 2019) with permission from Elsevier.

Table 7.3: EDX analysis of Fig. 7.17a (Summer et al., 2019).

Spectrum	C	O	P	S	Ti	Zn	Mo
1	100.0	–	–	–	–	–	–
2	36.4	3.4	–	37.5	–	–	22.7
3	24.7	–	–	49.7	–	–	25.6
4	45.6	29.8	0.6	2.3	20.8	0.2	0.8
5	9.2	57.8	0.5	1.9	30.3	0.2	–

Energy-dispersive X-ray (EDX) values obtained at 7.5 kV and given in at %.

is seen to accelerate running in events as well as to provide a low solid friction share and, hence, wear resistance.

7.5 Clutches and Brakes

One of the practical solutions for connecting two machine elements, usually shafts, are clutches and brakes. There are significant similarities between clutches and brakes; both of them are used in providing frictional, magnetic, or mechanical connection between two components.

7.5.1 Clutches

As mentioned above, the use of a clutch between an engine's crankshaft and the gearbox is one of the most familiar automotive settings. A clutch is a mechanical set which, in the simplest function, connects and disconnects power transmission especially from a driving shaft to a driven shaft. In such devices, the driving shaft is typically attached to an engine while the driven shaft provides output power for the gearbox. The connection/disconnection action can be achieved manually or automatically by various types of transmission means. This could be achieved by using a direct mechanical friction transmission (friction clutches) or an automatic transmission. In automatic transmissions, multiple plate clutches, band clutches and torque converter clutches are used either to transmit torque or to restrain a reaction member from rotating. The selection of clutch type and configuration depends on the application.

Friction clutches, the classic mechanical clutch, consist of two surfaces or two sets of surfaces that can be forced into frictional contact in order to allow gradual engagement action. Disc clutches may consist of single or multiple discs. An example of the design of disc clutches is illustrated in Fig. 7.18 (Childs, 2014; Childs, 2019).

The fundamental assumptions used in the design procedures for a friction disc clutch are based upon a uniform rate of wear at the mating surfaces and a uniform pressure distribution between the mating surfaces. Note that the uniform wear rate assumption gives a lower torque capacity clutch than the uniform pressure assumption. The preliminary design procedure for disc clutch design requires the determination of the torque and speed, specification of space limitations, selection of materials, i.e., the coefficient of friction and the maximum permissible pressure, and the selection of disc radii (Childs, 2014).

The friction material must have the required friction characteristics and it must withstand a wide range of operating temperatures, high shear forces and compressive loads. Commonly, gray cast iron or steel are the typical materials used for manufacturing clutch plates. While the friction contact surface is usually produced from a variety of fibers, particle fillers, and friction modifiers. These materials, with a properly blended composition, can provide the necessary material strength and bulk uniformity for a successful manufacturing, as well as providing a heat-resistant material with the required friction properties. The porosity of the friction material allows the transmission fluid to be stored near the friction interface as a lubricant reservoir, and the resilience of the friction material permits it to conform to the transmission mating surfaces (Kapoor et al., 2001).

Pantazopoulos et al. (Pantazopoulos et al., 2015) investigated the mechanisms that caused severe wear of an automotive metallic clutch pressure plate in contact with a consumable glass-fiber-reinforced-polymer-matrix composite disk. This tribosystem enables gear shifting in the drive train. Microscopic observations on the contact surface of the grey cast iron pressure plate revealed the activation of various wear modes, such as severe sliding wear, groove formation and oxidation, whereas observations on cross-sections indicated sub-surface damage. On the contact area, a ~ 10 mm–wide wear track was formed, as shown in Fig. 7.19, exhibiting a surface relief consisting of a sequence of deep valleys and crests. The stereo-microscopic observations of the wear track, shown in Figs. 7.19a and 7.19c, as well as the optical-microscopic ones, indicated severe dragging and deep grooving along the sliding direction. Also, the main wear modes were accompanied by micropolishing and microcutting, as shown in Fig. 7.19d. The deep, oblong grooves observed are consistent with the roughness measurements at various locations along the wear track, as shown in Fig. 7.19b. Within the wear track, further SEM observations showed a random

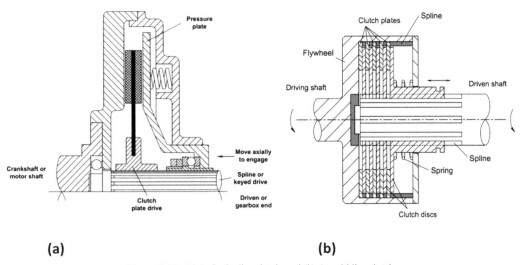

Figure 7.18: (a) A single disc clutch, and (b) A multidisc clutch.

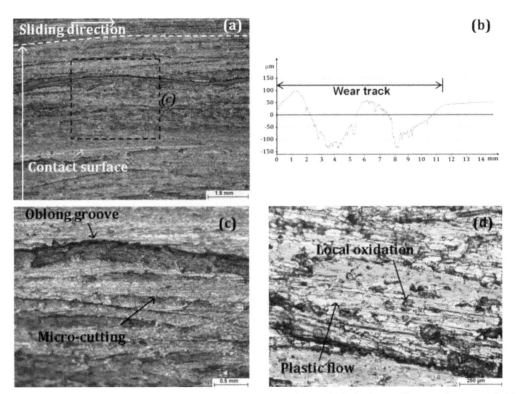

Figure 7.19: Worn surface of the pressure plate: (a) Top view stereo-micrograph, (b) Surface profile across the wear track, (c) Stereo- and (d) Optical-micrographs of the wear mechanisms. Reproduced after Pantazopoulos et al. (2015) with permission.

presence of voids with average dimensions of 30 μm, as well as a severe plastic flow of the metal along the sliding direction. The presence of voids suggests the occurrence of prior ductile tearing of the surface under a high stress state followed by crack opening and growth, leading to smearing under the operation conditions of the tribosystem.

Figure 7.20 shows a crack initiation at graphite flakes edges below the worn surface and its propagation within the ferrite-pearlite matrix in a direction parallel to the contact surface. The depth at which these cracks are observed was in all cases of the order of 20–30 μm. That is corresponding almost precisely to the depth at which maximum shear stress was established for the particular operating conditions and system configuration, according to Hertz contact stress theory (Vernon et al., 2004). Based of the above examinations, it was concluded that the graphite flakes' morphology facilitates crack initiation and propagation within the pearlite matrix.

Hydraulic clutches (or hydraulic automatic transmission) were invented in 1921. This type of clutch uses a fluid coupling in place of a friction clutch, and accomplishes gear changes by hydraulically locking and unlocking a system of planetary gears. Therefore, they are closely dependent on the transmission fluid characteristics as well as additives, including friction-modifying additives, to produce the desired transmission operation. Also, the transmission fluid can have a deep impact on static holding capacity and the friction interaction between the transmission fluid and clutch material.

From a friction standpoint, an automatic transmission clutch consists of two basic elements: The friction clutch plates and the steel reaction plates. The clutch plate assembly consists of three major components: A steel core, an adhesive coating, and the friction facings, as illustrated in Fig. 7.21. The friction facing is bonded to one or both sides of the steel core with an adhesive.

During clutch engagement, both transmission fluid and clutch friction material life can be affected by the amount of clutch energy. In a band or transmission power shifting clutch, the torque capacity must be great enough to produce a rapid clutch or band engagement. Otherwise, the energy dissipated as heat can raise clutch surface temperatures. This temperature rise will contribute to the deterioration of both clutch materials and transmission fluids. For wide-open-throttle conditions, typical clutch surface temperatures have been measured and found to be in the order of 95°C above the sump fluid temperature (Stebar et al., 1990).

Clutch energy dissipation for different transmission operating conditions is illustrated in Fig. 7.22. In the figure, the clutch horsepower (clutch friction torque × clutch sliding speed) is plotted vs. engagement time for three simulated heavy-throttle shifts (Kapoor et al., 2001). Curve A can be considered as a satisfactory shift, for which the clutch surface area is designed to provide a long clutch life. Curve B illustrates an unsatisfactory clutch engagement because of an extended engagement time, which results in a surface temperature as high as 230°C (Stebar et al., 1990). Curve C shows a clutch engagement for which the energy level is very different than for curve A, but the capacity of the clutch is high enough to produce a short engagement.

Characteristics of an Automatic Transmission Fluids (ATF) have an important role in controlling the transmission shift performance and long-term durability. The ATF must be able to endure the heat

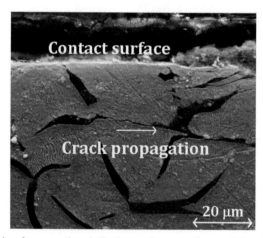

Figure 7.20: SEM micrographs of cross-sections indicating crack propagation parallel to the contact surface within the metallic matrix. Reproduced after Pantazopoulos et al. (2015) with permission.

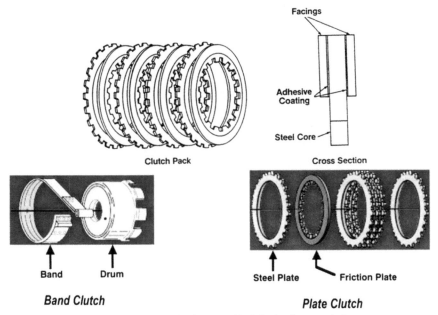

Facings

Adhesive Coating

Steel Core

Clutch Pack

Cross Section

Band   Drum

Steel Plate   Friction Plate

Band Clutch   *Plate Clutch*

Figure 7.21: The clutch plate assembly in hydraulic transmission.

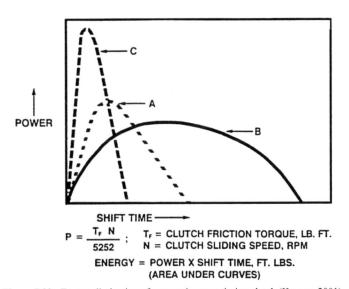

POWER

SHIFT TIME

$$P = \frac{T_f \ N}{5252} \ ; \quad \begin{array}{l} T_f = \text{CLUTCH FRICTION TORQUE, LB. FT.} \\ N = \text{CLUTCH SLIDING SPEED, RPM} \end{array}$$

ENERGY = POWER X SHIFT TIME, FT. LBS.
(AREA UNDER CURVES)

Figure 7.22: Energy dissipation of automatic transmission clutch (Kapoor, 2001).

and forces generated even under high load conditions. Also, they must resist chemical changes that are a consequence of oxidation and must minimize the adverse consequences of physical changes. This means that the ATF has the ability to flow into critical areas when cold, but not flow out excessively when the oil is hot. The ATF have to maintain the proper amount of friction and reduce wear and corrosion. This feature can be obtained either by fluid film formation between moving components or by providing a protective chemical film on the surface.

In general, ATF are formulated using a synthetic or mineral oil base stock plus a variety of additives. Such additives include friction modifiers to produce the desired clutch operation, corrosion inhibitors, viscosity index improvers, dispersants, seal swell agents, antioxidants and anti-wear agents, with a total additive treatment level typically in the range of 10 to 20% (Kemp and Linden, 1990; Lann et al., 2000). It is important to mention that friction modifiers have particular importance for proper transmission performance. Long-chain hydrocarbon molecules with a polar group on one end are usually employed to produce good clutch friction characteristics. Dilauryl phosphate and oleic acid are typical examples of friction-modifier additives.

7.5.2 Brakes

The basic objective of a brake in an automotive system is to slow down or stop an automobile movement. Braking systems are the most common method of braking in passenger cars, light and heavy trucks, buses and off-road vehicles. Physically, the braking action is achieved by absorbing kinetic energy and dissipating it in the form of heat by means of friction. There are numerous brake types, such as disc brakes, drum brakes, differential band brake, and others. The selection and configuration of a brake depends on the requirements. The brake factor of each type is the ratio of the frictional braking force generated to the actuating force applied.

Disc brakes are usually found in all automotive applications where they are employed widely for passenger cars, light and heavy trucks, buses, off-road vehicles and motorcycle wheels. Typically, these types of brakes involve a cast iron disc bolted to the wheel hub. This is sandwiched between two pads actuated by pistons supported in a caliper mounted on the stub shaft (Childs, 2014), as illustrated in Fig. 7.23.

In disc brakes, when the brake pedal is pressed, approximately 7 to 25% of the disc-rubbing surface is loaded by the brake shoe. Disk brakes allow heat dissipation because they have larger exposed surface areas and better cooling geometry. Furthermore, the use of a discrete pad provides relatively better heat transfer between the cooler disc and the hot brake pad.

Brake pads are steel backing plates where a friction material is bounded to the surface that faces the disc brake rotor.

Although a brake pad is considered as the consumable component in a braking system, the brake disk is also subjected to wear, even though in a significantly lower wear rate. Manufacturers specifications usually suggest that the used brake disc should be replaced when a specific reduction of its thickness is reached. Pantazopoulos et al. (Pantazopoulos et al., 2015) introduced a case study for investigating the premature failure of an AISI 410 grade stainless steel brake disk operating in a high performance motorcycle. Despite the fact that during a periodic maintenance, the disc thickness reduction was within acceptable levels (0,6 mm ~ 1,0 mm according to the manufacturer's specification), radial cracks starting only from the outermost ventilation holes were identified, as shown in Fig. 7.24. Fractographic evaluation of the conjugate surfaces indicated fatigue as the mechanism responsible for the crack initiation and propagation. Finite element analysis of the disc supported that fatigue cracks could start from the outermost ventilation holes of the brake disc, which are the most highly-loaded areas prone to fracture. The main disc surface wear mechanisms include micropolishing, ploughing and oxidative wear as a result of thermomechanical loading during sliding.

Drum brakes commonly refers to a system in which lining shoes press on the inner surface of the drum, as shown in Fig. 7.25. While, on the other hand, when the shoes press on the outside of the drum, it is often called a "clasp brake". Both types of brakes are applying friction, caused by a set of shoes or pads, to the circumference of a rotating cylinder (brake drum).

In these types of brakes, the shoe applies normal force on approximately 50% to 70% of the drum circumferential area (Kapoor et al., 2001). The drum lining is usually made of high-friction, low-wear materials, and the frictional loss at the interface provides dissipation of energy and the necessary braking action. Typically, the shoes are the consumable components and wear faster than the drum, thus requiring comparatively more frequent replacement.

Friction materials should satisfy a high friction coefficient along with a low wear rate. These materials consist of an optimized composition of different ingredients, including binder, friction additives, abrasive/ reinforcement, and filler/functional materials. It can be classified into one of four principal categories: Non-metallic materials, semi-metallic materials, fully metallic materials and ceramic materials. Another

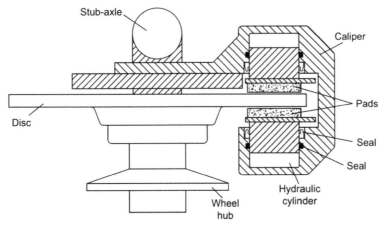

Figure 7.23: Disk brake configuration (Pantazopoulos, 2015).

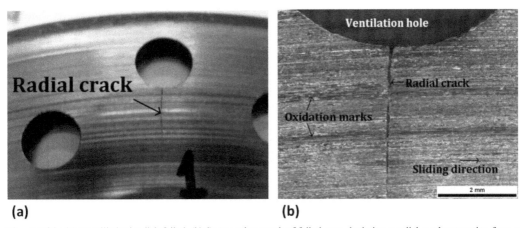

Figure 7.24: (a) Metallic brake disk failed, (b) Stereo-micrograph of failed area, depicting a radial crack emanating from a ventilation hole. Reproduced after Pantazopoulos et al. (2015) with permission.

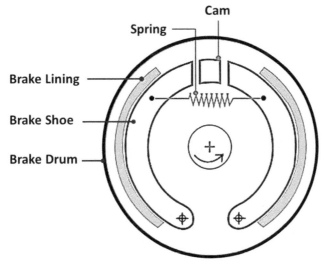

Figure 7.25: Illustration of internal drum brake.

classification was introduced by Anderson (Anderson, 1992) as organic, metallic and carbon. Carbon-based brakes, such as carbon-carbon composites, are generally used for aircraft and racing car applications, where weight is a critical design issue. Indeed, in the automotive industry, the selection of the friction material to be applied in the brake system is governed by the intended use of the vehicle. Friction materials with a higher coefficient of friction provide good braking with less brake pedal pressure requirement, but tend to lose efficiency at higher temperatures. Brake pads with a smaller and constant coefficient of friction do not lose efficiency at higher temperatures and are stable, but require a higher brake pedal pressure (Henderson and Haynes, 1994).

Because of the hazardous effects of some types of friction materials, e.g., asbestos, attempts were made to find alternative potential materials with eco-friendly behaviors. Many researches were performed in order to study the feasibility of using natural waste materials in the production of friction materials (Kukutschová et al., 2009; Kukutschová et al., 2010; Roubicek et al., 2008). Several agriculture wastes, such as palm slag, palm kernel fibers, palm kernel shells, banana peels, hyphaene thebaica kernel shell, cashew dust, and others were used to manufacture friction materials. Nano-alumina particles as an abrasive were also utilized along with banana peel (BP) and Bagasse (BG) fiber/particle to achieve an optimized friction materials composition (Amirjan, 2019).

Seven composite samples were prepared, in which the variables were BP and BG values and the amount of n-Al_2O_3 that was balanced with barite amounts in the overall composition, as shown in Table 7.4. Wear surfaces of composite samples showed the features of a good wear behavior and lower degree of peeling and wear debris, as shown in Fig. 7.26.

The wear mechanism present in braking systems is a complex issue because of the contact variations that exist in these specific systems. The connection between a brake pad surface topography and the occurrence of squeals could also have an important contribution on the wear behavior. Several researchers over the years were conducted in order to study the wear behavior of friction materials and its correlation to braking systems. Eriksson et al. (Eriksson et al., 1999; Eriksson and Rundgren, 2012), for instance, conducted investigations on the nature of the tribological contact in automotive breaks in disc pads. His results showed the presence of contact plateaus (flat areas rising a few micrometers over the rough surrounding) after the sliding action of the pads against the cast iron discs. The analysis of the wear mechanisms in automotive brakes produced by the contact surface variations suggests rapid deterioration

Table 7.4: The ingredients and composition of different friction composites (Amirjan, 2019).

Ingredient (wt.%)	Friction materials designation						
	BPG0	BG5	BG10	BP5	BP10	BPG5	BP10Al8
Rock wool	10	10	10	10	10	10	10
Carbon fiber	3	3	3	3	3	3	3
Glass fiber	5	5	5	5	5	5	5
Steel fiber	6	6	6	6	6	6	6
SiO_2	3	3	3	3	3	3	3
MgO	3	3	3	3	3	3	3
Graphite	8	8	8	8	8	8	8
Brass powder	8	8	8	8	8	8	8
Vermiculite	10	10	10	10	10	10	10
Phenolic Resin	15	15	15	15	15	15	15
Calcium carbonate	5	5	5	5	5	5	5
n-Al_2O_3	4	4	4	4	4	4	8
Bagasse	0	5	10	0	0	5	0
Banana peel	0	0	0	5	10	5	1
Barite	20	15	10	15	10	10	6

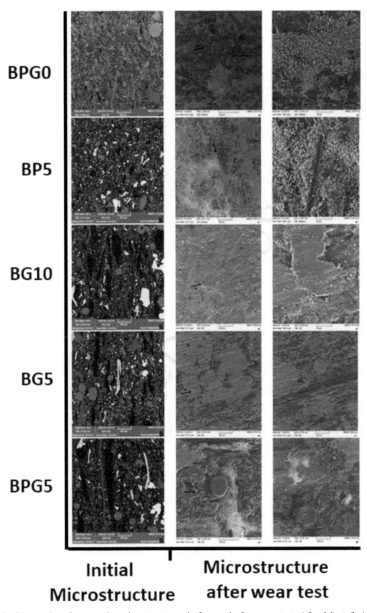

BPG0

BP5

BG10

BG5

BPG5

Initial
Microstructure

Microstructure
after wear test

Figure 7.26: SEM micrographs of composite microstructures before and after wear test. After Mostafa Amirjan (Amirjan, 2019) with permission from Elsevier.

processes. This was affected by braking force changes, vibrations and wave motions in the brake pad and cast iron disk, and slow degradation processes. Also, thermal deformation, shape adaptation and contamination and cleaning played important roles.

Laguna-Camacho et al. (Laguna-Camacho et al., 2015) investigated the wear mechanisms of disc and shoe pads subjected to real service. Figure 7.27 represents SEM images of a worn disc pad where it is possible to detect smeared wear debris on the surfaces that was caused by the deformation of the metallic particles and fibers, which first made contact with the steel disk. Some fillers such as Al_2O_3 and SiC did not appear in this specific worn zone, which means that these small particles were removed by an abrasion action. It is also possible to observe the appearance of fibers and hard particles, probably

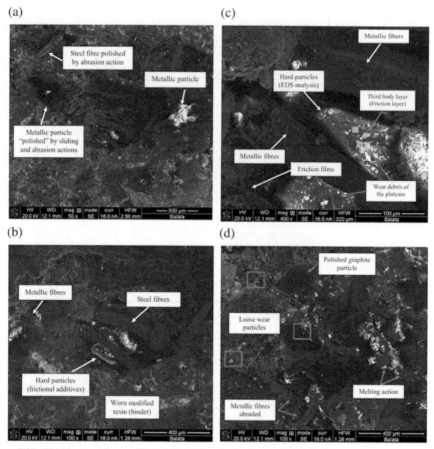

Figure 7.27: SEM images of worn disk pad. Reproduced after Laguna-Camacho et al. (2015) with permission.

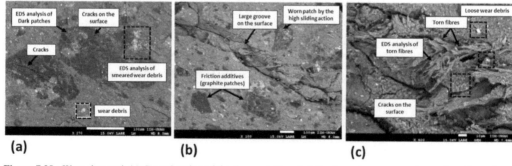

Figure 7.28: Worn shoe pad, (a) Central region of the shoe pad, (b and c) Cracks and grooves on the surface. Reproduced after Laguna-Camacho et al. (2015) with permission.

from the steel disc. The contact action of the steel disc with hard particles, as shown in Fig. 7.27c, could lead to an increase in the wear damage in this specific zone and a reduction in the friction coefficient. Figure 7.27d shows black graphite particles, in the upper part, that were "polished" by the sliding and abrasion actions.

On the other hand, The SEM images of the worn shoe pad, Fig. 7.28, showed cracks on the surface and smeared wear debris caused by the repetitive impact with the steel drum. Wear debris in random positions of the surface were also observed. In addition, large grooves generated by loose particles, inflicted the detachment of the threads of the asbestos as observed on the surfaces in Figs. 7.28b and 7.28c. These observations confirm the severity of the wear damage caused in these mechanical elements due to the high contact stresses generated during the service. In conclusion, the wear mechanisms identified in the disc pads were sliding and three-body abrasion wear. In the shoe pads, the dominant mechanism was fatigue cracking on the surfaces, caused by the high impact action of the pads against the drums during braking.

7.6 Automotive Lubricants

In general, any tribosystem consists of four elements: A contacting surface, the opposing contacting surface, the interface between the two, and the medium in the interface and the environment. There are also many variables in tribosystems, including the type of movement, the forces involved, temperature, speed and duration of the stress.

In regard to the lubrication of tribomotive subsystems, the lubricant must protect the automotive component against wear in different operating conditions. This protection maybe in the form of a fluid film to keep opposing surfaces separated or in the form of a chemical film on a surface to generate boundary lubrication protection. Moreover, anti-corrosion lubricants protect by virtue of alkaline agents in order to neutralize acids that form in hot spots. This variety of function implies that different engine components require substantially different kinds of lubricants, or protection. In particular, an engine oil must be effective under all driving conditions and must not evaporate or degrade excessively in high temperatures. Likewise, a transmission fluid must withstand high temperatures and loads. Among these diversities, three types of automotive lubricants will be discussed in this section.

7.6.1 Engine Oils

Since the development of the first internal combustion engine, engine oil has been a mandatory agent in any diesel or gasoline engine. The trivial tribological task of such oils is to guarantee the functional reliability of all friction points inside the engine in all operating conditions. In more details, they have much more complex requirements for providing wear resistance, ability to disperse wear and combustion products, and oxidation resistance.

Engine oils have to fulfil a wide range of functions in engines. That is, they should properly operate in different sliding friction speeds ranging from simple linear movement of a piston to extreme rotational movements (up to 200000 rpm with micron tolerances). They, also, should reduce friction and wear during extremely low-temperature start-ups as well as when the lubricating film is subject to high temperatures and pressures in bearings and around the piston rings. The typical operating temperature may range from −40°C, as an ambient temperature in the Arctic, to peak temperature values of over 300°C, as found under the piston crown (Tung and McMillan, 2004). However, there is another secondary task for the oils: It should burn off the cylinder wall without leaving any residue. In addition, as for the piston itself, the engine oil dissipates heat from the piston and, thus, cools it down.

Based on the above, the challenges in the development of engine oil arise because the environment in which engine oils operate is particularly extreme. Therefore, recently, lots of efforts were put in to improve the quality and efficiency of engine oils. One of the famous approaches is to incorporate particle additives into regular lubricants in order to reduce the friction and wear of frictional surfaces (Hemmat Esfe et al., 2019; Gaur et al., 2019; Laad and Jatti, 2018). A lot of investigations inferred that the addition of nano-particle additives to regular engine oils in order to improve the anti-frictional, chemical and physical properties of some regular oils is quite effective. However, one of the drawbacks of these nano oils is that, when nanoparticles are added to the base oil, there is a certain limit of volume fraction, after which an adverse effect may be obtained.

It was demonstrated that the addition of TiO_2 nanoparticles exhibited a good friction reduction and anti-wear behavior (Laad and Jatti, 2018). The addition of TiO_2 nanoparticles to the engine oil reduced the

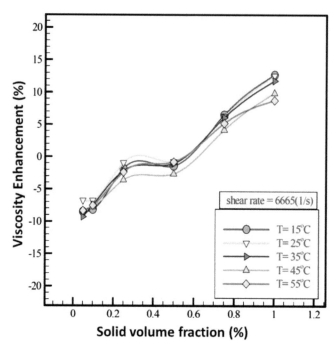

Figure 7.29: Viscosity enhancement versus solid volume fraction in different temperatures. After Esfe et al. (2019) with permission.

coefficient by 86% with 0.3% concentration by weight of the oil as compared to the oil without nanoparticles for load 4 kg. This reduction could be attributed to the rolling of the sphere like nanoparticles between the rubbing surfaces. Up to a certain limit, increasing the concentration of nanoparticles resulted in improvement in the coefficient of friction but not more than the coefficient of friction of oil without nanoparticles. This finding might be due to the agglomeration of TiO_2 nanoparticles. The anti-wear mechanism is attributed to the deposition of TiO_2 nanoparticles on the worn surface, which can decrease the shearing resistance, thus, improving the tribological properties. Gaur et al. (Gaur et al., 2019) also reported that the addition of CuO, FeO and CuO + FeO nanoparticles to the base oil (SAE 20W40) enhances the physical and tribological characteristics of the lubricating oil. Another study was conducted by Esfe et al. (Esfe et al., 2019) to evaluate the effect of introducing a mixture of ZnO and Carbon nanotubes MWCNT nanoparticles with 20:80 volume ratio in 5W50 base oil. A 10% reduction in oil viscosity was reached through addition of nanoparticles in solid volume fractions less than 0.25%, as shown in Fig. 7.29. This can be used to reduce heavy oil viscosity in the oil and prepare a good condition for enhancement of oil recovery. Thus, results of this research can be considered as an initial point in a feasibility study of using nanoparticles in enhanced oil recovery through injection of nanoparticles into the oil wells.

7.6.2 Transmission Fluids

By definition, Automatic Transmission Fluids ATFs are synthetic or mineral oils, formulated to meet the lubrication needs of Automatic Transmissions AT in automotive vehicles (Kemp and Linden, 1990). As in engine oils, various kinds of additives are applied to synthetic and mineral oils in order to help maintain the desired friction properties, minimize wear, and provide corrosion inhibition to the transmission components. Examples of additives used to give ATF its main properties are shown in Table 7.5.

Typically, one of the additives present in the fluid is a long-chain hydrocarbon that has a polar group on one end. The long chain hydrocarbon portion permits the additive to dissolve in oil, and the polar group on the end is attracted to a metal surface and provides the desired amount of friction. If too much friction

Table 7.5: Examples of additives applied to ATF. After Takanori Kugimiya et al. (Kugimiya et al., 1998) with permission from Springer Science Publishers.

Property	Additives	
	Classification	**Examples**
Anti-wear	Anti-wear agent	Organic sulfur compounds, organic phosphorus compounds, zinc dithiophosphates, etc.
Oxidation stability	Anti-oxidant	Aromatic amines, phenols, terpene sulfides, zinc dithiophosphates, etc.
	Detergent dispersant	Succinic imides, metallic sulfonates, metallic phenates, etc.
Friction	Friction modifier	Fatty acids, aliphatic ester, amines, amides, phosphate ester, etc.

is generated, the transmission may not shift smoothly. If too little friction occurs, the transmission may slip. Thus, great care must be taken to ensure that an appropriate additive is chosen for this important function (Tung and McMillan, 2004).

Sulfur compounds, phosphorus compounds and zinc dithiophosphates are typical anti-wear agents. Sulfur compounds act as extreme pressure agents. They prevent seizures by producing iron sulfide on the steel surface under extreme pressure conditions. Phosphorus compounds prevent the wear of the steel surface by producing iron phosphate on this surface under boundary lubrication conditions. These anti-wear agents also have anti-oxidation properties. However, amine type and/or phenol type anti-oxidants are generally used together with anti-wear agents (Kugimiya et al., 1998). In addition to the previous characteristics, viscosity index improvers, corrosion inhibitors and seal swellers are applied to ATF to satisfy each property (Tung and McMillan, 2004). Of all the ATF properties mentioned above, probably the friction property is the most important in the design of ATFs and also the most difficult to achieve all its requirements.

7.6.3 Gear Lubricants

Gear lubricants (commercially knows as Gearbox oils) are manufactured specifically for transmissions, transfer cases, and differentials in automotive as well as in other machinery. This type of lubricants is designed to withstand the harsh environment on gear mating surfaces. These are high pressure and consequently high temperature which are frequently generated at the point of contact.

Gear lubricating oils are categorized by the American Petroleum Institute (API) using GL ratings. The higher an oil's GL-rating, the more pressure can be sustained without any metal-to-metal contact taking place between transmission components. Differential (or separate differential) usually has higher GL-rating than gearboxes because of the higher pressure between metal parts. For example, GL-4 oil is recommended for most modern gearboxes while GL-5 oil is recommended for differentials (Herdan, 1997).

Gearbox oils should provide sufficient wear and extreme pressure protection to prevent material failure of the gear surfaces. Also, the properties of such lubricants must remain stable, including oxidative stability, anti-wear protection, maintenance of appropriate friction characteristics, low-temperature viscosity, corrosion protection, and resistance to foam formation (Herdan, 1997). Furthermore, the gear lubricant must remain functional for extended periods of time. For such reasons, most lubricants for gears and differentials contain extreme pressure additives (EP) as well as antiwear additives. The commonly used additives are dithiocarbamate derivatives and sulfur-treated organic compounds ("sulfurized hydrocarbons").

Ionic liquids (ILs) have been introduced as proposed additives in producing novel high-performance lubricants (Herdan, 1997; Totolin et al., 2013; Anand et al., 2015). They have been explored as lubricants for various device applications due to their excellent electrical conductivity as well as good thermal conductivity, where the latter allows frictional heating dissipation. The ILs are salts, formed by organic or inorganic cations and anions, with a melting point lower than 100°C. The special characteristics of ILs are their chemical and thermal stability, tunable properties, electrical conductivity, low vapor pressure and high thermal conductivity. These characteristics allow for the usage of ILs within different tribosystems, in vacuum applications, in harsh conditions, etc. (Herdan, 1997; Palacio and Bhushan, 2010).

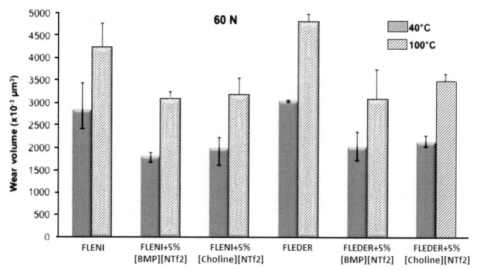

Figure 7.30: Wear behavior of the fully formulated gear oils and its mixtures with the ionic liquids. After Monge et al. (2015) with permission from Elsevier.

Monge et al. (Monge et al., 2015) investigated the tribological behavior of two fully formulated gearbox oils (polyalphaolefin FLENDER and mineral-based FLENI). Two ionic liquids ([Choline][NTf$_2$] and [BMP][NTf$_2$]) at 5 wt% concentration were used as additives. The addition of both ionic liquids resulted in improvement of the wear performance of the gear oils under all test conditions, as shown in Fig. 7.30. However, the results indicated that the addition of both ionic liquids did not significantly change the friction behavior.

References

Ali, M.K.A., Xianjun, H., Abdelkareem, M.A.A., Gulzar, M. and Elsheikh, A.H. 2018. Novel approach of the graphene nanolubricant for energy saving via anti-friction/wear in automobile engines. Tribology International 124: 209–229. Doi: https://doi.org/10.1016/j.triboint.2018.04.004.

Amirjan, M. 2019. Microstructure, wear and friction behavior of nanocomposite materials with natural ingredients. Tribology International 131: 184–190. Doi: https://doi.org/10.1016/j.triboint.2018.10.040.

Anand, M., Hadfield, M., Viesca, J.L., Thomas, B., Hernández Battez, A. and Austen, S. 2015. Ionic liquids as tribological performance improving additive for in-service and used fully-formulated diesel engine lubricants. Wear 334-335: 67–74. Doi: https://doi.org/10.1016/j.wear.2015.01.055.

Anderson, A.E. 1992. Friction and wear of automotive brakes. In ASM Handbook, 569–577. ASM International.

Andersson, P., Tamminen, J. and Sandström, C.-E. 2002. Piston ring tribology. A literature survey. In: Espoo 2002. Finland: JULKAISIJA – UTGIVARE – PUBLISHER.

Baek, H.K., Lee, S.W., Han, D., Kim, J., Lee, J. and Aino, H. 2014. Development of valvetrain system to improve knock characteristics for gasoline engine fuel economy.

Becker, E.P. 2004. Trends in tribological materials and engine technology. Tribology International 37(7): 569–575. Doi: https://doi.org/10.1016/j.triboint.2003.12.006.

Blau, P.J. 2013. Friction in internal combustion engines. pp. 1322–1325. *In*: Wang, Q.J. and Chung, Y.-W. (eds.). Encyclopedia of Tribology. Boston, MA: Springer US.

Childs, P.R.N. 2014. Chapter 13—Clutches and brakes. pp. 513–564. *In*: Childs, P.R.N. (ed.). Mechanical Design Engineering Handbook. Oxford: Butterworth-Heinemann.

Childs, P.R.N. 2019. 4—Machine elements. pp. 145–165. *In*: Childs, P.R.N. (ed.). Mechanical Design Engineering Handbook (Second Edition). Butterworth-Heinemann.

Chong, W.W.F. and De la Cruz, M. 2014. Elastoplastic contact of rough surfaces: A line contact model for boundary regime of lubrication. Meccanica 49(5): 1177–1191. Doi: 10.1007/s11012-013-9861-1.

Delprete, C. and Razavykia, A. 2017. Piston ring–liner lubrication and tribological performance evaluation: A review. Proceedings of the Institution of Mechanical Engineers, Part J: Journal of Engineering Tribology 232(2): 193–209. Doi: 10.1177/1350650117706269.

Eriksson, M., Bergman, F. and Jacobson, S. 1999. Surface characterisation of brake pads after running under silent and squealing conditions. Wear 232(2): 163–167. Doi: https://doi.org/10.1016/S0043-1648(99)00141-6.

Eriksson, M. and Rundgren, C.J. 2012. Vargfrågan-gymnasieelevers argumentation kring ett sociovetenskapligt dilemma. NorDiNa (Nordic Studies in Science Education) 8(1): 26–41.

Esfe, M.H., Arani, A.A.A., Esfandeh, S. and Afrand, M. 2019. Proposing new hybrid nano-engine oil for lubrication of internal combustion engines: Preventing cold start engine damages and saving energy. Energy 170: 228–238. Doi: https://doi.org/10.1016/j.energy.2018.12.127.

Fukuoka, S., Hara, N., Mori, A. and Ohtsubo, K. 1997. Friction loss reduction by new lighter valve train system. JSAE Review 18: 107–111.

Gangopadhyay, A., McWatt, D., Willermet, P., Crosbie, G.M. and Allor, R.L. 1999. Effects of composition and surface finish of silicon nitride tappet inserts on valvetrain friction. pp. 635–644. *In*: Dowson, D., Priest, M., Taylor, C.M., Ehret, P., Childs, T.H.C., Dalmaz, G., Berthier, Y., Flamand, L., Georges, J.M. and Lubrecht, A.A. (eds.). Tribology Series. Elsevier.

Gangopadhyay, A., Soltis, E. and Johnson, M.D. 2004. Valvetrain friction and wear: Influence of surface engineering and lubricants. Proceedings of the Institution of Mechanical Engineers, Part J: Journal of Engineering Tribology 218(3): 147–156. Doi: 10.1177/135065010421800302.

Gaur, M.K., Singh, S.K., Sood, A. and Chauhan, D.S. 2019. Experimental investigation of physical and tribological properties of engine oil with nano-particles additives. In: Advances in Design, Simulation and Manufacturing. Springer, Cham.

Gebauer, K. and Gavrilescu, A. 2004. Performance and Reliability of Deep Drawn Hollow Valves in the Automotive Engines.

Grützmacher, P.G., Rosenkranz, A., Szurdak, A., König, F., Jacobs, G., Hirt, G. and Mücklich, F. 2018. From lab to application—Improved frictional performance of journal bearings induced by single- and multi-scale surface patterns. Tribology International 127: 500–508. Doi: https://doi.org/10.1016/j.triboint.2018.06.036.

Henderson, B. and Haynes, J.H. 1994. Disc brakes. In: The Haynes Automotive Brake Manual, 1–20. Haynes North America.

Herdan, J.M. 1997. Lubricating oil additives and the environment—an overview. Lubrication Science 9(2): 161–172. Doi: 10.1002/ls.3010090205.

Jang, J.Y. and Khonsari, M.M. 2019. Performance and characterization of dynamically-loaded engine bearings with provision for misalignment. Tribology International 130: 387–399. Doi: https://doi.org/10.1016/j.triboint.2018.10.003.

Jog, S., Anthony, K., Bhoinkar, M., Kadam, K. and Patil, M.M. 2020. Modelling and analysis of IC engine piston with composite material (AlSi17Cu5MgNi). *In*: Gunjan, V., Singh, S., Duc-Tan, T., Rincon Aponte, G. and Kumar, A. (eds.). ICRRM 2019 – System Reliability, Quality Control, Safety, Maintenance and Management. ICRRM 2019. Springer, Singapore. Doi: https://doi.org/10.1007/978-981-13-8507-0_25.

Kapoor, A., Tung, S.C., Schwartz, S.E., Priest, M. and Dwyer-Joyce, R.M. 2001. Automotive tribology. *In*: Bhushan, B. (ed.). Modern Tribology Handbook, Michigan, USA: CRC Press.

Kemp, S.P. and Linden, J.L. 1990. Physical and Chemical Properties of a Typical Automatic Transmission Fluid. SAE.

Khurram, M. 2016. Tribological analysis of engine valve train performance considering effects of lubricant formulation. Ph.D. School of Mechanical and Manufacturing Engineering, National University of Sciences and Technology (2010-NUST-DirPhD-ME-61).

Kugimiya, T., Yoshimura, N. and Mitsui, J. 1998. Tribology of automatic transmission fluid. Tribology Letters 5(1): 49–56. Doi: 10.1023/A:1019156716891.

Kukutschová, J., Roubíček, V., Malachová, K., Pavlíčková, Z., Holuša, R., Kubačková, J., Mička, V., MacCrimmon, D. and Filip, P. 2009. Wear mechanism in automotive brake materials, wear debris and its potential environmental impact. Wear 267(5): 807–817. Doi: https://doi.org/10.1016/j.wear.2009.01.034.

Kukutschová, J., Roubíček, V., Mašláň, M., Jančík, D., Slovák, V., Malachová, K., Pavlíčková, Z. and Peter Filip. 2010. Wear performance and wear debris of semimetallic automotive brake materials. Wear 268(1): 86–93. Doi: https://doi.org/10.1016/j.wear.2009.06.039.

Kumar, V. and Agarwal, A.K. 2019. Tribological aspects of automotive engines. *In*: Katiyar, J., Bhattacharya, S., Patel, V. and Kumar, V. (eds.). Automotive Tribology. Energy, Environment, and Sustainability. Springer, Singapore. Doi: https://doi.org/10.1007/978-981-15-0434-1_2.

Laad, M. and Jatti, V.K.S. 2018. Titanium oxide nanoparticles as additives in engine oil. Journal of King Saud University—Engineering Sciences 30(2): 116–122. Doi: https://doi.org/10.1016/j.jksues.2016.01.008.

Laguna-Camacho, J.R., Juárez-Morales, G., Calderón-Ramón, C., Velázquez-Martínez, V., Hernández-Romero, I., Méndez-Méndez, J.V. and Vite-Torres, M. 2015. A study of the wear mechanisms of disk and shoe brake pads. Engineering Failure Analysis 56: 348–359. Doi: https://doi.org/10.1016/j.engfailanal.2015.01.004.

Lai, F., Qu, S., Duan, Y., Lewis, R., Slatter, T., Yin, L., Li, X., Luo, H. and Sun, G. 2018. The wear and fatigue behaviours of hollow head & sodium filled engine valve. Tribology International 128: 75–88. Doi: https://doi.org/10.1016/j.triboint.2018.07.015.

Lann, P.L., Derevjanik, T.S., Snyder, J.W. and Ward, W.C. 2000. Nylon degradation with automatic transmission fluid. Thermochimica Acta 357-358: 225–230. Doi: https://doi.org/10.1016/S0040-6031(00)00392-0.

Lee, J. 1993. Dynamic modeling and experimental verification of a valve train including lubrication and friction. Ph.D., University of Michigan.

Ma, W., Biboulet, N. and Lubrecht, B.B. 2018. Performance evolution of a worn piston ring. Tribology International 126: 317–323. Doi: https://doi.org/10.1016/j.triboint.2018.05.028.

Martínez, D.L., Valverde, R.R. and Ferrarese, A. 2010. Ring packs for friction optimised engines. MTZ Worldwide 71(7): 26–29. Doi: 10.1007/BF03227030.

Monge, R., González, R., Hernández Battez, A., Fernández-González, A., Viesca, J.L., García, A. and Hadfield, M. 2015. Ionic liquids as an additive in fully formulated wind turbine gearbox oils. Wear 328-329: 50–63. Doi: https://doi.org/10.1016/j.wear.2015.01.041.

Muzakkir, S.M., Patil, M.G. and Hirani, H. 2015. Design of innovative engine valve: Background and need. International Journal of Scientific Engineering and Technology 4(3): 178–181.

Notay, R.S., Priest, M. and Fox, M.F. 2019. The influence of lubricant degradation on measured piston ring film thickness in a fired gasoline reciprocating engine. Tribology International 129: 112–123. Doi: https://doi.org/10.1016/j.triboint.2018.07.002.

Palacio, M. and Bhushan, B. 2010. A review of ionic liquids for green molecular lubrication in nanotechnology. Tribology Letters 40(2): 247–268. Doi: 10.1007/s11249-010-9671-8.

Pantazopoulos, G., Tsolakis, A., Psyllaki, P. and Vazdirvanidis, A. 2015. Wear and degradation modes in selected vehicle tribosystems. Tribology in Industry 37(1): 72–80.

Pashneh-Tala, S., Lewis, R. and Malins, A. 2014. Tribological Desigen Guide, Part 4: The wear analysis prosess.

Priest, M. and Taylor, C.M. 2000. Automobile engine tribology—approaching the surface. Wear 241(2): 193–203. Doi: https://doi.org/10.1016/S0043-1648(00)00375-6.

Roubíček, V., Raclavska, H., Juchelkova, D. and Filip, P. 2008. Wear and environmental aspects of composite materials for automotive braking industry. Wear 265(1): 167–175. Doi: https://doi.org/10.1016/j.wear.2007.09.006.

Ruddy, B.L., Dowson, D. and Economou, P.N. 1982. A review of studies of piston ring lubrication. 9th Leeds-Lyon Symp. on Tribology: Tribology of Reciprocating Engines, Leeds, UK.

Rycroft, J.E., Taylor, R.I. and Scales, L.E. 1996. Elastohydrodynamic effects in piston ring lubrication in modern gasoline and diesel engines. In: Elastohydrodynamics—Fundamentals and Applications in Lubrication and Traction. 23rd Leeds-Lyon Symposium on Tribology.

Röhrle, M.D. 1995. Pistons for internal combustion engines—fundamentals of piston technology. In: MAHLE GmbH. Verlag Moderne Industrie. Landsberg/Lech, Germany.

Schommers, J., Scheib, H., Hartweg, M. and Bosler, A. 2013. Reibungsminimierung bei Verbrennungsmotoren. MTZ—Motortechnische Zeitschrift 74(7): 566–573. Doi: 10.1007/s35146-013-0170-y.

Stebar, R.F., Davison, E.D. and Linden, J.L. 1990. Determining Frictional Performance of Automatic Transmission Fluids in a Band Clutch. SAE.

Summer, F., Grün, F., Schiffer, J., Gódor, I. and Papadimitriou, I. 2015. Tribological study of crankshaft bearing systems: Comparison of forged steel and cast iron counterparts under start–stop operation. Wear 338-339: 232–241. Doi: https://doi.org/10.1016/j.wear.2015.06.022.

Summer, F., Grün, F., Offenbecher, M. and Taylor, S. 2019. Challenges of friction reduction of engine plain bearings— Tackling the problem with novel bearing materials. Tribology International 131: 238–250. Doi: https://doi.org/10.1016/j.triboint.2018.10.042.

Taylor, C.M. 1994. Fluid film lubrication in automobile valve trains. Proceedings of the Institution of Mechanical Engineers, Part J: Journal of Engineering Tribology 208(4): 221–234. Doi: 10.1243/PIME_PROC_1994_208_377_02.

Ting, L.L. 1980. Development of a laser fluorescence technique for measuring piston ring oil film thickness. Journal of Lubrication Technology 102(2): 165–170. Doi: 10.1115/1.3251458.

Tomanik, E. and Ferrarese, A. 2006. Low Friction Ring Pack for Gasoline Engines. (42606): 449–455. Doi: 10.1115/ICEF2006-1566.

Torabi, A., Akbarzadeh, S. and Salimpour, M. 2017. Comparison of tribological performance of roller follower and flat follower under mixed elastohydrodynamic lubrication regime. Proceedings of the Institution of Mechanical Engineers, Part J: Journal of Engineering Tribology 231(8): 986–996. Doi: 10.1177/1350650116684403.

Torabi, A., Akbarzadeh, S., Salimpour, M. and Khonsari, M.M. 2018. On the running-in behavior of cam-follower mechanism. Tribology International 118: 301–313. Doi: https://doi.org/10.1016/j.triboint.2017.09.034.

Totolin, V., Minami, I., Gabler, C. and Dörr, N. 2013. Halogen-free borate ionic liquids as novel lubricants for tribological applications. Tribology International 67: 191–198. Doi: https://doi.org/10.1016/j.triboint.2013.08.002.

Tung, S.C. and McMillan, M.L. 2004. Automotive tribology overview of current advances and challenges for the future. Tribology International 37(7): 517–536. Doi: https://doi.org/10.1016/j.triboint.2004.01.013.

Turturro, A., Rahmani, R., Rahnejat, H., Delprete, C. and Magro, L. 2012. Assessment of friction for cam-roller follower valve train system subjected to mixed non-newtonian regime of lubrication. Proceedings of the Spring Technical Conference of the Internal Combustion Division, Torino, Italy.

Vatavuk, J. and Ferrarese, A. 2013. Design of wear-resistant coatings for engine components. pp. 711–719. *In*: Wang, Q.J. and Chung, Y.-W. (eds.). Encyclopedia of Tribology, Boston, MA: Springer US.

Vernon, E.E., Stevenson, M.E., McDougall, J.L. and McCall, L. 2004. Failure analysis of gray iron pump housings. Journal of Failure Analysis and Prevention 4(6): 15–18. Doi: 10.1361/15477020421791.

von Kaenel, A., Grahle, P. and Abele, M. 2000. A new concept for steel-composite lightweight valves. SAE Technical Paper 2000-01-0906, 2000, https://doi.org/10.4271/2000-01-0906.

Xiuxu Zhao, Xiangyu He, Liting Wang and Peng Chen. 2020. Research on pressure compensation and friction characteristics of piston rod seals with different degrees of wear. Tribology International. Vol. 142. Doi: https://doi.org/10.1016/j.triboint.2019.105999.

Chapter 8

Friction and Wear in Extreme Conditions

8.1 Introduction

The primary task in addressing this chapter is to establish a definition for the term "Extreme Conditions". Indeed, there are many operating conditions that can be classified as extreme. Extreme operating conditions could be related to high loads and/or temperatures (e.g., cutting tools used in machining processes or ball valves used for cutting off oil/gas flow), or severe environments, such as in space. Also, they may be related to high transitory contact conditions (e.g., wheel/rail interface). In addition, extreme situations with near-impossible monitoring and maintenance opportunities (e.g., sea bed/sea installed ball valves for cutting off oil/gas flow, mechanical sub-sea oil pipe repair connectors, oil risers, off-shore wind turbines, and nuclear power components).

In general, extreme conditions can typically be categorized into those involving abnormally high or excessive exposure to cold, heat, pressure, vacuum, voltage, corrosive chemicals, particle and electromagnetic radiation, vibration, shock, moisture, contamination, or dust, or extreme fluctuations in operating temperature range (Clatterbaugh et al., 2011). Likewise, Pinchuk et al. (Pinchuk et al., 2002) suggested that the extreme conditions should include any device or system requiring a lubricant operating under any of the following conditions:

- Beyond the original machinery design specifications.
- Beyond the original machinery ambient parameters.
- Application in an environmentally sensitive location.
- Beyond the original lubricant design specification.

Operation in such extreme conditions requires tribologists to develop tribosystems that could meet these extreme requirements, which is often a challenge. This chapter will discuss selected examples of friction and wear challenges in extreme operating situations.

8.2 Challenges in Tribology

Tribology is the science of interacting surfaces in relative motion. This usually includes the study of friction, wear and, in some cases, lubrication. Since 1960, the work of tribologists has primarily been focused on solving the immediate industrial problems of wear and friction through a thorough understanding of lubrication mechanisms. The progression of humanity has introduced new technologies, devices, materials and surface treatments which require novel lubricants and lubrication systems. Likewise, the development of high-speed trains, aircrafts, space stations, computer hard discs, artificial implants, bio-medical systems and many other engineering systems, have only been possible through the advances in tribology.

Currently, we face new challenges in tribology, such as sustainability, climate change and gradual degradation of the environment. These problems require new solutions and innovative approaches. As humanity progresses, tribology continues to make vital contributions by addressing the demands for advanced technological developments, resulting in, for example, a reduction in fuel consumption and greenhouse gas emissions, an increase in machine durability and improvements to quality of life through artificial implants (Stachowiak, 2017; Hintze et al., 2020).

Also, the scientific development in the previous century changed the subjects of tribological studies. For example, it aimed at expanding mechanisms of friction and wear on the molecular level and at creating tribological materials capable of serving under the most adverse external impact conditions. The most popular tendencies in tribology development and areas of application for the last three decades of the previous century could be summarized in the following groups (Sviridenok et al., 2015; Barber, 2020):

- Structural, mechanical and tribological properties of prospective materials and coatings.
- Mechanisms, types and peculiarities of wear of tribo-couplings.
- Modelling the mechanisms of wear and frictional interaction.
- Wear resistance and durability of materials.
- Quantitative analysis of friction and wear processes.
- Instruments and equipment used in tribology.
- Tribological and rheological properties of lubricants.
- Surface parameters and properties with regard to tribological properties of materials.
- Tribological properties of polymeric materials and composites.
- Space tribology.
- Friction and wear in MEMS/NEMS.

8.3 Friction and Wear Under Extreme Loads

8.3.1 Wheel-rail Interactions

The principal feature that distinguishes a railway is steel wheels rolling on steel rails, as shown in Fig. 8.1. This interaction is extremely complex and is absolutely critical to the safe and efficient operation of railways. The tribological behavior of such a contact is affected basically by the vehicle speed, wheel slip, contact pressure, temperature rise, and environmental conditions. In fact, the contact area of a wheel-rail interaction is typically about 100 mm^2 and should support not less than 3 t (lightweight passenger coach), in some cases, however, it may support more than 17 t (heavy freight). Also, relative slip between the wheel and rail results in rapid temperature rises that can reach several hundred degrees Celsius in routine operation, and over 1000°C in extreme conditions (Kapoor et al., 2000).

Practically, a wheel and a rail meet at a contact patch that sees a severe stress and working environment. These conditions inevitably lead to wear, deformation and damage to the wheels and rails. Therefore, the effect of solid or liquid contaminants on the tribological response in the wheel/rail contact is of a particular relevance to a number of works (Arias-Cuevas et al., 2010; Wang et al., 2011; Lyu et al., 2015).

Khalladi et al. (2016) focus in their work on the identification of the various causes that contribute to the premature wear of the wheels in the presence of four contaminants: Sand, phosphate, sulfur and cement. Experimental friction tests were conducted using a pin-on-disk tribometer, maintaining the same sliding velocity and Hertzian pressure, respectively, at 0.1 m/s and 1000 MPa. The most interesting finding is that sand provides a higher adhesion between the wheel and the rail than sulfur, phosphate and cement contaminants. Silica grains that are found in sand, cement and phosphate penetrate within the wheel/rail contact and cause severe surface damage to the wheel and the rail materials and increase the wear rate of the wheel with an abrasive process.

Generally, due to the load (weight) of the wheel on the rail, a force exists normal to the plane of the contact within the wheel-rail contact patch. Tractions are also aroused in the plane of contact by the vehicle steering forces. This force system results in complex hydrostatic and shear stresses in the rail and wheel. Of greatest interest is the compressive contact stress normal to the plane of contact, which has a generally elliptical distribution and affects both wheel-rail wear and rolling contact fatigue (RCF), as shown in Fig. 8.2 (Kapoor et al., 2000).

Figure 8.1: Wheel-rail interaction.

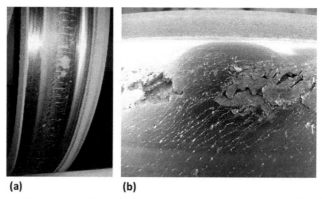

(a)   **(b)**

Figure 8.2: Surface fatigue cracks on the railway, (a) wheel tread; (b) rail head.

The formation of RCF damage of a wheel-rail is often associated with many factors, such as the wheel and rail materials, the contact stress, the lubricant and other factors. A greater problem may be an increased RCF, manifested as cracks, spalls, and shells on the rail surface. An increase in adhesion leads to an increase in the shear forces that produce ratcheting of the surface layers of the rail, leading to an increase in surface cracks. Therefore, many researchers have focused on the initiation, damage mechanism and evolvement characteristics of RCF surface cracks of a wheel-rail. Huang et al. explored the formation and damage mechanism of RCF surface cracks under dry conditions using a rolling-sliding wear testing apparatus (Huang et al., 2018). The cyclic vertical force ranged from 0 to 2000 N while applying a compressed spring. Firstly, tests were run for various cycles (1.5, 2.0, 3.0 and 4.0×10^5 Cycles) in order to explore the evolvement of RCF of wheel and rail materials. Then, the vertical force was changed in the middle of the tests in order to investigate the effect of a vertical force on the propagation mechanism of surface cracks. The first finding of this work is that the wear rate is affected by changing the vertical force. Under 400 N vertical force, the wear rates of wheel and rail rollers slowly increase. When the vertical force increases to 500 N in the later 1.5×10^5 cycles, the wear rate of rollers sharply increases and the hardness of rollers decreases. It is assumed that when the vertical force increases, the removal rates of materials become larger and exceed the hardening rate.

It is also demonstrated that the dependence of surface damage of rail and wheel rollers changes according to the number of loading cycles, as shown in Figs. 8.3 and 8.4. Cracks were observed after 1.5×10^5 cycles, and became much denser after 2×10^5 cycles. After that, the surface cracks damage relatively lessens due to the wear with an increase in the number of cycles.

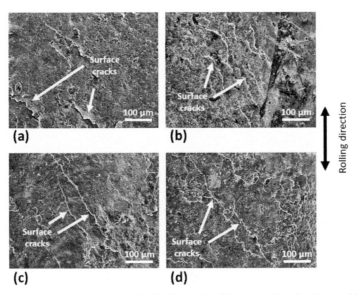

Figure 8.3: SEM micrographs of surface damage of rail rollers under different number of cycles conditions, (a) 1.5 × 105; (b) 2.0 × 105; (c) 3.0 × 105; (d) 4.0 × 105. After Huang et al. (2018) with permission from Elsevier.

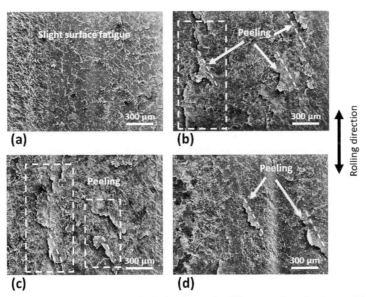

Figure 8.4: SEM micrographs of surface damage of wheel rollers under different number of cycles conditions, (a) 1.5 × 105; (b) 2.0 × 105; (c) 3.0 × 105; (d) 4.0 × 105. After Huang et al. (2018) with permission from Elsevier.

SEM micrographs show that fatigue cracks grow along a certain angle to depth on the rail rollers, but are easy to become parallel to the surface or turn towards the surface on the wheel rollers, which results in removing large pieces of material from the surface. Subsurface cracks can be also observed, which initiate on both sides of the principal cracks as well as near the crack tips, as shown in Fig. 8.5.

It should be assumed that when the vertical force of 400 N is applied, the contact stress of wheel-rail rollers is about 567 MPa, which exceeds the yield strength of both the wheel and the rail materials. In this case, each cycle of rolling contact load will bring a plastic deformation and material hardening. With the increase of cumulative plastic deformation, tiny fatigue cracks may initiate on the roller surface.

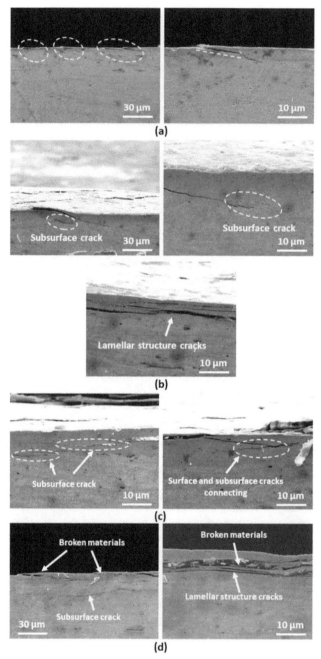

Figure 8.5: SEM micrographs of fatigue cracks of rail rollers under different number of cycles conditions, (a) 1.5×105; (b) 2.0×105; (c) 3.0×105; (d) 4.0×105. After Huang et al. (2018) with permission from Elsevier.

8.3.2 *Effect of Cyclic Loading on Wear of Polymers*

Under cyclic loading, the polymeric material suffers damage and deformation resulting from imperfect recovery in successive cycles, creep and stress relaxation effects and rise in temperature owing to the dissipation of energy in cyclic deformation. The effect of the cyclic parameters on the fatigue behavior of polymeric materials, in classical fatigue tests, is not far from its effect on the wear behavior under cyclic loading conditions. Working under cyclic loading conditions, as in many mechanical and biomedical applications, resulted in subsurface cracking in the highly strained regions (Cooper et al., 1994). During sliding or rolling of polymers, fatigue wear occurs when the polymer undergoes cyclic deformation (due to asperity interaction or as a result of cyclic loading). This form of wear causes separation of fragments from the surface by fatigue cracks, and also due to micro-cracking of the polymer surface because of intensive contact, mechanical and thermal effects. The subsurface fatigue cracks start below the polymer rubbing surface at a point where the shear stress has a maximum value. This would explain the occurrence of high wear factors in acetabular cups under cyclic loading, which are more than two orders of magnitude greater than these under constant load in equivalent wear tests (Barbour et al., 1995).

The effect of cyclic loading on occlusal contact wear and the possible presence of fatigue wear mechanisms in four composite resins (Silux, Z100, Ariston and Surefil) were also investigated by Yap et al. (Yap et al., 2002). Cyclic loading resulted in deep and wide microcracks in Silux composite, which may lead to catastrophic failure. Fine microcracks were also observed on the surface of the tested composites. Furthermore, it was realized that there is a correlation between the number of cycles and wear.

Since the polymer fatigue behavior is related to the number of loading cycles and sliding parameters, variation of one or both of them may affect the wear behavior of the polymer. This indicates the need for more understanding of the effect of the load frequency on mechanical and fatigue properties of polymers, and its effect on wear behavior. At high frequencies, increasing the frequency produces an increase in temperature, leading to a decrease in mechanical properties and fatigue safe-life, while relatively low frequencies have a limited effect on fatigue properties of the polymer. In contrast to the wear process, relevant fatigue properties are those of a surface layer of the polymer that may have been drastically modified during sliding.

The formation and propagation of cracks would be expected to depend on the frequency and waveform of the fatigue cycle (Abdelbary, 2014). Under cyclic loading, the macroscopic polymer asperity is cyclically deformed at the frequency of the loading cycle, and this may produce crack propagation and increase the macroscopic polymer asperity wear and surface fatigue wear. Results of wear factors (*WF*) at different frequencies of cyclic loads show a clear increase in the wear volume with a frequency increase, as presented in Fig. 8.6 (Abdelbary, 2004). On the other hand, the propagation of subsurface microcracks was accelerated by heating the crack tip, which directly leads to an increase in the wear of the polymer. A useful observation is that there is a significant increase in the total worn volume due to the applied frequency. The result of

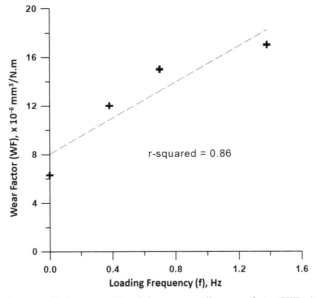

Figure 8.6: Relation between cyclic frequency (f) and the corresponding wear factor (WF). Adopted from Abdelbary (Abdelbary, 2014) with permission from Elsevier.

such an observation is that, in many mechanical and medical polymeric applications, the total volume loss may have a superior priority in calculating the life time of a component, rather than the wear factor.

8.4 Micro/Nano-Scale Friction and Wear

The fundamental difference that distinguishes micro/nano tribology from classical macro tribology is that micro/nano tribology considers the friction and wear of two objects in relative sliding whose dimensions range from micro-scales down to molecular and atomic scales. MEMS refer to micro-electromechanical systems that have a characteristic length of 100 nm to 1 mm, while NEMS are the nano-electromechanical systems that have a characteristic length of less than 100 nm (Mylvaganam and Zhang, 2011). Possible applications of MEMS that would allow contact between surfaces include gears and transmissions, as presented in Fig. 8.7. The NEMS field, in addition to fabrication of nanosystems, has provided impetus to development of experimental and computation tools. As an example, Fig. 8.8 shows an atomic force microscopy (AFM) based nanoscale data storage system for ultrahigh density magnetic recording which experiences tribological problems (Vettiger et al., 1999). It is clear that tribology is an important factor affecting the performance and reliability of MEMS and NEMS devices.

There are great challenges in the development of a fundamental understanding of tribology, surface contamination and environment in MEMS/NEMS. One of these challenges in such extreme tribological situations is the adhesion force, which can be up to a million times greater than the force of gravity. This is due to the fact that the adhesion force decreases linearly with size, whereas the gravitational force decreases with the size cubed.

In MEMS devices, when the length of the machine decreases from millimeters to microns, the area decreases by a factor of a million and the volume decreases by a factor of a billion. Therefore, micro objects adhere to their neighbors or surfaces, which is an obstacle in the miniaturization of components (Kendall, 1994). Furthermore, surface forces, such as friction, forces, viscous drag and surface tension that are proportional to area, become a thousand times larger than the forces proportional to volume, such as inertial and electromagnetic forces (Sundararajan and Bhushan, 1998). Therefore, tribologists must understand the origins of friction, adhesion and wear over a broad range of length scales, from the macroscopic to the molecular scales, in order to guarantee the function and reliability of MEMS devices.

Regarding adhesion problems, they are often expected in micro- and nanoelectromechanical systems. In order to overcome the adhesion in these systems, reduction in adhesion and friction were realized by applying the principles of surface chemistry (Bhushan, 2005). Low-surface-energy, hydrophobic coatings applied to oxide surfaces are promising for minimizing adhesion and static-charge accumulation (de Boer, 2001). Typically, these are very thin or monolayer organic coatings, either physisorbed or covalently bound to the surface. Nonpolar organic groups on the surface exhibit low adhesive energy and low static-charge accumulation. In addition, coatings must be compatible with subsequent device processing, including packaging processes involving thermal treatments of 400 to 500°C. Examples of coatings that have been successfully introduced to commercial MEMS products is the perfluorodecanoic acid coating.

(a)   (b)

Figure 8.7: (a) Advanced MEMS have numerous contacting and sliding interfaces that can be subject to a range of loads. (b) SEM image of wear debris in the ~ 10 μm receiver hole for a failed drive gear of a Si-MEMS device. After Tanner et al. (1999), with permission from Elsevier publishing.

In mid-1980, advances in microfabricating Polycrystalline silicon (polysilicon) technology led to the development of MEMS. Mechanical applications for MEMS include acceleration, pressure, flow, and gas sensors, linear and rotary actuators, and other microstructures of micro-components, such as electric motors, gear trains, gas turbine engines, nozzles, fluid pumps, fluid valves, switches, grippers, and tweezers (Bhushan, 2005). Data recording head, as illustrated in Fig. 8.9, and moving mechanical arrays (MMA) are also practical applications of MEMS. Nevertheless, further development of MMA has been restricted due to the poor tribological properties of polysilicon. Today, MEMS are fabricated using non-lithographic micromachining processes such as LIGA (a German term for lithography, electroforming, and plastic moulding) and other laser machining processes.

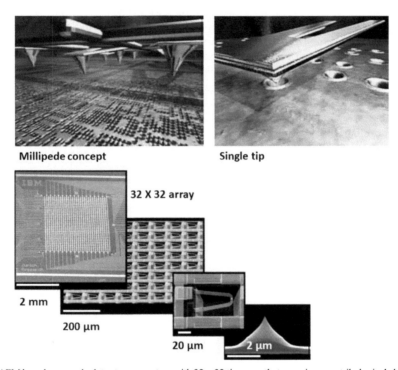

Figure 8.8: AFM based nanoscale data storage system with 32 × 32 tip array-that experiences a tribological challenges. After Vettiger et al. (1999), with permission from Elsevier publishing.

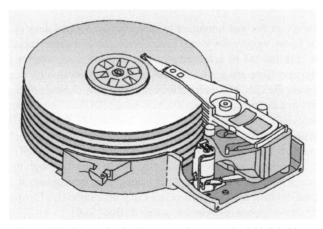

Figure 8.9: Schematic of a data-processing magnetic rigid disk drive.

On the other hand, NEMS are produced by nanomachining in a typical top-down approach, by removing material atom by atom, and bottom-up approach by building blocks to produce three dimensional nanostructures (nanocrystals) (Bhushan, 2004). Examples of NEMS include nanocomponents, nanodevices, nanosystems and nanomaterials such as microcantilever with integrated sharp nanotips for data storage. Examples also include, tips for nanolithography, dippen nanolithography for printing molecules, biological (DNA) motors, molecular gears, and molecularly-thick films.

Friction and wear of ultra-small moving components are highly reliant on the interaction between moving surfaces. Until recently, little work has been devoted to studying the friction and wear problems of ultra-small moving components (MEMS/NEMS). The reason is possibly due to a lack of effective test methods that are relevant to the operational conditions of proposed MEMS devices, in which case loads of a few micro/milli-Newtons are typical at contact areas and moving distances of a few microns, contact pressures of tens to hundreds of MPa with wear depths limited to nm (Charitidis, 2013). Now, the study of tribology at micron and nanometer scales has become experimentally possible with the invention of surface force apparatus (SFA), scanning tunnelling microscope (STM), and atomic force microscopy (AFM) as well as friction force microscopy (FFM) (a subsequent modification of AFM).

Subhash et al. (Subhash et al., 2011) investigated the tribological behavior of MEMS devices made of polycrystalline silicon using a surface-micromachined nanotractor device. In this study, twenty-five devices operating for several days were tested with different test parameters. The first device was subjected to loaded wear cycling with 35 μN drag force and friction tests at large intervals. This device operated for 700,000 wear cycles. A second device was subjected to unloaded wear testing and friction tests at more frequent intervals. The first observation was that, after 252,000 cycles, the device became adhered during the friction test (device failure). Moreover, during the wear testing, evidence of changes in the evolving surface conditions could be inferred from the travel macroscopic characteristics of the nanotractor.

Surface features that correlate to the above macroscopic travel were examined using a high-resolution scanning electron microscope (SEM). The collective remark is that the wear intensity is highly nonuniform along the contact surfaces. However, the overall features were similar in all devices irrespective of the test conditions. Based on this observation, the mechanisms responsible for material removal during the wear process are classified into:

(i) adhesion-dominated wear, as shown in Fig. 8.10, in which the contact regions and wear tracks appear polished on the surface of the friction pad. The possible reason for this observation is that the asperities were sheared off or were plastically deformed during early wear cycling.

(ii) third-body wear (associated with device failure), where the agglomerates formed in the previous regime are trapped between the friction pad and the wear track. The applied normal force causes these agglomerates to stick to one of the contact surfaces and form debris patches along the wear track, as shown in Fig. 8.11. The severe wear in this localized region caused the debris patch on one of the clamps to get lodged in the trench of the counter surface. Finally, the clamps undergo "severe wear" characterized by a trench formation where the contact surfaces get locked and cause device failure.

Friction and wear issues become important in sophisticated MEMS devices which employ sliding contacts in rotary- or linear motion devices. These devices have a great appeal because of the high functionality at a low cost that can be achieved with gears, linear racks, rotating platforms, and pop-up mirrors. Organic monolayer films are again the focus of efforts to develop lubricants for MEMS. These films have low friction coefficients, and their performance naturally depends on the chemical nature and/ or structure of the films. For example, Octadecyltrichlorosilane (ODTS) films show a lower initial friction coefficient than shorter-chain fluorinated (FDTS) films, presumably due to the higher packing density and ordering of the hydrogenated alkyl groups. The friction coefficient of organic film-modified surfaces increases after moderate periods of applied shear force, and the surfaces exhibit significant wear. These observations suggest that organic monolayer films may not be robust enough to serve as lubricants in devices requiring extended sliding contact. Other hard coatings may be necessary to prevent extensive wear in sliding contacts. A self-limiting conformal chemical vapor deposition (CVD) of tungsten on polysilicon to form a hard coating shows superb resistance to wear (de Boer, 2001).

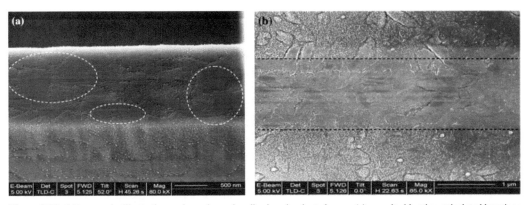

Figure 8.10: Micrographs illustrating various phases in adhesion-dominated wear: (a) asperity blunting at isolated locations of a clamp and (b) thin film formation on wear track. After Subhash et al. (2011) with permission from Elsevier.

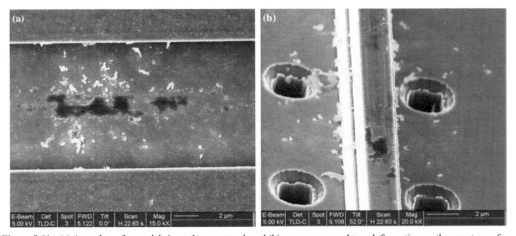

Figure 8.11: (a) A patches of wear debris on the wear track and (b) severe wear and trench formation on the counter surface of the friction pad. After Subhash et al. (2011) with permission from Elsevier.

8.5 Friction and Wear at Extreme Temperatures

In many tribological applications, the system components are exposed to harsh operating conditions, such as very high or ultra-low temperatures. Examples of such applications can be found in the aerospace, mining, power generation, and metalworking industries. The distinguishing line between what is a high or low temperature is rather vague and markedly related to the application, but most importantly to the materials involved. In this regard, while certain ceramics exhibited good tribological and mechanical properties at up to 2000°C, this temperature is already above the decomposition or melting point for most polymers and metallic alloys. Thus, the idea of homologous temperature, T_{hom}, was introduced in order to categorize hot and cold working operations. This can be defined as the working temperature of the material normalized by its melting point in an absolute temperature scale (K). If the result fraction is less than 0.4, then it is considered that thermally activated mechanisms do not take place and the operation is classified as cold working (Hernandez et al., 2015; Wheeler et al., 2015).

In tribology, an application can be considered to operate at elevated temperatures when the use of conventional lubricants, i.e., oils and greases is no longer effective due to their rapid decomposition at around 300°C. In the case of metals and alloys, the introduction of elevated temperatures into a tribological system involves several chemical and physical phenomena occurring sequentially or even simultaneously, as presented in Fig. 8.12.

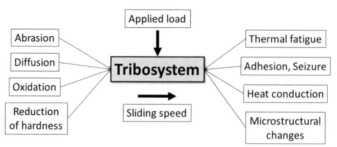

Figure 8.12: Phenomena occurring in a tribological system at elevated temperature. After Hernández (Hernández, 2016).

Many components for aerospace, electronics and defense industries are made out of superalloys (nickel-, titanium- or cobalt-based alloys) mainly due to their good mechanical strength and resistance to surface degradation under severe operating conditions (Opris et al., 2007; Xu et al., 2003). Nevertheless, they can show a relatively poor abrasive wear resistance at elevated temperatures due to chemical reaction with abrasive particles.

Steel remains the first choice for several applications due to good mechanical and tribological properties, abundant resources and relatively low cost of processing. Tool steels are iron based alloys which can exhibit high hardness and good wear resistance even at temperatures around 600°C by a controlled chemical composition and proper heat treatment. Accordingly, this allows for their implementation in high temperature applications, such as hot forming processes, cutting tools, forging and plastic moulding. These steels are treated, by quenching and tempering processes, in order to obtain a microstructure containing a tempered martensite matrix with evenly distributed carbides. Primary undissolved carbides form during the solidification process and can increase hardness and wear resistance, however they can also act as crack initiation sites, reducing toughness (Hojerslev, 2001). Tempering of steel in the range of 400 and 600°C allows for the diffusion of substitutional elements such as Mo, Cr or V through the iron lattice, resulting in the formation of secondary fine carbides. The formation of such carbides is accompanied by an increase in hardness and wear resistance. This process is called secondary hardening (Bhadeshia and Honeycombe, 2006).

8.5.1 Mechanisms of Wear at High Temperatures

Tribological systems operating in severe conditions, like high temperatures, are subjected to different phenomena, such as friction, wear, oxidation and changes in mechanical properties. Usually, there are several mechanisms occurring simultaneously. The prevailing wear mechanism present will depend on the materials in contact, surrounding media, and operating conditions. It is worth noting that abrasive and adhesive wear were identified as the most important types of wear occurring in industry. The presence of high temperatures in the tribosystems can increase the severity of these wear mechanisms. In the investigation of the tribological behavior of surface-treated and post-oxidized tool steels, Syed et al. (Syed et al., 2010) suggested that the wear mechanisms are predominantly adhesive at room temperature and a combination of abrasive and adhesive at higher temperature.

Adhesive wear is generally caused by the formation of adhesive bonds between asperities of mating surfaces. At high temperatures, the strength of metals decreases and their ductility increases, which, in turn, increases the number of bonds and the severity of adhesion. In addition, the detached wear debris may accumulate and mix, forming built-up particles. These can adhere or embed into one of the surfaces and cause grooves by ploughing into the counter surface (Stachowiak and Batchelor, 2006). As well as to the abrasion caused by built up particles, three body abrasive wear at high temperatures can also occur due to the presence of hard oxides or oxidized wear debris. In many industries, such as in mineral handling, the abrasive particles can also come from an external source. Therefore, the performance of the tribological system can be affected by the mechanical properties of the abrasive particles, their shape and how they interact with the surfaces in contact (Hardell et al., 2014). On the other hand, since metals soften at high temperatures, the abrasive particles can get embedded in the surface and form a tribolayer, which can increase the wear resistance of the bulk material. It was also demonstrated that the thickness and appearance of tribolayers formed at high temperatures depend on the microstructure and hardness of the materials involved (Varga et al., 2013; Soemantri et al., 1985). Thus, soft materials are prompted to form more uniform tribolayers in abrasive environments, increasing wear resistance at high temperatures.

A typical application for the above discussion is the influence of working conditions to the mechanical properties of the bulldozer track B1 (30MnB) steel pads used for the slag removal in the steel plant. Brkovski et al. (Brkovski et al., 2013) observed that, over a period of 6 months, almost 8 mm of track pads were removed (worn off). While for the same period of time, only 3 mm of the track pads were removed if the same bulldozer worked at fine grain sand separation. In fact, working in a steel plant is a very special case and extreme working conditions are present. The elevated temperature is higher than 400°C (sometimes much higher) and the media is slag with different chemical composition and mechanical properties compared with earth. Nonetheless, the red color of slag shown in Fig. 8.13 confirms that in some periods of work the temperature was much higher. Such conditions contribute in more intensive wearing of bulldozer parts mainly track pads which are in direct contact with the hot slag.

In the study by Brkovski et al., metallorgraphic investigation was performed to old and new track pads for microstructure comparison. Also, nine hardness measurements were taken from each specimen (old and new) at different positions, as can be seen in Fig. 8.14. Tensile strength testing was performed on both old and new track pads at room temperature and at an elevated temperature of 300°C.

Figure 8.13: Working conditions in steel plant. After Brkovski et al. (2013).

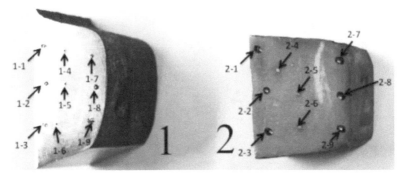

Figure 8.14: Metallographic specimens of track pads and hardness measuring points, sample (1) new and sample (2) old. After Brkovski et al. (2013).

Metallographic investigation of new and old parts of the track pads confirmed that in both cases oxide inclusions were present. Hardness measurement of the used track didn't show any significant difference at different locations of the track pad. Besides, there was no significant difference between hardness of the used and the new track pads. So this high strength steel poses excellent wear resistance at ambient temperature. At elevated temperatures, however, the strength and wear resistance are lowered. This means that heating and consequently softening happen only at contact surface.

8.5.2 Effect of Surface Oxidation

Due to exposure to the oxygen in air, surface oxide layers are present in almost all metals. The rate of oxidation is governed by the interaction of oxygen and metal ions which is, in turn, controlled by temperature. In general, chemical reaction rates are increased at high temperatures. A general theory of oxidational wear was developed by Quinn (Quinn and Winer, 1985). The theory does not only take into account oxidation caused by frictional heating through asperity contacts, but also the one occurring when the sliding surfaces are exposed to a high temperature. It was postulated that the tribological activation energy for oxidation was about half of that needed for the static oxidation of ferrous materials.

The formation of oxide layers during sliding of tribo-surfaces is not only dependent on the natural oxides but also on the *in situ* formed ones. Nevertheless, the presence of oxides at the interface of tribo-surfaces does not necessarily need to be beneficial in terms of wear and friction behavior. The detachment of a hard and brittle layer may result in increased wear through abrading particles, whereas a ductile oxide layer can accommodate more plastic deformation in the presence of tangential stresses and normal loads. Finally, the formation of an oxide layer may be beneficial. If the strength of the matrix is high enough to support the oxides, then these can provide protection against wear. But, on the other hand, if the matrix is soft or has been softened, the presense of tribo-oxides does not reduce wear and they may even act as three body abrasives.

The role of oxide layers at elevated temperatures was discussed by Hernandez et al. (Hardell et al., 2014; Hernandez et al., 2015). In this work, high temperature tribological studies of boron steel sliding against tool steel were conducted using a pin-on-disc machine. The tests were performed under unlubricated conditions at temperatures ranging from 25 to 400°C, under three different loads: 25, 50 and 75 N and a sliding speed of 0.2 m/s. Figure 8.15 shows the overall influence of different loads and temperatures on friction where the average value of the friction coefficient at the steady state region was considered. It is clear that, at a given load, the average friction coefficient decreases as the temperature is increased. A maximum friction coefficient value of about 1.2 is observed at room temperature and 25 N, whereas the minimum value is found at 75 N and 400°C. It is demonstrated that the decrease of the friction coefficient is related to the development of oxidized protective layers on the worn surfaces of the materials in contact.

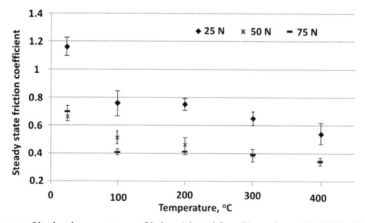

Figure 8.15: Influence of load and temperature on friction. Adopted from Hernandez et al. (2015) with permission from Elsevier Publishing.

Throughout high temperature tests, the drop in friction can be explained by the formation of a protective oxidized layer. As the temperature is raised, more particles can get oxidized and sintered, thus forming a glaze layer that further decreases friction. In addition, the effectiveness of such layers against wear and high friction is governed by the composition and type of oxides formed, which at the same time depend on several factors such as temperature, load, sliding speed and the materials in contact.

Scanning electron micrographs (SEM) of the boron steel surface revealed adhesion as the main wear mechanism from the test carried at 25°C. As a result of the lower hardness, the boron steel suffered more plastic deformation. Consequently, the deformed asperities became detached and mixed with the wear particles from the tool steel material, forming agglomerates or transfer particles. These built-up particles produced grooves caused by ploughing on both surfaces, as shown in Fig. 8.16.

At 200°C, the wear track of the boron steel pin was protected by a more continuous and smooth oxide layer compared to that exposed to 100°C, resulting in a decrease in the wear rate. At 300°C, the presence of grooves caused by adhesive wear on the tool steel worn surface decreased compared to the ones seen at 200°C. As a consequence of the temperature rise, the sintering of the wear debris increased. A common observation at 100 to 300°C, is that the increase in the load to 75 N facilitated the detachment rate of the oxide layers, causing a slight increase in the wear rate. At 400°C, the existence of a continuous and smooth oxidized layer covering most of the wear track of the boron steel pin was observed under 25 and 50 N, as shown in Fig. 8.17. Also, an increase in the load to 75 N increased the rupture of the oxide layers, thus generating more oxidized wear debris. This, in turn, acted as three body abrasive particles and increased the wear rate.

The obtained data were used to develop the friction and wear mechanisms map for boron steel and tool steel, as shown in Fig. 8.18. The map provides an overview of the tribological behavior by presenting friction and observed mechanisms as a function of the temperature and contact pressure. The predominating tribological mechanisms observed during certain conditions are represented in different colors. The boundaries between the wear mechanisms cannot be well determined, hence, more experiments with intermediate temperatures and loads are necessary in order to define them with more accuracy.

It is worth considering that plasma nitriding and post-oxidation of nitrided and nitrocarburized steels is an economical way of improving the tribological and mechanical properties of tool steel. Experimental studies were carried out in order to reveal the friction and wear behavior of different tool steels (plasma nitrided and post-oxidized at different temperatures) sliding against ultra-high strength boron steel at room

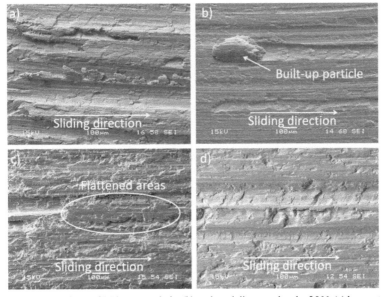

Figure 8.16: SEM of worn surfaces of (a) boron steel pin, (b) tool steel disc tested under 25 N, (c) boron steel pin and d tool steel disc tested under 50 N, at 25°C. After Hernandez et al. (2015) with permission from Elsevier Publishing.

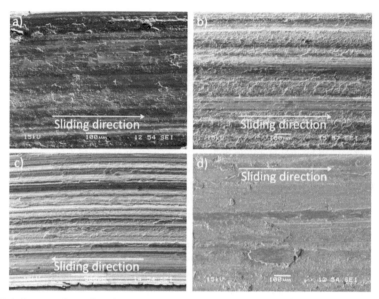

Figure 8.17: SEM of worn surfaces of specimens tested at 400°C: boron steel pin under (a) 25 N and (b) 75 N; tool steel disc under (a) 25 N and (b) 75 N. After Hernandez et al. (2015) with permission from Elsevier Publishing.

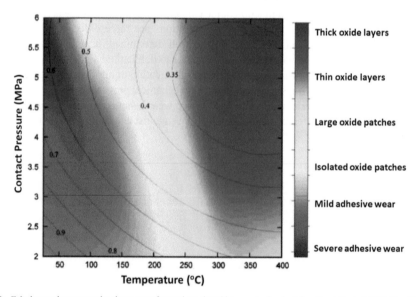

Figure 8.18: Friction and wear mechanisms map for tool steel and boron steel pair. After Hardell et al. (2014) with permission from Taylor & Francis Publishing.

and elevated temperatures (Syed et al., 2010). It was reported that the tool steel, post-oxidized at 500°C, resulted in a better friction and wear performance at room temperature and also improved wear resistance at an elevated temperature. Results indicate that the operating temperature, composition and post-oxidation temperatures have significant influence on the tribological behavior of the tool steel during sliding against ultra high strength boron steel.

The role of an oxide layer is also distinguished in metal forming processes at high temperatures. In fact, the main problems occurring during hot forming processes are high friction and wear, and surface initiated fatigue. These create a great challenge to achieve a high quality product and an extended tool life. Friction and wear mechanisms of oxide layers in hot rolling and hot stamping in different processes and conditions on high strength steel were investigated by Dohda et al. (Dohda et al., 2015). It was found that the development of layers at the interface of two materials contributed to the reduction of friction and wear rate. Surface coatings on both tool and workpiece helped in improving friction and wear at elevated temperatures. However, heat treatment, surface roughness, and coating layers had influences on the performance of tool surface coatings. Al–Si and Zn–Ni coatings were commonly applied to high strength steel sheets for applications at elevated temperatures.

8.5.3 Effect of Applied Load

In fact, most of the literatures covered the effect of a load on the wear of metallic materials are restricted to room temperature (RT) testing. However, Wang et al. (Wang et al., 2010) introduced a pioneer work to study the wear behavior of cast steel for several different microstructures under varying normal loads up to 400°C. Recently, Torres et al. (Torres et al., 2016) conducted a further study to investigate the role exerted on wear behavior by changing contact loads and stresses during a prolonged high temperature sliding. To this end, a series of wear tests were performed at different loads using several iron-based materials. Three different iron-based alloys were chosen for testing: Ferritic/pearlitic steel, martensitic hardfacing and chromium carbide-rich hardfacing. The results at room temperature indicated no discernible effects of normal load on the calculated wear rates for any of the chosen materials. However, significant differences were found in some cases at high temperatures, as shown in Fig. 8.19. In particular, the wear rates for the ferritic/pearlitic steel at the lower load of 45 N showed a minimum at 500°C, compared to the value measured at room temperature. A subsequent increase was found at 700°C, reaching the maximum measured value. At 130 N, the ferritic/pearlitic steel grade performed differently, as the normalized wear rates remained almost constant within the entire temperature range. In the case of the martensitic hardfacing, it was observed that the wear rates clearly decreased at high temperatures compared to room temperature for both applied loads. Wear rates for the hypereutectic hardfacing were found to be almost temperature independent at 130 N. However, at the lower load of 45 N, a significant increase in the wear resistance was found after testing at high temperatures. It was concluded that, at high temperatures, the wear rates were observed to be load-dependent, with lower values corresponding in general to lower loads. Also, high temperatures in combination with higher applied forces led to higher wear rates for the ferritic/pearlitic steel and the martensitic hardfacing.

    An et al. (An et al., 2017) also introduced a study regarding the effect of an applied load on the tribological behavior of $Mg_{97}Zn_1Y_2$ alloy at high temperatures. In addition, the roles of friction-induced microstructural evolution and hardness change in subsurfaces on the wear transition were also investigated. It was demonstrated that, at temperatures of 50–150°C, the friction coefficient went down rapidly, then decreased slowly with an increasing load in the mild wear regime, and finally decreased modestly until the largest applied load in the severe wear regime. At a temperature of 200°C, the coefficient of friction

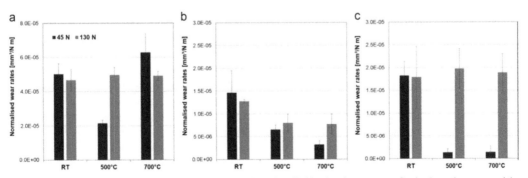

Figure 8.19: High temperature sliding wear rates as a function of applied load and temperature for the three chosen materials: (a) ferritic/pearlitic steel, (b) martensitic and (c) hypereutectic hardfacings. After Torres et al. (2016), with permission from Elsevier publishing.

decreased rapidly while increasing the load in the mild wear regime, and further decreased to the lowest value in the severe wear regime before rising up to a high value and decreasing again. The main wear mechanisms were identified as abrasion plus oxidation and delamination plus surface oxidation in the mild wear regime and severe plastic deformation plus spallation of oxide layer and surface melting in the severe wear regime. At 200°C, the main wear mechanisms were delamination in the mild wear regime, severe plastic deformation and surface melting in the severe wear regime. The reason for the transition from mild to severe wear of the tested alloy at high temperatures of 50–200°C was softening in the subsurface.

8.5.4 Effect of Surface Texture

Surface texturing is a surface modification approach used to enhance the tribological performance of mechanical components with artificial topography. It is well known in many applications, such as cutting tools, cylinders, cylinder liners, bearing bushes, mechanical seals, piston rings and others. Surface texture would reduce both the friction coefficient and the degree of wear under oil lubrication (Li et al., 2015). In unlubricated sliding, Sugihara et al. (2012) reported that a groove depth of 5 μm denotes the lowest adhesion and cutting temperature of cutting aluminum alloys. The highest temperature reduction of 103°C was observed with a proper design of the groove texture in comparison with the smooth surface. In addition, the friction coefficient and wear performance improved with a proper groove design. It is generally concluded that the groove surface texture improves the tribological properties under the extreme tribological conditions.

Under high temperature conditions, the groove surface textures have a positive effect on the performance of thermal fatigue, but the magnitudes of the improvements are different due to both the parameter of textures and the material properties (Zhou et al., 2006). The main mechanisms are the release of thermal stresses, and a reduction in the probability of a crack initiation and propagation.

Wu et al. (Wu et al., 2016) investigated the effect of surface textures on the tribological behavior of nitrided 316 stainless steel. In this study, three types of groove surface textures with different parameters were machined by a laser machine, as represented in Fig. 8.20 and Table 8.1. The sliding friction and wear performance of the textured and untextured surfaces were conducted under different high temperature and other working conditions. The tribological characteristics were evaluated in terms of the friction coefficient, wear scar width, microphotography of the wear surface, and their chemical composition analysis.

According to the experimental results, texture (1) shows the lowest friction coefficient under the C1 condition, Table 8.2, and the content reduction reached approximately 14% compared to the untextured one. Similarly, texture (1), texture (2) and texture (3) show the lowest friction coefficients under the C2, C3, and C4 experimental conditions, respectively. Meanwhile, the highest reductions in the friction coefficients were approximately 14%, 6%, and 10% under the C2, C3, and C4 conditions, respectively. Generally, the groove surface texture decreases the friction coefficient compared to the untextured surface to some extent. The reason is that a surface texture releases the high heat stress under the high temperature conditions. The thermal stress at high temperatures easily causes cracking and flaking of the surface, which rapidly increases further serious friction and wear. Simultaneously, different surface textures showed a disparate ability to release the heat stress under distinctive experimental conditions. Therefore, the surface texture demonstrated a dissimilar influence of the friction coefficient under different high-temperature friction conditions.

In fact, a friction reduction is more easily achieved by proper design of the groove surface texture, at a temperature of 500°C and a high oscillation frequency, than at 300°C and a low oscillation frequency. The mechanism is the ability to vary the heat stress release and wear debris storage by different groove surface textures under distinct experimental conditions. It is further postulated that more stress was released with an increasing groove width. Therefore, it can be considered that a wider groove width may be beneficial for a greater release of thermal stress under high-temperature friction conditions. On the other hand, the ability of wear debris storage is enhanced as the grooved texture width increases, and it has a positive effect on reducing serious wear. Also, the textured area can store wear debris in order to avoid further abrasive wear in the frictional contact area. However, it can be seen in Fig. 8.21 that there is more serious wear at 300°C than at 500°C.

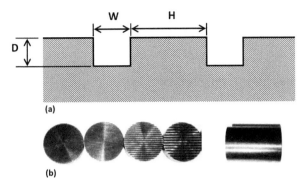

(a)

(b)

Figure 8.20: (a) Schematic diagram of the surface texture and (b) Photograph of some specimens. After Wu et al. (Wu et al., 2016), with permission.

Table 8.1: Geometric parameter of the groove textured surface. After Wu et al. (Wu et al., 2016).

Specimen	D (μm)	W (μm)	H (μm)
Texture (1)	90 ± 5	200 ± 10	370 ± 10
Texture (2)	90 ± 5	400 ± 10	740 ± 10
Texture (3)	90 ± 5	600 ± 10	1110 ± 10
Untextured	0	0	0

Table 8.2: Parameter of experimental arrangement. After Wu et al. (Wu et al., 2016).

Number	Load, (N)	Temperature, (°C)	Frequency, (Hz)	Sliding speed, (mm/s)	Stroke, (mm)
C1	40	500	20	20	0.5
C2	40	300	20	20	0.5
C3	40	500	10	20	1
C4	40	300	10	20	1

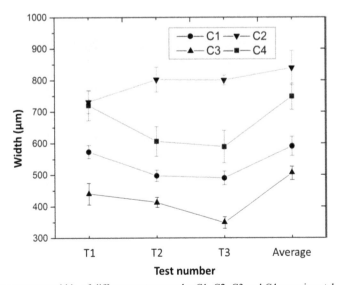

Figure 8.21: Average wear scar width, of different textures under C1, C2, C3 and C4 experimental conditions. After Wu et al. (Wu et al., 2016), with permission.

It is concluded that under the extreme conditions of a high temperature and a dry friction, it is helpful to increase the groove texture width by the mechanisms of thermal stress release and wear debris storage in order to improve wear performance.

8.5.5 High Temperature Tribometers

In general, tribometers (or tribotesters) are used as instruments that perform tribological measurements, such as the coefficient of friction, friction force, wear volume, and contact temperature between two rubbing surfaces. The investigation of friction and wear behavior of materials at elevated temperatures has become increasingly important, especially for the development of power plants and internal combustion engine industries. High temperature tribometers are used for studying friction and wear properties at elevated temperatures, up to 1500°C. Such types of tribometers generally obtain high temperatures by using insulated chamber or furnace equipped with heating elements.

8.6 Friction and Wear at Extremely Low Temperatures

Low temperature is an important factor in the tribo-mechanical and thermodynamic characteristics of materials. This includes slipping behaviors, resistance to wear or abrasion, and other tribo-characteristics. A practical example for these cases is the viscosity change of the gearbox oil in a diesel engine during the winter season, which can influence the cold-start abilities of engines at sub-zero temperatures. Also, the condition of a road in freezing winter, where the weather creates slippery ice, snow, and slush on the road. Today, with the remarkable development of the railway industry, many trains work in extremely low temperatures, as low as –45°C (Wang et al., 2012). Consequently, a consistent investigation of the friction and wear behavior in such temperatures is required in order to enhance the understanding of the tribological materials performance for low temperature applications.

Many researches on friction and wear at low temperatures, in connection with space applications, have been conducted in the last decade (Gradt et al., 1998). In cryogenic conditions, components with interacting surfaces in relative motion, like bearings, seals and valves, cannot be lubricated conventionally using oils and greases, and thus, they are critical with respect to wear and frictional heat generation. Furthermore, in a hydrogen environment, other problems for tribosystems arise because of its ability to reduce protective oxide layers as well as the ductility of the materials of interest.

8.6.1 The Challenge in Wheel/Rail Interaction

In cold regions, tribology of the railway transport, as an open system, can be challenging in terms of friction, adhesion and wear of the contact surfaces. In addition, a safety issue due to a varying coefficient of friction that prolongs the stopping distance should be considered.

In the winter of 2017, several incidents regarding the stopping distance of freight wagons equipped with composite brake blocks occurred in the north of Sweden at temperatures as low as –10°C. It seems as if the friction coefficient of composite brake blocks decreases considerably at low temperatures, resulting in a prolonged braking distance and near misses. A similar scenario has also been noticed in Finland. A study by Lyu et al. (Lyu et al., 2019) addresses the difference in friction and wear from cast iron, sinter and composite railway brake blocks at low ambient temperatures. In this study, three different railway brake block materials, cast iron, sinter and composite, were investigated. Brake specimens were tested at five different temperature levels: 10°C, 3°C, –10°C, –20°C and –30°C. Figure 8.22 indicates that the cast iron brake block has a transition in friction coefficient with the decreasing temperature from 10°C to –30°C. At –10°C and –20°C, the friction coefficient of a cast iron brake block is much lower than at the other three temperatures. The composite brake block has the lowest friction coefficient at all five temperatures. It may also be noted that the friction coefficient of the composite brake block decreases rapidly in subzero temperatures. For the sinter brake block, the friction coefficient remains stable at all five temperatures and seems not to be sensitive to the change of temperature. It is demonstrated that the composite brake

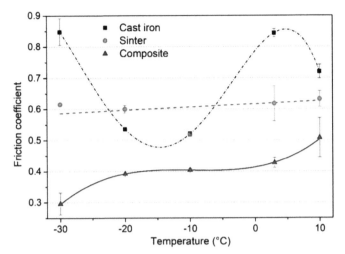

Figure 8.22: Friction coefficient (mean value and standard deviation) of three brake blocks as a function of temperature. After Lyu et al. (2019), with permission from Elsevier.

material contains a large portion of a phenolic resin, which is prone to adsorbing water vapor. The adsorbed water vapor forms an ice condensation layer on the composite brake block, which will act as a lubricant and reduce the friction coefficient.

Results of wear tests show that at –10°C and –20°C, the cast iron specimens and counterface yielded extremely high wear (one magnitude greater than at other temperatures), showing an opposite behavior to the friction coefficient (at –10°C and –20°C, cast iron has a lower friction coefficient than at the other three temperatures). This could be attributed to the graphite removal from the cast iron brake block into the interface acting as a solid lubricant and somewhat relieving the friction between the specimen and the counterface. The transferred graphite is only observed at –10°C and –20°C, and not at –30°C. With a further decrease of temperature, there is a pronounced tendency to have an ice condensation layer on the metal surface. This condensed ice layer encourages the generation of oxide layers, protecting the contacting bodies from severe wear (Cornell and Schwertmann, 2003).

Relatively stable wear at different temperatures is detected in sinter brake block specimens and its mating disc, similar to its performance in the friction coefficient. Finally, brake specimens of the composite block demonstrate a very minor wear at all five temperatures. It is important to mention that, at –30°C, the disc sample tested against composite brake block has a negative wear, indicating that material was transferred to the disc surface rather than being worn off.

On the other hand, a low temperature has a large effect on adhesion in the wheel-rail interaction. It is found that the wheel/rail adhesion coefficient in low temperatures is higher than that at room temperature due to the decline of oxidation. The iron oxides formed in the rolling contact generally play an important role in the wheel/rail adhesion. As a third-body medium in the rolling contact, the increasing oxidation would result in the decline of friction. When the temperature decreases below –20°C, the decrease of metal ductility slightly reduces the adhesion coefficient (Shi et al., 2018). The adhesion force is generally produced by the shear of micro cold-welding (bonding of micro asperities), so the adhesion depends on the anti-shearing of asperities. Due to the decrease in ductility, these bonds are broken more easily for the brittle fracture. Therefore, as the temperature drops from –20 to –40°C, the anti-shearing of asperities would be weakened, resulting in a slight decline of the adhesion coefficient.

The presence of oil, water, or antifreeze markedly decreases the adhesion coefficient, even at a low temperature. The wheel/rail adhesion coefficient under antifreeze conditions is lower than that under water conditions but higher than that observed under oil conditions, as presented in Fig. 8.23. The possible reason is that when a small amount of liquid is applied, a portion of the added liquid would remain on the sideways of the rolling track on the rollers. At room temperature, the liquid is easily supplemented into

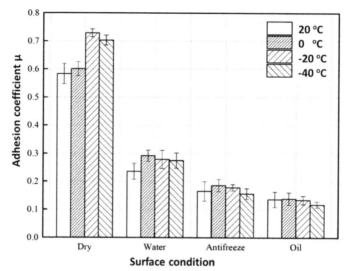

Figure 8.23: The effect of low temperature on the adhesion coefficient under water, oil and antifreeze conditions. After Shi et al. (2018), with permission from Elsevier.

the contact interface, while the amount of liquid in the wheel/rail interface at a low temperature would be considerably less than that at room temperature due to the freezing or poor liquidity of the liquid.

8.6.2 Sliding on Snow and Ice

Sliding on snow and ice has been the subject of scientific enquiry since over 100 years ago. It is recognized that the kinetic friction coefficient (μ) for ice is strongly dependent on the temperature of the ice/slider system and the sliding velocity. At a temperature greater than $-10°C$ and a velocity greater than 0.01 m/s, frictional heating is sufficiently high to melt the ice surface and provide a lubricating film of liquid water.

New technologies in manufacturing polymers and composites, as alternatives to metals, led to a growing interest in material investigations for low temperature applications. The material properties of polymers and composites are strongly temperature-dependent. Young's modulus and hardness of polymers are much higher at low temperatures than at room temperature. Therefore, while adhesive wear dominates at ambient temperatures, abrasive wear is observed at low temperatures (Marmo et al., 2005). Tribological performances of polymers depend markedly on the temperature at the friction contact. Studies at a cryogenic environment showed that both friction and wear are reduced due to the change in adhesion and deformation characteristics of the polymer materials (Hübner et al., 1998).

It is important to mention that the thermal properties of the cryogenic medium have a significant influence on the tribological performance of the sliding polymers and composites. The tribological behavior of two PTFE polymer-composites was investigated at low temperatures, with a special attention given to the elementary processes at the friction contact (Theiler, 2007; Theiler and Gradt, 2018). Tribological tests were carried out with a pin-on-disc configuration at room temperature (RT) in air, in liquid nitrogen LN_2 ($T = 77$ K) and in liquid helium LHe ($T = 4.2$ K). Other tests were also performed at $T = 77$ K in a Helium gas (He-gas) medium.

Figure 8.24 presents the results of the friction measurements at two sliding velocities (0.2 and 1 m/s). It is concluded that the coefficient of friction of PTFE composites decreases when cooled from room temperature to $T = 77$ K. This can be explained by the increased hardness of polymers at low temperature and decreased in friction due to deformation. Nevertheless, the temperature of the cryogen is not the significant parameter to understand and explain the tribological behavior of these composites at low temperature. The determinant factor is the physical, especially thermal, properties of the cryogen (liquid nitrogen or liquid helium). The decrease of friction with the temperature of the cryogen is not observed down to extreme

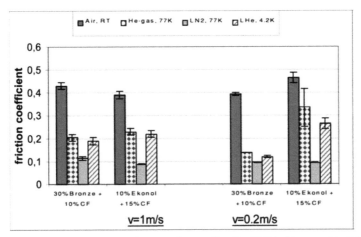

Figure 8.24: Coefficient of friction of PTFE matrix composites at RT, at T = 77 K in He-gas and in LN2, and at T = 4.2 K in LHe (F = 50 N; s = 2000 m). After Theiler et al. (2007), with permission from Elsevier.

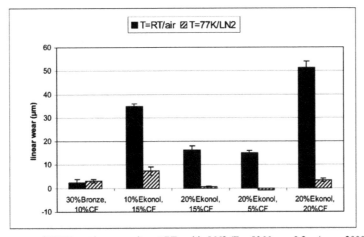

Figure 8.25: Linear wear of PTFE matrix composites at RT and in LN2 (F = 50 N; v = 0.2 m/s; s = 2000 m). After Theiler et al. (2007) with permission from Elsevier.

low temperature. Indeed, a higher coefficient of friction is measured at $T = 4.2$ K compared to the one at $T = 77$ K. Thermal properties of the cooling medium are, in this case, responsible for this increase.

The linear wear of PTFE matrix composites measured at RT and in LN_2, with $F = 50$ N and $v = 0.2$ m/s are illustrated in Fig. 8.25. The figure clearly shows the good performance of the polymer composites at low temperatures, with low linear wear. Opposing to the friction coefficient, linear wear depends on the material composition. A material with 30 wt% bronze and 10 wt% carbon fibre (CF) gives the lowest wear at RT. The materials with 20 wt% Ekonol, however, have a very small linear wear in LN_2, and are quite difficult to detect after 2000 m. The analyses of the metallic counterface surface show that PTFE matrix composites transfer down to very low temperatures. The transfer film at $T = 4.2$ K is characterized by small cracks and wear particles. The transfer film formed during the test is affected by the low temperature, possibly after the test or between the passes.

Marmo et al. (Marmo et al., 2005) determined the friction processes for ice samples sliding on steel by examining the wear and debris morphology with a low-temperature scanning electron microscopy and relating the processes to the velocity and temperature of formation. Friction experiments were carried out over a temperature range of –27 to –0.5°C and a velocity range of 0.008–0.37 m/s. Two distinguished levels of friction were observed:

i) Low friction (μ < 0.1) both at high velocity–low temperature and at low velocity–high temperature is characterized by the development of a cohesive mass of debris with well-defined grain boundaries and spheroidal air bubbles indicative of the deposition and freezing of liquid water.

ii) High friction (μ > 0.15) at low velocity–low temperature is characterized by the scuffing of the wear surface and sheets of solid debris that accumulate at the trailing edge of the wear surface, it is indicative of a brittle failure of ice at the sliding interface.

An ice friction map was produced by plotting each of the friction experiments in a temperature-velocity space and contouring the coefficient-of-friction data, as shown in Fig. 8.26. The development of such maps is of great use to engineers working on structures for ice-prone environments, winter sports and traction control systems for automobiles.

In the case of rubbers, static and dynamic coefficients of friction COF of a rubber ball against an ice surface were studied at different temperatures, from 0 to –50°C (Duanjie, 2016). The coefficient of friction of a rubber ball sliding on an ice surface at different temperatures below the ice point was evaluated. The speed increased at an exponential rate in a step-less fashion from 0.01 to 10 rpm, corresponding to sliding speeds of 2.6×10^{-5} m/s and 2.6×10^{-2} m/s, respectively. The test parameters are summarized in Table 8.3.

The evolution of COFs of a rubber ball against the ice at different temperatures is shown in Fig. 8.27, and the static and average dynamic COFs are summarized and compared in Fig. 8.28. In Fig. 8.27, as a certain threshold is reached, the relative motion at the interface of the rubber specimen and the ice surface

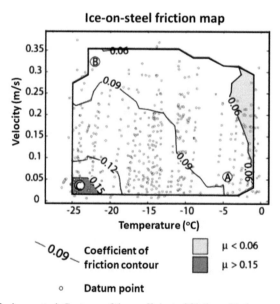

Figure 8.26: Friction map for ice on steel. Contours of the coefficient of friction with the temperature and velocity of each friction test identified. After Marmo et al. (Marmo et al., 2005), with permission from International Glaciological Society.

Table 8.3: Test parameters of the COF measurement. After Duanjie (Duanjie, 2016), with permission from Nanovea Inc.

Parameter	Value
Temperature	–50, –40, –30, –20, –10, 0°C
Normal force	2 N
Speed	0.01 to 10 rpm continuous, changed at an exponential rate
Duration of test	0.33 min

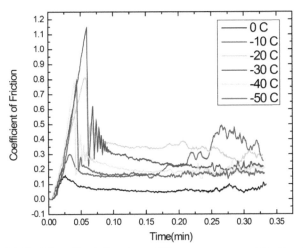

Figure 8.27: Evolution of COF of the rubber ball on ice at different temperatures. After Duanjie (Duanjie, 2016), with permission from Nanovea Inc.

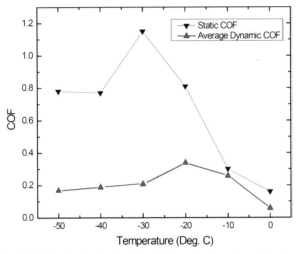

Figure 8.28: Static and dynamic COFs of rubber ball on ice at different temperatures. After Duanjie (Duanjie, 2016), with permission from Nanovea Inc.

takes place, and the subsequent measured COF decreases. This internal spike of the force at the beginning of the COF test arises from the static friction. It may result from several sources, such as reactions between surfaces, interlocking surface features, or other subtler phenomena, such as van der Waals forces, cohesion of surface films, and even microscopic solid static junctions between surfaces.

Figure 8.28 indicates that rubber has the lowest static and dynamic COF at the ice point of 0°C. This is due to the formation of a thin water film present between the rubber ball and the ice surface at this temperature, which may result from pressure melting, pre-melting or heat generated from friction at the interface. This liquid film acts as a lubricant at the sliding interface and substantially decreases friction. This experiment illustrates the risk of driving in a freezing rain. It generates ice pellets of an extremely smooth surface and low friction, which is notorious for causing accidents on roadways (Duanjie, 2016).

Nevertheless, the initial static COF progressively increases as the temperature decreases to –30°C, as a result of the freezing of the surface water film. Moreover, higher COF above 0.3 was observed at speeds higher than 1 rpm during the wear test at –10°C. However, as the temperature further drops below –40°C, the static and dynamic COFs decrease to values of ~ 0.8 and ~ 0.2, respectively, for the test carried out at –40 and –50°C. This may be related to a change in the mechanical properties of the rubber ball at this temperature, such as "glass transition."

8.6.3 Low Temperature Tribometers (Cryotribometers)

These types of tribometers are used for measuring tribological properties of materials at low temperatures, down to –50°C or less. Most of these tribometers are used for investigating materials used for very low temperature regions or in space applications. Some kinds of these tribometers allow one to perform tribological studies in vapors of cryogenic liquids. The sample chambers of all cryotribometers are thermally insulated by vacuum superinsulation and cooled directly by a bath of liquid cryogen or by a heat exchanger (continuous flow cryostat). These types of tribometers generally use liquid nitrogen (LN2), liquid hydrogen (LH2) or liquid helium (LHe) to cool the interface. In a bath cryostat operation, the liquid coolant is filled directly into the sample chamber. Thus, the complete friction sample is immersed into the coolant and the environmental temperature is equal to the boiling temperature of the coolant. The advantage of this method is a very effective cooling of the sample. The frictional heat is not only removed by heat conduction and convection, but also by evaporation of the liquid (Gradt et al., 2001). Two examples of cryotribometers setup are shown in Figs. 8.29 and 8.30.

Bäurle (Bäurle, 2007) designed and built a large-scale tribometer (diameter 1.80 m, pin-on-disc geometry) for friction measurements on ice, as shown in Fig. 8.31. Due to difficulties in conducting experiments with snow, the tribometer focuses on friction between polyethylene—the principal component at the snow-contacting face of skis and ice. The apparatus consists of a rotating table, carrying the ice annulus, and two arms for holding the slider and the ice surface preparation tools.

Both comparative studies of ski bases at a laboratory scale and experiments to explain the process of snow and ice friction have been conducted on the tribometer. On the tribometer, the slider is heated constantly, as in real skiing, while the ice is heated periodically. Results from laboratory measurements have to be interpreted carefully when the aim is to draw conclusions for real situations. Scaling up can be problematic, especially in the ski friction case, where water films are expected to build up at the front of the ski and increase in thickness and area covered towards the end.

It is finally concluded that the tribometer allows for reproducible measurements of friction on ice. It has been noticed that trends seen in the laboratory measurements are not always obeyed in real skiing. This is thought to be due to the difference between sliding on snow as compared to ice. Laboratory measurements on snow are therefore desired. An approach to producing a snow or snow-like surface on the tribometer is to sieve snow onto the track during a friction measurement, with the instrument running.

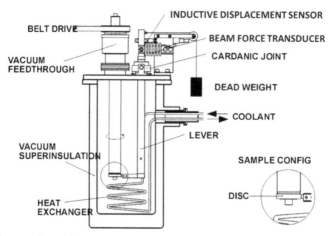

Figure 8.29: Cryotribometer CT1 which can be operated a flow cryostat as well as a bath cryostat using LN2 or LHe as coolants. After Gradt et al. (Gradt et al., 2001), with permission from Elsevier.

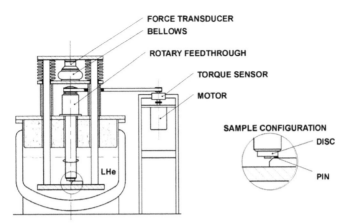

Figure 8.30: Cryotribometer CT2 which designed for higher loads, forces and higher environmental pressures respectively. After Gradt et al. (Gradt et al., 2001), with permission from Elsevier.

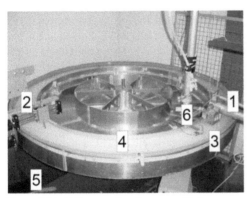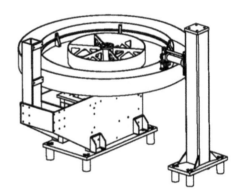

Figure 8.31: Tribometer. Loading arm with force sensor (1), ice surface preparation arm (2), slider (3), ice annulus (4), foundation and motor (5), ice temperature measurement (6). After Bäurle (Bäurle, 2006).

8.7 Friction and Wear at High Speeds

Friction and wear are complex processes and attributed to several operating parameters. One of the most effective parameters is the sliding speed. However, other parameters, such as the applied force, material properties, and roughness of the contact surfaces, lubricated or dry conditions and thermal effects, also have pronounced contributions.

In high speed applications, very severe contact conditions are obtained at the contact interface, with high pressures and large sliding speeds (up to several tens of meters per second). This was the reason for the development of several experimental techniques and devices for determining friction and wear at high speeds. For instance, due to the complex operating conditions during hot rolling, it is quite difficult to evaluate tribological properties of work rolls on-line. As a result, pilot testers have been used in order to simulate the operating conditions in hot rolling (Hao et al., 2017).

Earlier, Bowden (Bowden, 1958) conducted very high speed tests using a steel ball rapidly spinning around its vertical axis at a speed of around 700 m/s. The test was performed in order to study the effect of frictional heating on the deformation and surface melting of solid in contact. However, the accuracy of the results is notably affected by the time of contact, which varied between 30 μs and 140 μs. Montgomery (Montgomery, 1976) also performed very high speed tests (up to 560 m/s) in order to simulate the friction of a metallic projectile sliding down a cannon bore. A very important conclusion from this work is that the mechanism of wear at high sliding speeds is almost certainly surface melting followed by subsequent removal of a portion of the melted surface layer. The surfaces are probably not actually in contact at all but are separated by a lubricating film of melted material. The wear rate, excluding metallurgical reactions which have been shown to be important in some cases, is almost entirely a function of the melting point. Thermal conductivity, no doubt, has effects but they are probably minor as long as the values are not grossly different. Of the materials studied, with the possible exception of copper, there was no indication that any property other than melting point had a significant influence on wear rate at high sliding speeds.

Regarding polymer-metal contact, the sliding speed in some applications can reach values of up to 50 m/s. Many of those applications have relatively large contact surfaces in a continuous and discontinuous contact. In such applications, very high temperatures are developed in the contact area and surface melting is frequently observed on the polymer surface. Therefore, the *pv* limit, contact temperature, area of contact, and glass transient temperature are important factors in studying the friction and wear of polymers at high speeds. To investigate such high speed contact, the parameters contributing to heat generation and the condition of heat transfer should be maintained. Such parameters include applied load, contact speed, area of contact and ambient temperature.

El Fahham (El Fahham, 1996) designed and constructed a large scale tribometer to produce a high speed and insure a relatively high load and large contact surface in both continuous and discontinuous contacts, as shown in Fig. 8.32. The tribometer has the capacity to run under lubricated conditions with and without abrasive particles. Four polymers were tested in order to investigate their friction and wear properties at a high sliding speed and load. The polymers tested are Polytetrafluoroethylene PTFE, Polyoxymethylene Acetal, Ultra high molecular weight polyethylene UHMWPE, and Nylon 6/6. Two main groups of tests were conducted; first, variable speed tests were conducted, in which the load is maintained constant and the speed is increased to a very high wear rate or until severe surface melting occurs. The purpose of such tests is to obtain the *pv* limit of the material. Table 8.4 presents different variable speed tests conducted on the polymers and their speed limits.

Figure 8.33 shows the variation of the friction coefficient and contact temperature with time and speed in variable speed tests of polymers at a 200 N load. It was recognized that PTFE and Acetal have a relatively low coefficient of friction (< 0.2) and the material maintains its stability up to the maximum tests speed of 50–56 m/s. UHMWPE has a moderate coefficient of friction (~ 0.2) but it increases dramatically with the contact temperature when approaching its speed limit (34 m/s). Nylon 6/6 has both the highest coefficient of friction (> 0.2) and the lowest speed limit in all tests (10 m/s). In the second group of tests,

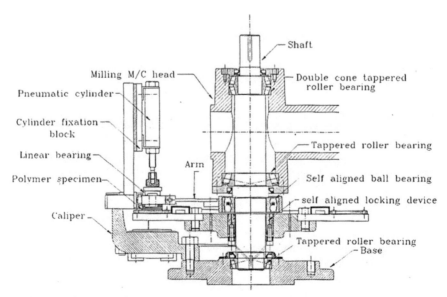

Figure 8.32: General assembly of the high speed tribometer. After El Fahham (El Fahham, 1996), with permission.

Table 8.4: Variable speed tests in continuous contact. After El Fahham (El Fahham, 1996), with permission.

Material	Load N	Speed m/s	Duration min	Speed limit m/s	Notes
PTFE	120	54	170	> 54	↑
PTFE	200	50	185	> 50	↑
Acetal	200	24	115	22	↑↑
UHMWPE	120	34	165	32	↑↑↑
UHMWPE	200	10	50	8	↑↑↑
Nylon 6/6	120	10	45	8	↑↑↑
Nylon 6/6	200	8	32	6	↑↑↑

Area of contact A = 2380 mm²
Speed increment of 2 m/s each 10 minutes
↑   high wear rate
↑↑   very high wear rate leading to test termination
↑↑↑   severe wear with surface melt leading to test termination

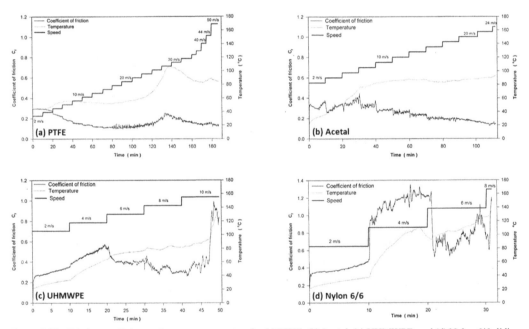

Figure 8.33: Friction coefficient and contact temperature for (a) PTFE, (b) Acetal, (c) UHMWPE, and (d) Nylon 6/6 sliding against steel at variable speed tests under 200 N load. After El Fahham (El Fahham, 1996), with permission.

the polymers were tested under constant load and speed in order to determine the coefficient of friction and specific wear rate of each material for a specific speed and load, and to analyze the wear mechanism of different materials at different speeds.

It is suggested that the wear mechanism of polymers when exceeding their *pv* limits would greatly affect the performance of the machine component involved. For PTFE and Acetal, the material may wear rapidly with relatively small debris or melt without depositing a film on the counterface surface; this would not affect the counterface materials and might take some time before contact failure occurs. For UHMWPE, the material maintains its stability for a certain time, after which a very high wear rate occurs and large particles are smeared out from the surface. The particles adhere to the counterface surface and both surfaces may weld together after the machine is stopped. As for Nylon 6/6, the large film with high polymer surface distortion deposited on the counterface surface induces high vibration in the system, which could cause mechanical problems. The large film deposited may also be projected from the surface, which may cause certain damage to other components of the machine (El Fahham, 1996).

In general, at high sliding speeds in polymer-metal tribosystems, the coefficient of friction and the wear rate decrease with increasing speed up to the material *pv* limit. Nevertheless, when exceeding the polymer *pv* limit, it is found that the coefficient of friction does not decrease and a very high wear rate occurs. This contradiction may be attributed to the surface contact area and the non-uniform wear distribution along the rubbing surface which limits the lubrication effect of the molten surface.

8.8 Friction and Wear Under Vacuum Conditions

The advances in aerospace industries in the past decades around the world were the trigger for an urgent attention to investigate friction and wear under extreme conditions, such as vacuum environments. Under such a harsh space environment, it is often a problem to achieve acceptable endurance of tribological components. These severe conditions also include atomic oxygen, ultraviolet irradiation, proton and electron irradiation, high vacuum and thermal cycling.

8.8.1 Friction in Vacuum

As previously presented in Chapter 3, when two solid surfaces are placed in contact, the contact occurs across the interface between surface asperities. A plastic flow of these metal asperities will then occur, and a certain amount of metal-to-metal contact will take place through any surface films present. According to the adhesion theory of friction, adhesion occurs at the contacting asperities, and the force required to shear the adhered junctions will be the shear strength in the surficial region. The force required to initiate motion between these surfaces will represent the static friction force. Once one of the surfaces is in motion, the friction force measured is the dynamic friction.

It is widely believed that the reduction in the ambient pressure increases friction forces of tribological components. This is probably due to the removal of residual physically adsorbed films. Also, sliding at reduced pressures will produce frictional heat at the interface between the two contacting surfaces. This frictional heat will help to promote the desorption of physically adsorbed species, such as the atmospheric gases and water vapor. Furthermore, at a low ambient pressure, the concentration of oxygen in the environment is reduced, and the temperature for the thermodynamic stability of the metal oxide is also reduced. These reductions in the adsorbed films and metal oxides increase the tendency of metal surfaces to adhere one to another.

In the particular case of vacuum environments, destructive adhesion occurs in solid-state contacts, causing a high friction and heavy surface damage. It is demonstrated that the vacuum environment and the surface cleanliness strongly influence the adhesion behavior of solids (Miyoshi, 1999). An analogy between adhesion and friction in vacuum can exist. Figure 8.34 shows two comparisons of adhesion and friction results obtained in vacuum: Adhesion (pull-off force) versus the coefficient of friction and atomically clean surface versus as-received, contaminated surface.

The relation between adhesion and friction in a vacuum is also observed at high temperatures. Figure 8.35 shows the pull-off forces and the static and dynamic coefficients of friction for as-received ceramic flat surfaces in contact with clean ceramic pins as a function of temperature. On the figure, the low adhesion and friction at temperatures up to 300°C can be attributed to the presence of contaminants on the as-received flat surfaces. On the other hand, at temperatures between 400 and 700°C, the high adhesion and friction can be associated with the absence of adsorbed contaminants, such as carbon and water vapor.

In fact, there are many material properties contributing to the friction in vacuum. For example, the elastic properties of metals are related to their cohesive energy. Usually, the stronger the cohesive bonds in the metal, the higher the elastic modulus. With increasing cohesive energy and modulus of elasticity, metals tend to become less and less ductile. For example, in the body-centered-cubic system, iron is fairly ductile. Tungsten, on the other hand, which has a markedly higher cohesive energy and elastic modulus, is relatively brittle and fractures very readily. With clean iron surfaces in contact, strong adhesion occurs

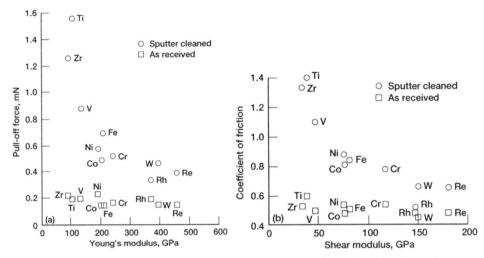

Figure 8.34: Pull-off force (adhesion) as function of Young's modulus (a) and coefficient of friction as function of shear modulus (b) of metals in contact with polycrystalline manganese zinc ferrite in ultrahigh vacuum. After Miyoshi (Miyoshi, 1999), with permission from Elsevier publishing.

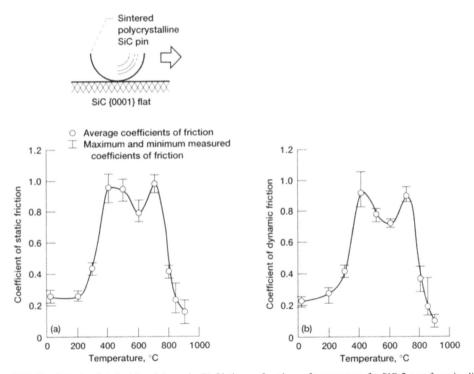

Figure 8.35: Coefficients of static (a) and dynamic (b) friction as functions of temperature for SiC flat surfaces in sliding contact with sintered polycrystalline SiC pins in ultrahigh vacuum. After Miyoshi (Miyoshi, 1999), with permission from Elsevier publishing.

across the interface. With sliding, a continuous increase in true contact area occurs, resulting in junction growth, and friction coefficients continue to increase until complete seizure of the iron surfaces occurs (Buckley, 1971; Buckley, 1974).

Ductility of metals influences the real area of contact and, accordingly, the adhesion and friction at the rubbing surfaces. Friction measurements have been made in a vacuum environment with iron, molybdenum and tantalum at temperatures below 300 K (Buckley, 1971; Miyoshi, 1999). At sufficiently low temperatures, decreases in friction coefficients for iron and molybdenum have been observed. These decreases are attributed to a transformation from the ductile to the brittle states in these metals. Thus, in a vacuum environment, ductility is an important factor in determining friction characteristics. It was suggested that both the real area of contact and the surface energy must be minimized in order to reduce friction in a vacuum.

The coupled materials have a dominant role in the coefficient of friction for interacting surfaces in ultra-high vacuum. Consequently, the judicious selection of counterpart materials can reduce the coefficient of friction. Figure 8.36 presents examples of coefficients of friction for clean metal–metal couples, clean metal–nonmetal couples, and clean nonmetal–nonmetal couples measured in ultrahigh vacuum.

Table 8.5 presents data about friction coefficients in a standard atmosphere and in a vacuum for traditional materials. It can be noted that there are no specific values of friction coefficient in the low, middle, high and ultrahigh vacuum. Indeed, vacuum degree, temperature, sliding speed, and contacting load have a great influence on the behavior of a friction coefficient of metals and nonmetals (Deulin et al., 2010).

MoS$_2$ filled PEEK composite was investigated in a vacuum at room temperature (RT) under 1 MPa and 7 MPa (Theiler and Gradt, 2010). The effect of the sliding speed on the coefficient of friction is shown in Fig. 8.37. Particularly between 0.01 m/s and 0.1 m/s, increasing speed reduces the friction. When the contact pressure increases at moderate sliding velocities, friction decreases slightly, but increases at 1 m/s at higher contact pressure. Also, at room temperature, the friction coefficient of PEEK composite in a vacuum is lower than in air. The better tribological properties of PEEK composites in a vacuum compared to air can suggest that PEEK is sensitive to the amount of water vapor in the environment. It is demonstrated that the lower coefficient of friction of the MoS$_2$ filled composite at –80°C is associated with a thin polymer transfer film on the counterface. In addition, there is a significant relation between sliding speed and the tribological performance. The transition from severe wear to mild wear is found between 0.01 and 0.1 MPa, as shown in Fig. 8.38. The optimal performance was obtained by p.v. = ~ 1 MPa m/s.

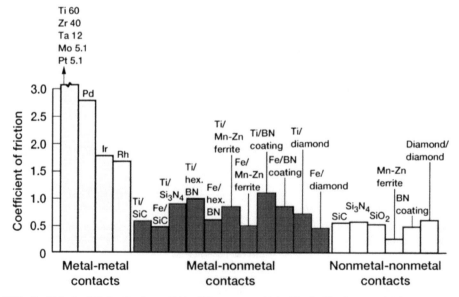

Figure 8.36: Coefficients of friction for clean solid in sliding contact with itself and with other materials in ultrahigh vacuum. After Miyoshi (Miyoshi, 1999), with permission from Elsevier publishing.

Table 8.5: Friction coefficients of the metals, solid lubricants and antifriction materials. After Deulin et al. (Deulin et al., 2010) with permission from Springer publishing.

Material	Friction coefficient in air	Friction coefficient in vacuum
Fe-Fe	0.3	1.9
Fe-Mg	1.0	0.6
Fe-Cd	1.5	0.4
Fe-Pb	0.9	0.4
Stainless steel-Stainless steel	0.5	2.9
Stainless steel-Cu	–	0.3
Stainless steel-Al	0.4	0.3
Stainless steel-Kovar	–	0.4
Stainless steel-brass	0.4	0.8
Stainless steel-Mo	–	0.8
Stainless steel-lanthanum	–	0.8
Stainless steel-Ni	–	0.8
Stainless steel-Dimolit	–	0.05
Stainless steel-Teflon	–	0.2
Stainless steel-Cu+MoS_2 (antifriction film)	–	0.2
Stainless steel-Si (antifriction film)	–	0.2
Stainless steel-Ge (plished)	–	0.2
Stainless steel-ground ceramic "22XC"	–	0.2
Stainless steel-polished ceramic "22XC"	–	0.3
Stainless steel-polished glass ceramic "CT-50-1"	–	0.4
Stainless steel-ground glass ceramic "CT-50-1"	–	0.5
Stainless steel-polished glass "C48-3"	–	0.3
Stainless steel-ground glass "C48-3"	–	0.5
Chromium steel-Chromium steel	–	0.3
Chromium steel-Cu	0.5	0.5
Cu-Cu	0.7	1.0
Cu-Ni	0.5–1.0	4.8–21.0
Cu-Ta	0.6	1.5–2.0
Cu-W	0.4	0.4
Al-Al	0.3	0.6
Al-Ni	0.8	1.6–2.2
Al-Cu	–	2.4
Al-Ag	–	1.5
Brass-brass	–	2.2
Beryllium bronze-beryllium bronz	0.4	0.7
Beryllium bronze-brass	0.7	1.1
Ni-Ni	0.4	0.9
Ta-Ta	–	4.9
W-W	0.2	4.7
Ag-Ag	0.2	2.5

Table 8.5 contd. ...

...Table 8.5 contd.

Material	Friction coefficient in air	Friction coefficient in vacuum
Co-Co	–	3.9
Cr-Cr	–	0.3–0.5
Mo-Mo	0.6	3.0
Au-Au	1.0	2.5
Zn-Zn	0.6	4.5
Sn-Sn	1.0	3.0
Ti-Ti	1.0	1.0
Pt-Pt	–	4.2
Nb-Nb	–	3.5
Zr-Zr	–	4.2
Chromium steel-MoS_2 film (vacuum deposition)	–	1.5
Chromium steel-MoS_2 film (friction deposition)	–	0.06
Chromium steel-MoS_2 film+epoxy compound	–	0.06
Chromium steel-MoS_2 film+organosilicon compound	–	0.1
Chromium steel-MoS_2 film+sodium silicate	–	0.2
Chromium steel-MoS_2 film+TSO_2	–	0.1
Cu-MoS_2	0.2	0.07
Cu-CdS_2	0.5	0.2
Cu-ViJ	0.4	0.3–0.5
Cu-phthalocyanine	0.4	0.4
Chromium steel-MoSe	0.2	–
Chromium steel-WSe_2	0.2	0.1
Chromium steel-TaSe	0.2	0.1

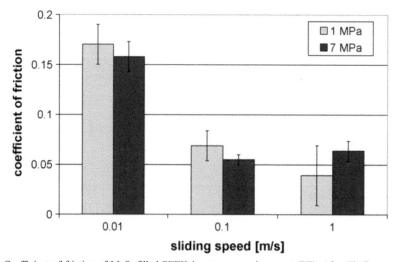

Figure 8.37: Coefficient of friction of MoS_2 filled PEEK in vacuum environment (RT). After Theiler et al. (2010) with permission from Elsevier.

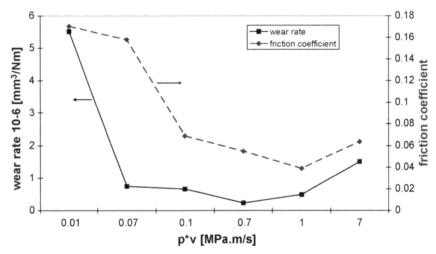

Figure 8.38: Wear rate and friction coefficient versus p.v. values of MoS_2 filled PEEK in vacuum environment (RT). After Theiler et al. (2010) with permission from Elsevier.

8.8.2 Wear in Vacuum

As already discussed, tribological properties of materials exhibit different characteristics in the vacuum of space as compared to under atmospheric pressure. It is widely accepted that adhesive and fatigue wear are the two important types of wear encountered in a vacuum environment. In particular, adhesive wear is the most detrimental and most frequently encountered.

In an ordinary environment, adhesion can occur between contacting surfaces. It can also occur in a vacuum with the simple approach of two clean surfaces. When two smooth and clean solid surfaces are in sliding, rubbing or rolling contact in a vacuum environment, adhesive wear will take place because of adhesion across the interface. In the case of similar materials, adhesion occurs across the interface, and with tangential motion, fractures may occur at the interface or subsurface in one of the two mechanical components in a manner described earlier with reference to the zones of greatest mechanical weakness. When a subsurface fracture occurs, material may be transferred from one surface to another, from each surface to the other, or back and forth from one surface to another (Buckley, 1974). This transfer will frequently result in the destruction of the operating clearance between the mechanical components and in the failure of that component.

On the other hand, when dissimilar metals are in contact with sliding, or rolling, contact in the absence of a contaminating film, the adhesion process will generally result in the transfer of particles of the cohesively weaker of the two materials to the cohesively stronger. The adhesive junctions generally being stronger than the cohesive bonds of the cohesively weaker of the two materials. This transfer results in adhesive wear. If repeated contact is made over the same general area, sufficient adhesive wear particles will transfer from the cohesively weaker to the cohesively stronger material, such that, after a period of time, the surface of the cohesively stronger material will be well populated with transferred wear particles.

In the case of metallic alloys, the absence of adhesion or complete seizure in many instances is due to the presence of minor alloying constituents which can, on heating of the surfaces by the frictional energy put into the surfaces during sliding, rolling, or rubbing contact, diffuse to the surface and provide a protective surface film (Buckley, 1971). Finally, a vacuum not only radically affects the wear behavior of metals and alloys in contact, but also has a pronounced influence on nonmetals as well.

Fatigue wear can occur in a vacuum for materials that do not adhere strongly because of the presence of surface films or where there is a lubricant film. When two surfaces are in rubbing contact, the surface micro-contacts are subjected to both compressive and tensile forces (Radchik and Radchik, 1958). Thus, with repeated passes, surface and subsurface cracks can develop and may lead to the generation of free wear particles.

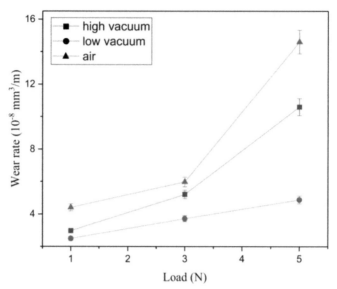

Figure 8.39: Wear rate of the alloy at different load under different vacuum degree. After Niu et al. (2019), with permission from Elsevier publishing.

In the previous discussion, it is clear that wear in a vacuum environment is highly associated with the materials in contact. Therefore, it is important to research and develop some new kinds of materials, which can potentially operate in vacuum environments. For instance, an investigation of the wear behavior of copper alloy $CuZn_{39}Pb_3$ against steel under both atmospheric and vacuum conditions showed that the wear rates of $CuZn_{39}Pb_3$ were lower under vacuum conditions (Küçükömeroğlu and Kara, 2014).

Studying tribological behavior of Ni_3Si-based composites postulated that it had superb wear properties under dry sliding, sulfuric acid solution and seawater conditions. Furthermore, $Ni_3(Si,Ti)$ alloy has good ductility and an elongation value of 30% or more under vacuum conditions (Liu et al., 1996). Therefore, it is interesting to study the tribological property of the alloy in vacuum environments. Recently, Niu et al. (Niu et al., 2019) investigated the impact of different vacuum degrees on tribological behaviors of Ni_3Si alloy. The tests were conducted using ball-on-disk vacuum tribometer, in which the alloy disk was rotated against a commercial GCr15 steel ball. The tests were performed under high vacuum (4.0×10^{-4} Pa), low vacuum (4.0 Pa) as well as atmospheric pressure. It is observed that Ni_3Si alloy has excellent anti-wear properties in all the vacuum conditions, and the wear rate was in the magnitude of 10^{-8} mm^3/m, as shown in Fig. 8.39. The alloy showed the optimal tribological property in low vacuum as compared to high vacuum and air conditions. Studying the morphologies of the worn surface suggested that the wear mechanism in air condition was oxidation wear. But the wear mechanisms were adhesive wear in high vacuum, and was slight abrasive wear in low vacuum. Finally, it is suggested that the Ni_3Si alloy had great application potential in extreme environments, such as vacuum and high load.

8.8.3 Vacuum Tribometers

Commercial or in-house vacuum tribometers are usually utilized to investigate tribological behavior under vacuum conditions (down to 10^{-7} mbar). These types of tribometers are designed to provide controlled vacuum conditions for friction and wear studies. Some of the tribometers are equipped with heating devices in order to be able to perform tests at a high temperature. These devices are required for the testing of critical equipment with moving parts that work under cryogenic and/or vacuum conditions (machines operating in harsh conditions, such as space, tools in the semiconductor industry, scanning electron microscopy or cryo pumps, etc.). A simple design of such a tribometer is shown in Fig. 8.40 (Niu et al., 2019). The specimen disk with dimensions of $18 \times 18 \times 3$ mm^3 was rotated against a counterface

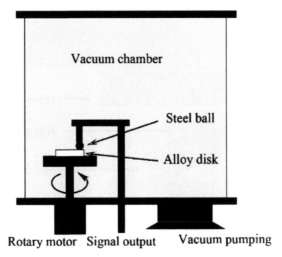

Figure 8.40: Schematic of vacuum tribometer. After Niu et al. (2019).

ball (diameter of 3.18 mm). To minimize the effect of the residual gas under high vacuum, the chamber is vacuumed to 4×10^{-4} Pa, then refilled with Ar gas, and then evacuated again. The wear rate is calculated by wear track perimeter times cross-sectional area divided by the sliding distance.

8.9 Wear in Mining and Mineral Processing

There are numerous influencing factors in a tribological system and handling most of them is very challenging. A practical example of such tribological systems is the mining and mineral processing field.

Mining and mineral processing production moves volumes of solids that abrade and erode equipment. In this industry, the most important and cost-effective types of wear are impact, abrasion and their combinations (Wills and Napier-Munn, 2005). Over the equipment lifetime, ongoing maintenance and replacement can result in extra expenses and an increased downtime. Therefore, the chosen material should have a good resistance against impact-abrasive wear and adequate other mechanical properties. Tribo-simulative tests are frequently used to evaluate the tribological behavior of the materials involved in mining and mineral processing. These types of tests mimic the real process, as far as possible, and imitate the conditions and environment of a particular machine component. However, tribologists often criticize tribo-simulation, because the results of laboratory tests have turned out to be useless in many practical applications. This criticism is often due to an overestimation and misinterpretation of published wear rates and friction coefficients.

In studying the wear of powder-metallurgical (PM) materials, a hammer-mill device is used to demonstrate severe impact conditions and impact conditions combined with abrasive particles (Teeria et al., 2006). Such conditions take place in rotating grinding mills in ore processing.

All experiments of impact wear tests were performed with the same impact angle and impact energy. The impact angle of hammers was chosen to be 90° between the chamfer of the hammer and the specimen surface, as shown in Fig. 8.41. The impact energy of a single impact was about 31 J. The tests were designed to last for 690 min per specimen, provided that the specimen would bear the whole test. The impact frequency was 130 impacts per minute, which lead to approximately 90000 impacts per test. In the second phase of the experimental work, abrasive particles were added in the tests to simulate, as accurately as possible, a real mineral grinding process.

The impact wear tests proved that high hardness is not necessarily enough to give good resistance against impacts. Impact toughness plays a key role in avoiding brittle behavior. The best wear resistance results were obtained when a suitably treated material had high hardness combined with high impact toughness and homogeneous microstructure.

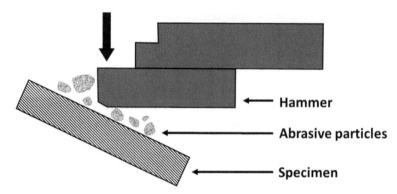

Figure 8.41: The contact between the specimen and the hammer.

8.10 Geotribology

It is believed that the term "geotribology" was first used by Blok (Blok, 1963) with no significant discussion. Later, geotribology framework was employed to analyze the flow mechanics of granular sand (Dove and Jarrett, 2002). In fact, even though tribological concepts can be utilized to many geosciences phenomena, the two research communities are separated. In earth science, many tribological concepts were applied successively, particularly in rock friction analyses. Actually, many components of tribology were used in earth sciences, particularly in rock friction analyses. For instance, the asperity-asperity contact mechanism (Bowden and Tabor, 1950) was applied to rock friction experiments that led to the rate-state friction law that prevails in earthquake analyses.

Recently, Boneh et al. (Boneh et al., 2018) followed previous studies to develop 'geotribology' as an interactive framework for the analysis of geologic processes of bodies in relative motion involving friction, wear and lubrication. In this study, a critical review of wear and friction along brittle faults is presented, considering the effective asperity model. The compilation of wear-rates along experimental rock faults indicates a strong dependency on slip-velocity that was not considered in the Archard model (Archard, 1953).

Wear mechanisms of faults

An example of geotribological systems is a fault which is defined as a planar fracture or discontinuity in a volume of rock. This phenomenon results in a slip, defined as the relative movement of geological features present on either side of a fault plane. The slip is associated with friction resistance and wear. Due to friction of the constituent rigid rocks, both sides of a fault cannot always slip or flow past each other easily, and so, occasionally, all movement stops. The regions of higher friction along a fault plane, where it becomes locked, are called asperities. When a fault is locked stress builds up, and when it reaches a level that exceeds the strength threshold, the fault ruptures and the accumulated strain energy is released in part as seismic waves, forming an earthquake (Allaby, 2013).

Friction and wear are related to each other and both have complex correlations to the fault system. This complexity is demonstrated by the different wear mechanisms of the interacting blocks: Adhesive, abrasive, delamination, fatigue and corrosive wear. These wear mechanisms can act simultaneously, and the global wear, which is the sum of all mechanisms, is measured.

Experimental examination of rock faults exposes two general modes of wear features (Boneh and Reches, 2018). The first one is the two-body mode where the wear is localized at isolated asperities. This mode is associated with the slip distance. Therefore, the two-body wear mode is characterized mostly by interactions at isolated asperities, as shown in Fig. 8.42a.

The second one is the three-body mode. It occurs while the fault is covered by a continuous gouge layer of finite thickness, as shown in Fig. 8.42b. The gouge layer separates the two fault locks and prevents direct contact between the isolated asperities of the fault blocks. In this case, the wear occurs at the contacts

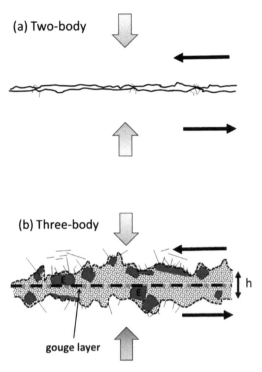

Figure 8.42: Simplified view of the contact between two loaded fault blocks. (a) Two-body fault with two bare blocks in direct contact only at touching isolated asperities. (b) Three-body fault zone in which the two are separated by a gouge layer composed of fine grains at the nanometer to micron scale (dotted line). After Boneh et al. (2018) with permission from Elsevier publishing.

between the gouge and the host blocks, and the gouge thickens by fracturing and plucking the host block. The plucked particles integrate into the gouge and can be much larger than the initial surface roughness. This process modifies the nature of contact between the gouge layer and the host rock with multiple-scale roughness (Sagy et al., 2017) referred to as "effective roughness." Thus, the effective roughness is different from the original surface roughness by its failure mechanism and scale. Consequently, the effective roughness is a time-dependent feature of a fault that controls the wear and damage of the fault blocks.

8.11 Abrasion of Concrete Paver Blocks

Currently, the interlocking concrete block pavement has been extensively used as an alternative to concrete and asphalt pavements. Most concrete is composed mainly of Portland cement, coarse and fine aggregates and water. Deterioration of pavements, due to traffic loads, usually occurs in form of wear, such as erosion, cavitation and simple abrasion. The abrasion resistance of a concrete pavement may be defined as its ability to resist being worn away by friction and rubbing. The abrasion resistance is influenced by several factors, such as cement content, water-cement ratio (w/c), cement type, and aggregate type influence. For concrete with high abrasion resistance, it is desirable to use a hard surface material, aggregate and cement with low porosity and high strength (Yazıcı and İnan, 2006). Hardness of coarse aggregate is important here. It was reported that the service life of concrete blocks can be extended by using harder aggregate types for most modes of wear. From this point of view, several research studies have been conducted in order to study the viability of replacing limestone aggregate with alternative coarse aggregate in the production of concrete pavement blocks (Gencel et al., 2011; Thomas et al., 2016; Pacheco-Torgal and Jalali, 2010; Yüksel, 2017). A pioneer work was conducted by El-Haggar et al. (Korany and El-Haggar, 2001), in which different slag types were proposed as replacements for coarse aggregates in producing cement masonry bricks and paving interlock units.

Recently, Abdelbary et al. (2016; 2018) investigated the impact of replacing coarse aggregate with steel slag in order to produce high wear resistance pavement blocks. The study was evaluated in the light of the ASTM C936 standard.

Five groups of mixtures were prepared in order to evaluate the impact of the slag replacement level and Cement/Slag mix ratio, as presented in Table 8.6.

Table 8.6: Mix proportions of paving interlock groups. After Abdelbary et al. (2018).

Group	Cement pcf (Kg/m³)	Sand pcf (Kg/m³)	Limestone pcf (Kg/m³)	Slag pcf (Kg/m³)	w/c ratio	Rep. %	Mix ratio (Cement: F.A. : C.A.)
1	24.16 (387)	72.5 (1162)	36.27 (581)	–	0.35	–	1 : 3 : 1.5
2	24.47 (392)	73.35 (1175)	18.35 (294)	18.35 (294)	0.40	50	1 : 3 : 1.5
3	24.60 (394)	73.80 (1182)	–	36.83 (590)	0.40	100	1 : 3 : 1.5
4	20.41 (327)	61.30 (982)	–	61.30 (982)	0.40	100	1 : 3 : 3
5	24.35 (390)	48.69 (780)	–	73.04 (1170)	0.40	100	1 : 2 : 3

F.A. fine aggregate, C.A. coarse aggregate.

8.11.1 *Effect of Slag Replacement Level*

Experimental results of abrasion tests are presented in Table 8.7. It is indicated that test groups (2 and 3) have higher abrasion resistance than the control group. These groups showed much lower abrasion coefficient than the ASTM C936 limit of 15 cm³/50 cm². Mixes 2 and 3 (containing 50% and 100% of slag aggregate) gave abrasion resistance that was higher by 11% and 58%, respectively, than that of the control mix. It is clear that there is a direct proportionality between slag replacement level and abrasion behavior of paving units, as shown in Fig. 8.43.

The average abrasion resistance was estimated based on ASTM C241-90 (ASTM International, 2005).

On the figure, increasing the slag replacement level resulted in increasing the abrasion resistance. This may be attributed to the hardness that slag imparts to concrete; hardness is believed to be the most important factor that controls the wear of the aggregate in concrete. The hard aggregate should protect the softer paste, provided that there is an adequate aggregate/paste bonding strong enough to hold the aggregate securely in the face of the 'attacking' abrasion load (Gencel et al., 2011). Investigation of the specimen abraded surfaces indicated that the hard slag aggregates in the matrix prevent abrasive particles from penetrating more into the concrete, as shown in Fig. 8.44. While on the other hand, abrasion of limestone in the control mix was observed. This finding agrees with the physical properties of slag aggregates where the measured hardness of limestone is about 57% of the hardness of EAFS. Yi et al. (Yi et al., 2012) also reported that the typical hardness of slag is about 50% compared to the hardness of limestone.

Table 8.7: Results of abrasion tests. After Abdelbary et al. (2018).

Group	Original wt. lb (gm)	Final wt. lb (gm)	Weight loss lb (gm)	Thickness loss ft (mm)	Abrasion resistance H_a	Abrasion coefficient (cm³/50 cm²)
1	0.91 (412)	0.87 (394)	0.040 (18)	0.0051 (1.57)	1.50	7.85
2	0.92 (419)	0.89 (403)	0.035 (16)	0.0047 (1.46)	1.66	7.30
3	0.99 (448)	0.96 (436)	0.026 (12)	0.0035 (1.07)	2.37	5.34
4	0.98 (444)	0.95 (432)	0.026 (12)	0.0034 (1.06)	2.51	5.31
5	1.06 (482)	1.04 (472)	0.022 (10)	0.0025 (0.77)	3.29	3.86

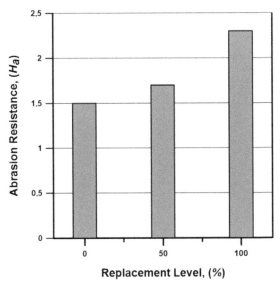

Figure 8.43: Effect of slag replacement level on abrasion resistance (for Mix 1:3:1.5). After Abdelbary et al. (2016; 2018).

(a)   **(b)**

Figure 8.44: Photography of abraded surface; (a) limestone specimen (Mix group 1), and (b) slag specimen (Mix group 3). After Abdelbary et al. (2016).

8.11.2 Effect of Slag Content (Mix Ratio)

It is well recognized that coarse aggregate plays an important role in concrete pavers since it occupies at least one-quarter of the total volume of concrete. Results signified that changes in mixing ratio and slag aggregate content can change the abrasion resistance of paving units, as tabulated in Table 8.8. Also,

Table 8.8: Abrasion resistance for mix designs. After Abdelbary et al. (2018).

Group	Mix ratio	Slag content (%)	Abrasion resistance (H_a)
1	1 : 3 : 1.5	0	1.50
2	1 : 3 : 1.5	13	1.66
3	1 : 3 : 1.5	27	2.37
4	1 : 3 : 3	43	2.51
5	1 : 2 : 3	0.50	3.29

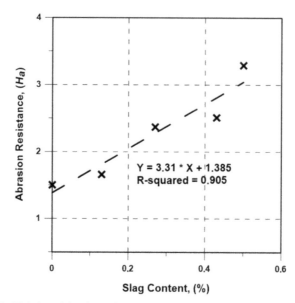

Figure 8.45: Variation of abrasion resistance with slag content. After Abdelbary et al. (2016).

Fig. 8.45 indicates that there is a clear dependency between slag content and abrasion resistance. As in many particulate composite materials, properties and content of filler particles play a controlling role in determining its wear and abrasion behavior. Hence, it was expected to find the highest abrasion resistance at the highest slag content.

References

Abdelbary, A., Abouelwafa, M.N., El Fahham, I.M. and Gomaa, A.I. 2004. A new reciprocating tribometer for wear testing under different fluctuating loading conditions. Alexandria Engineering Journal 43(5): 615–619.

Abdelbary, A. 2014. 3—Fatigue wear of unfilled polymers. In: Wear of Polymers and Composites, edited by Ahmed Abdelbary, 67–93. Oxford: Woodhead Publishing.

Abdelbary, A. and Ragab Mohamed, A. 2016. The impact of incorporating slag aggregates on the abrasion behavior of concrete paver blocks. Landscape Architecture and Regional Planning 1(1): 18–24. Doi: 10.11648/j.larp.20160101.13.

Abdelbary, A. and Ragab Mohamed, A. 2018. Investigating abrasion resistance of interlocking blocks incorporating steel slag aggregate. Materials Journal 115(01). Doi: 10.14359/51700898.

Allaby, M. 2013. Strike-slip fault.

An, J., Feng, J.H., Yan, X.H. and Li, R.G. 2017. Tribological behavior of Mg97Zn1Y2 alloy at elevated temperatures of 50–200°C. Journal of Materials Engineering and Performance 26(10): 4940–4952. Doi: 10.1007/s11665-017-2926-x.

Archard, J.F. 1953. Contact and rubbing of flat surfaces. Journal of Applied Physics 24: 981–988.

Arias-Cuevas, O., Li, Z., Lewis, R. and Gallardo-Hernández, E.A. 2010. Rolling–sliding laboratory tests of friction modifiers in dry and wet wheel–rail contacts. Wear 268(3): 543–551. Doi: https://doi.org/10.1016/j.wear.2009.09.015.

ASTM International. 2005. Standard Test Method for Abrasion Resistance of Stone Subjected to Foot Traffic. In ASTM C241-90. West Conshohocken, USA.

Barber, J.R. 2020. Contact problems involving friction. In: Paggi, M. and Hills, D. (eds.). Modeling and Simulation of Tribological Problems in Technology. CISM International Centre for Mechanical Sciences. Vol. 593. Springer. Cham. Doi: https://doi.org/10.1007/978-3-030-20377-1_2.

Barbour, P.S.M., Barton, D.C. and Fisher, J. 1995. The influence of contact stress on the wear of UHMWPE for total replacement hip prostheses. Wear 181-183: 250–257. Doi: https://doi.org/10.1016/0043-1648(95)90031-4.

Bäurle, L., Szabó, D., Fauve, M., Rhyner, H. and Spencer, N.D. 2006. Sliding friction of polyethylene on snow and ice: Tribometer Measurements. In: Tribology Letters 24: 77. https://doi.org/10.1007/s11249-006-9147-z.

Bäurle, L., Szabó, D., Spencer, N.D. and Kaempfer, T.U. 2007. Sliding friction of polyethylene on snow and ice: Contact area and modeling. In: Cold Regions Science and Technology 47(3): 276–289. Doi: 10.1016/j.coldregions.2006.10.005.

Bhadeshia, H.K.D.H. and Honeycombe, Sir R. 2006. 8—The heat treatment of steels: Hardenability. pp. 167–181. *In*: Bhadeshia, H.K.D.H. and Honeycombe, Sir R. (eds.). Steels (Third Edition), Oxford: Butterworth-Heinemann.

Bhushan, B. 2004. Springer Handbook of Nanotechnology. New York: Springer.

Bhushan, B. 2005. Nanotribology and Nanomechanics. New York: Springer.

Blok, H. 1963. The flash temperature concept. Wear 6(6): 483–494. Doi: https://doi.org/10.1016/0043-1648(63)90283-7.

Boneh, Y. and Reches, Z. 2018. Geotribology—Friction, wear, and lubrication of faults. Tectonophysics 733: 171–181. Doi: https://doi.org/10.1016/j.tecto.2017.11.022.

Bowden, F.P. and Tabor, D. 1950. The Friction and Lubrication of Solids. U.K.: Clarendon Press Oxford.

Bowden, F.P. 1958. A review of the friction of solids. Wear 1(4): 333–346. Doi: https://doi.org/10.1016/0043-1648(58)90005-X.

Brkovski, D., Cvetkovski, S. and Jancevski, J. 2013. Investigations of extreme working conditions to the mechanical properties of crawler vehicle. Machines Technologies Materials 12(12/2013): 18–20.

Buckley, D.H. 1971. Friction, Wear, and Lubrication in Vacuum. Washington, D.C.: NASA Lewis Research Center.

Buckley, D.H. 1974. Adhesion, friction, wear and lubrication in vacuum. Japanese Journal of Applied Physics 13(S1): 297. Doi: 10.7567/jjaps.2s1.297.

Charitidis, C.A. 2013. Nanotribological behavior of carbon-based thin films. pp. 2381–2396. *In*: Wang, Q.J. and Chung, Y.-W. (eds.). Encyclopedia of Tribology, Boston, MA: Springer US.

Clatterbaugh, G.V., Trethewey Jr., B.R., Roberts, J.C., Ling, S.X. and Dehghani, M.M. 2011. Engineering Systems for Extreme Environments. Johns Hopkins APL Technical Digest 29.

Cooper, J.R., Dowson, D., Fisher, J., Isaac, G.H. and Wroblewski, B.M. 1994. Observations of residual sub-surface shear strain in the ultrahigh molecular weight polyethylene acetabular cups of hip prostheses. Journal of Materials Science: Materials in Medicine 5(1): 52–57. Doi: 10.1007/BF00121154.

Cornell, R.M. and Schwertmann, U. 2003. The Iron Oxides: Structure, Properties, Reactions, Occurrences and Uses: John Wiley & Sons.

de Boer, M.P. and Mayer, T.M. 2001. Tribology of MEMS. Doi: 10.1557/mrs2001.65.

Deulin, E.A., Mikhailov, V.P., Panfilov, Y.V. and Nevshupa, R.A. 2010. Friction in vacuum. pp. 33–67. *In*: Deulin, E.A., Mikhailov, V.P., Panfilov, Y.V. and Nevshupa, R.A. (eds.). Mechanics and Physics of Precise Vacuum Mechanisms, Dordrecht: Springer Netherlands.

Dohda, K., Boher, C., Rezai-Aria, F. and Mahayotsanun, N. 2015. Tribology in metal forming at elevated temperatures. Friction 3(1): 1–27. Doi: 10.1007/s40544-015-0077-3.

Dove, J.E. and Jarrett, J.B. 2002. Behavior of dilative sand interfaces in a geotribology framework. Journal of Geotechnical and Geoenvironmental Engineering 128(1): 25–37.

Duanjie, L. 2016. Low Temperature Tribology of Rubber Using Tribometer.

El Fahham, I.M. 1996. Friction and Wear of High Speed Polymer-Metal Contact. Ph.D., University De Montreal.

Gencel, O., Sabri Gok, M. and Brostow, W. 2011. Effect of metallic aggregate and cement content on abrasion resistance behaviour of concrete. Material Research Innovations 15(2): 116–123.

Gradt, T., Schneider, T., Hübner, W. and Börner, H. 1998. Friction and wear at low temperatures. International Journal of Hydrogen Energy 23(5): 397–403. Doi: https://doi.org/10.1016/S0360-3199(97)00070-0.

Gradt, T., Börner, H. and Schneider, T. 2001. Low temperature tribometers and the behaviour of ADLC coatings in cryogenic environment. Tribology International 34(4): 225–230. Doi: https://doi.org/10.1016/S0301-679X(01)00005-6.

Hao, L., Wu, H., Wei, D., Cheng, X., Zhao, J., Luo, S., Jiang, L. and Jiang, Z. 2017. Wear and friction behaviour of high-speed steel and indefinite chill material for rolling ferritic stainless steels. Wear 376-377: 1580–1585. Doi: https://doi.org/10.1016/j.wear.2017.02.037.

Hardell, J., Courbon, C., Winkelmann, H. and Prakash, B. 2014. High temperature friction and wear mechanism map for tool steel and boron steel tribopair AU - Hernandez, S. Tribology—Materials, Surfaces & Interfaces 8(2): 74–84. Doi: 10.1179/1751584X13Y.0000000049.

Hernandez, S., Hardell, J., Winkelmann, H., Rodriguez Ripoll, M. and Prakash, B. 2015. Influence of temperature on abrasive wear of boron steel and hot forming tool steels. Wear 338-339: 27–35. Doi: https://doi.org/10.1016/j.wear.2015.05.010.

Hintze, V., Bierbaum, S. and Scharnweber, D. 2020. Implant surface modifications and new development in surface coatings. *In*: Alghamdi, H. and Jansen, J. (eds.). Dental Implants and Bone Grafts. Woodhead Publishing. Doi: https://doi.org/10.1016/B978-0-08-102478-2.00005-2.

Hojerslev, C. 2001. Tool steels. Riso National Laboratory, Tool Steels (Denmark): 5–25.

Huang, Y.B., Shi, L.B., Zhao, X.J., Cai, Z.B., Liu, Q.Y. and Wang, W.J. 2018. On the formation and damage mechanism of rolling contact fatigue surface cracks of wheel/rail under the dry condition. Wear 400-401: 62–73. Doi: https://doi.org/10.1016/j.wear.2017.12.020.

Hübner, W., Gradt, T., Schneider, T. and Börner, H. 1998. Tribological behaviour of materials at cryogenic temperatures. Wear 216(2): 150–159. Doi: https://doi.org/10.1016/S0043-1648(97)00187-7.

Kapoor, A., Fletcher, D.I., Schmid, E., Sawley, K.J. and Ishida, M. 2000. Tribology of rail transport. In: Modern Tribology Handbook, Bhushan, B., Ed. CRC Press.

Kendall, K. 1994. Adhesion: Molecules and mechanics. Science 263(5154): 1720. Doi: 10.1126/science.263.5154.1720.

Khalladi, A. and Elleuch, K. 2016. Tribological behavior of wheel–rail contact under different contaminants using pin-on-disk methodology. Journal of Tribology 139(1): 011102-011102-9. Doi: 10.1115/1.4033051.

Korany, Y. and El-Haggar, S. 2001. Using slag in manufacturing masonry bricks and paving units. TMS Journal 19(1): 97–106.

Küçükömeroğlu, T. and Kara, L. 2014. The friction and wear properties of CuZn39Pb3 alloys under atmospheric and vacuum conditions. Wear 309(1): 21–28. Doi: https://doi.org/10.1016/j.wear.2013.10.003.

Li, G., Shen, M., Meng, X., Li, J., Li, X. and Peng, X. 2015. An experimental study on tribological properties of groove-textured surface of 316L stainless steel. Journal of Function Materials 44: 2033–2037.

Liu, C.T., George, E.P. and Oliver, W.C. 1996. Grain-boundary fracture and boron effect in Ni3Si alloys. Intermetallics 4(1): 77–83. Doi: https://doi.org/10.1016/0966-9795(95)96901-5.

Lyu, Y., Zhu, Y. and Olofsson, U. 2015. Wear between wheel and rail: A pin-on-disc study of environmental conditions and iron oxides. Wear 328-329: 277–285. Doi: https://doi.org/10.1016/j.wear.2015.02.057.

Lyu, Y., Bergseth, E., Wahlström, J. and Olofsson, U. 2019. A pin-on-disc study on the tribology of cast iron, sinter and composite railway brake blocks at low temperatures. Wear. Doi: https://doi.org/10.1016/j.wear.2019.01.110.

Marmo, B.A., Blackford, J.R. and Jeffree, C.E. 2005. Ice friction, wear features and their dependence on sliding velocity and temperature. Journal of Glaciology 51(174): 391–398. Doi: 10.3189/172756505781829304.

Miyoshi, K. 1999. Considerations in vacuum tribology (adhesion, friction, wear, and solid lubrication in vacuum). Tribology International 32(11): 605–616. Doi: https://doi.org/10.1016/S0301-679X(99)00093-6.

Montgomery, R.S. 1976. Friction and wear at high sliding speeds. Wear 36(3): 275–298. Doi: https://doi.org/10.1016/0043-1648(76)90108-3.

Mylvaganam, K. and Zhang, L.C. 2011. 4—Micro/nano Tribology. pp. 121–160. In: Paulo Davim, J. (ed.). Tribology for Engineers, Woodhead Publishing.

Niu, M., Zhang, X., Chen, J. and Yang, X. 2019. Friction and wear properties of Ni3Si alloy under different vacuum conditions. Vacuum 161: 443–449. Doi: https://doi.org/10.1016/j.vacuum.2019.01.015.

Opris, CD., Liu, R., Yao, M.X. and Wu, X.J. 2007. Development of Stellite alloy composites with sintering/HIPing technique for wear-resistant applications. Materials & Design 28(2): 581–591. Doi: https://doi.org/10.1016/j.matdes.2005.08.004.

Pacheco-Torgal, F. and Jalali, S. 2010. Reusing ceramic wastes in concrete. Construction and Building Materials 24(5): 832–838. Doi: https://doi.org/10.1016/j.conbuildmat.2009.10.023.

Pinchuk, D., Pinchuk, J., Akochi-Koblé, E., Ismail, A. and van de Voort, F.R. 2002. Tribology and Lubrication in Extreme Environments (Two Case Studies).

Quinn, T.F.J. and Winer, W.O. 1985. The thermal aspects of oxidational wear. Wear 102(1): 67–80. Doi: https://doi.org/10.1016/0043-1648(85)90092-4.

Radchik, V.S. and Radchik, A.S. 1958. About the deformations on the surface layer in sliding friction. Rep. Acad. of Sciences, USSR 119(5): 933–935.

Sagy, A., Tesei, T. and Collettini, C. 2017. Fault-surface geometry controlled by faulting mechanisms: Experimental observations in limestone faults. Geology 45(9): 851–854. Doi: 10.1130/G39076.1.

Shi, L.B., Ma, L., Guo, J., Liu, Q.Y., Zhou, Z.R. and Wang, W.J. 2018. Influence of low temperature environment on the adhesion characteristics of wheel-rail contact. Tribology International 127: 59–68. Doi: https://doi.org/10.1016/j.triboint.2018.05.037.

Soemantri, S., McGee, A.C. and Finnie, I. 1985. Some aspects of abrasive wear at elevated temperatures. Wear 104(1): 77–91. Doi: https://doi.org/10.1016/0043-1648(85)90247-9.

Stachowiak, G.W. and Batchelor, A.W. 2006. 12—Adhesion and adhesive wear. In: Engineering Tribology (Third ed.). Burlington: Butterworth-Heinemann. pp. 553-572. Doi: https://doi.org/10.1016/B978-075067836-0/50013-4.

Stachowiak, G.W. 2017. How tribology has been helping us to advance and to survive. Friction 5(3): 233–247. Doi: 10.1007/s40544-017-0173-7.

Subhash, G., Corwin, A.D. and de Boer, M.P. 2011. Evolution of wear characteristics and frictional behavior in MEMS devices. Tribology Letters 41(1): 177–189. Doi: 10.1007/s11249-010-9696-z.

Sugihara, T. and Enomoto, T. 2012. Improving anti-adhesion in aluminum alloy cutting by micro stripe texture. Precision Engineering 36: 229–237.

Sundararajan, S. and Bhushan, B. 1998. Micro/nanotribological studies of polysilicon and SiC films for MEMS applications. Wear 217(2): 251–261. Doi: https://doi.org/10.1016/S0043-1648(98)00169-0.

Sviridenok, A.I., Myshkin, N.K. and Kovaleva, I.N. 2015. Latest developments in tribology in the journal Friction and Wear. Journal of Friction and Wear 36(6): 449–453. doi: 10.3103/S106836661506015X.

Syed, A.K., Hardell, J. and Prakash, B. 2010. Tribological behaviour of surface-treated and post-oxidized tool steels at room temperature and 400°C. Estonian Journal of Engineering 16(2): 123–134. Doi: 10.3176/eng.2010.2.01.

Teeria, T., Kuokkalaa, V., Siitonen, P., Kivikytö-Reponen, P. and Liimatainen, J. 2006. Impact wear in mineral crushing. Proceedings of the Estonian Academy of Sciences 12(4): 408–418.

Theiler, G. and Gradt, T. 2007. Polymer composites for tribological applications in hydrogen environment. International Conference on Hydrogen Safety. S. Sebastian, Spain.

Theiler, G. and Gradt, T. 2010. Friction and wear of PEEK composites in vacuum environment. Wear 269(3): 278–284. Doi: https://doi.org/10.1016/j.wear.2010.04.007.

Theiler, G. and Gradt, T. 2018. Friction and wear behaviour of polymers in liquid hydrogen. Cryogenics 93: 1–6. Doi: https://doi.org/10.1016/j.cryogenics.2018.05.002.

Thomas, B.S., Kumar, S., Mehra, P., Gupta, R.C., Joseph, M. and Csetenyi, L.J. 2016. Abrasion resistance of sustainable green concrete containing waste tire rubber particles. Construction and Building Materials 124: 906–909. Doi: https://doi.org/10.1016/j.conbuildmat.2016.07.110.

Torres, H., Varga, M., Adam, K. and Ripoll, M.R. 2016. The role of load on wear mechanisms in high temperature sliding contacts. Wear 364-365: 73–83. Doi: https://doi.org/10.1016/j.wear.2016.06.025.

Varga, M., Rojacz, H., Winkelmann, H., Mayer, H. and Badisch, E. 2013. Wear reducing effects and temperature dependence of tribolayer formation in harsh environment. Tribology International 65: 190–199. Doi: https://doi.org/10.1016/j.triboint.2013.03.003.

Vettiger, P., Brugger, J., Despont, M., Drechsler, U., Dürig, U., Häberle, W., Lutwyche, M., Rothuizen, H., Stutz, R., Widmer, R. and Binnig, G. 1999. Ultrahigh density, high-data-rate NEMS-based AFM data storage system. Microelectronic Engineering 46(1): 11–17. Doi: https://doi.org/10.1016/S0167-9317(99)00006-4.

Wang, S.Q., Wei, M.X. and Zhao, Y.T. 2010. Effects of the tribo-oxide and matrix on dry sliding wear characteristics and mechanisms of a cast steel. Wear 269(5): 424–434. Doi: https://doi.org/10.1016/j.wear.2010.04.028.

Wang, W.J., Shen, P., Song, J.H., Guo, J., Liu, Q.Y. and Jin, X.S. 2011. Experimental study on adhesion behavior of wheel/rail under dry and water conditions. Wear 271(9): 2699–2705. Doi: https://doi.org/10.1016/j.wear.2011.01.070.

Wang, Y.-Q., Zhou, H., Shi, Y.-J. and Feng, B.-R. 2012. Mechanical properties and fracture toughness of rail steels and thermite welds at low temperature. International Journal of Minerals, Metallurgy, and Materials 19(5): 409–420. Doi: 10.1007/s12613-012-0572-8.

Wheeler, J.M., Armstrong, D.E.J., Heinz, W. and Schwaiger, R. 2015. High temperature nanoindentation: The state of the art and future challenges. Current Opinion in Solid State and Materials Science 19(6): 354–366. Doi: https://doi.org/10.1016/j.cossms.2015.02.002.

Wills, B.A. and Napier-Munn, T. 2005. 7—Grinding mills. pp. 146–185. *In*: Wills, B.A. and Napier-Munn, T. (eds.). Wills' Mineral Processing Technology (Seventh Edition), Oxford: Butterworth-Heinemann.

Wu, W., Chen, G., Fan, B. and Liu, J. 2016. Effect of Groove surface texture on tribological characteristics and energy consumption under high temperature friction. PLOS ONE 11(4): 1–20. Doi: 10.1371/journal.pone.0152100.

Xu, X., Yu, Y. and Huang, H. 2003. Mechanisms of abrasive wear in the grinding of titanium (TC4) and nickel (K417) alloys. Wear 255(7): 1421–1426. Doi: https://doi.org/10.1016/S0043-1648(03)00163-7.

Yap, A.U.J., Teoh, S.H. and Chew, C.L. 2002. Effects of cyclic loading on occlusal contact area wear of composite restoratives. Dental Materials 18(2): 149–158. Doi: https://doi.org/10.1016/S0109-5641(01)00034-3.

Yazıcı, Ş. and İnan, G. 2006. An investigation on the wear resistance of high strength concretes. Wear 260(6): 615–618. Doi: https://doi.org/10.1016/j.wear.2005.03.028.

Yi, H., Xu, G., Cheng, H., Wang, J., Wan, Y. and Chen, H. 2012. An overview of utilization of steel slag. Procedia Environmental Sciences 16: 791–801. Doi: https://doi.org/10.1016/j.proenv.2012.10.108.

Yüksel, İ. 2017. A review of steel slag usage in construction industry for sustainable development. Environment, Development and Sustainability 19(2): 369–384. Doi: 10.1007/s10668-016-9759-x.

Zhou, H., Zhang, Z.H., Ren, L.Q., Song, Q.F. and Chen, L. 2006. Thermal fatigue behavior of medium carbon steel with striated non-smooth surface. Surface and Coatings Technology 200: 6758–6764.

Chapter 9

Lubrication and Coating Challenges in Extreme Conditions

9.1 Introduction

The challenges associated with extreme conditions cover a wide variety of environments, from the human body to space engineering. These include high temperatures, cryogenic temperatures, vacuum (such as in space), radiation, dust environments, corrosive environment, and mechanical/tribological stresses. Devices and equipment used in such environments are abundantly found in automotive, nuclear, aerospace, heavy machinery, power and thermal industries.

Lubrication is the introduction of a substance between the contact surfaces of moving parts to reduce friction and to dissipate heat. The selection of a suitable lubricant and understanding of the mechanism by which it acts to separate surfaces in a bearing or other machine components is a major area for study in tribology. Tribological research increased significantly in response to a technological demand for advanced lubrication techniques with high performance levels in extreme working environments.

Coatings of special materials on the existing components is a highly practiced method. Coating techniques are favorable in several applications where liquid lubrication is not suitable or recommended. For example, liquid lubrication is not appropriate in extreme environments, such as high temperatures in the chemical and power generation industry and vacuum in space applications. Also, in the food processing industry, liquid lubrication poses a very high risk of contamination. Thus, there is a need for surface coatings in such extreme environments.

9.2 Lubrication Challenges

The environment, or operating conditions, can play a significant role in the longevity of a lubrication system. Temperature abnormalities, such as ultra-high and cryogenic temperatures, extremely wet or dusty surroundings and excess vibration, can all have detrimental effects on lubrication. Indeed, it is usual to experience more than one of these conditions at a time in the same tribosystem.

9.2.1 Lubrication at High Temperatures

In the field of tribology, lubrication at high temperatures is one of the challenges encountered. In fact, at high temperatures, liquid lubricants can decompose, degrade or oxidize. However, suitable solid lubricants can extend the operating temperatures of systems beyond 500°C while maintaining relatively low coefficients of friction (Bhushan, 2000). On the basis of the classical theory of adhesion and solid lubrication, if chemical reaction film or physical adsorption film with low shear stress covers on the contact surface, it can achieve a low friction coefficient in boundary lubrication. During high temperature friction, the solid-lubricating film generated on the worn surface by physical or chemical reaction can effectively reduce the friction. If the sliding bodies differ chemically or if there is a third or fourth body at the sliding interface, two or more oxides may form on the sliding surface and control friction and wear.

High temperature lubricant is a kind of lubricant that can be used to reduce friction and minimize wear between tribo-surfaces at high temperatures. Some oxides, fluorides, and sulfates become soft at elevated temperatures and, therefore, can be used as lubricants.

In particular, traditional solid lubricants with a layered structure, like graphite and MoS_2, can provide lubrication below 400°C due to undesirable oxidation at elevated temperatures, while fluoride graphite and WS_2 can withstand temperatures of up to 500°C. Moreover, hexagonal boron nitride (HBN) can maintain lubrication at temperatures of up to 1000°C (Zhu et al., 2019). Indeed, there is no unique lubricant that can operate at a broad temperature range.

Recently, a mixture of two or more solid lubricants is one of the more promising strategies to achieve high temperature solid-lubricating materials. For instance, Ag and BaF_2/CaF_2 eutectic combination can provide favorable lubricity from room temperature to 1000°C.

Various high temperature solid-lubricating materials have been developed to meet different requirements. According to the composition, a high temperature solid-lubricating material is divided into metal matrix, intermetallic matrix and ceramic matrix high temperature solid-lubricating material.

Metal matrix composites mostly consist of Al matrix, Fe matrix, Cu matrix, Ni matrix and Co matrix solid-lubricating composites. Al matrix and Cu matrix composites are viable candidates for use below 400–500°C, while the permissible temperature of Fe matrix solid-lubricating composites is at most 700°C. Ni matrix and Co matrix solid-lubricating composites have high usable temperature, but not exceeding 900°C for long term applications. Table 9.1 presents tribological properties of selected metal matrix composites used as high temperature solid-lubricating materials (Zhu et al., 2019).

On the other hand, intermetallic matrix materials often offer a compromise between ceramic and metallic properties. Lately, studying intermetallic compounds has lead to various novel high temperature solid-lubricating materials developments, such as titanium aluminum matrix, nickel aluminum matrix, ferric aluminum matrix, nickel silicon matrix and other high temperature solid-lubricating materials (Zhu et al., 2019). Table 9.2 displays the tribological properties of selected intermetallic matrix composites used as solid-lubricating materials.

Ceramic matrix composites could be considered as a serious solution for controlling friction and wear at operating temperatures of up to 1000°C, with the added advantages of corrosion and oxidation resistance. The main reason for this is their hardness stability at elevated temperatures. Typical advanced ceramics have a hardness of about 15–30 GPa at room temperature. Nevertheless, the main drawback of such materials is their low toughness, which results in microfractures at contact surfaces during the sliding process, leading to a low wear resistance and a strong plowing effect. Therefore, a major challenge in advanced ceramic matrix solid-lubricating composites is to develop long-life tribo-components for use in high temperature environments. The most famous examples of ceramic matrix composites are: Zirconia matrix $ZrO_2(Y_2O_3)$, Alumina matrix Al_2O_3, Silicon nitride matrix Si_3N_4 and Sialon matrix $Si_4Al_2O_2N_6$.

Currently, smart lubricating materials and multifunctional lubricating materials are developed as new class materials with increased safety, long-term durability and the minimum repairing costs possible. Such materials are designed to be self-diagnosing, self-repairing and self-adjusting. These materials include structural/lubricating integrated material, anti-radiation lubricating material, conductive or insulation lubricating material, etc.

9.2.2 Lubrication at Low Temperatures and in Cryogenic Environments

First, we should define the cryogenic environment as those temperatures below –153°C (120 K) (Gradt, 2013). An example of cryogenic conditions can be found in the mechanical structure of large superconducting magnets for magnetic plasma suspension in fusion reactors. The environmental temperature for these spacers and supports is below 4.2 K. Other examples are valves and fuel pumps for liquid hydrogen and liquid oxygen in cryogenic rocket engines. These pumps reach speeds of up to 60.000 min^{-1} and are exposed to temperatures between 20 K at rest and 900 K during operation.

Table 9.1: Tribological properties of selected metal matrix composites used as high temperature solid-lubricating materials. After Zhu et al. (Zhu et al., 2019) with permission from Elsevier publishing.

Materials	Synthetic method and materials composition	Tested conditions and tribological results
Al matrix composites	Spark plasma sintering Al-Si-Fe-Ni-Graphite	Al_2O_3 ball; 3 N; 0.2 m/s; RT-500°C, μ: 0.3–0.4, W: ~ 10^{-5} mm³/Nm
	Spark plasma sintering Al-Fe-V-Si-Graphite	Al_2O_3 ball; 3 N; 0.2 m/s; RT-350°C; μ: 0.34–0.46; W: 4.36–9.75 × 10^{-4} mm³/Nm
Cu matrix composites	Hot pressing sintering Cu-Sn-Zn-Pb-Graphite	45 # steel; 39.2 N; 1 m/s; RT, μ: ~ 0.14, W: 4 × 10^{-6} mm³/Nm 450°C, μ: ~ 0.15, W: 7 × 10^{-4} mm³/Nm
	Hot pressing sintering Cu-Ni-Sn-Fe-Graphite	$Cr1_2$ steel; 26 N; 0.51 m/s; RT ~ 500°C; μ: ~ 0.2; W: ~ 10–10 mm³
Fe matrix composites	Hot pressing sintering Fe-Mo-CaF_2	5140 steel pin; 39.2 N; 1 m/s; RT, μ: 0.36–0.38, W: 8 × 10^{-5} mm³/Nm; 600°C, μ: 0.28–0.30, W: 1.8 × 10^{-6} mm³/Nm
	Cold pressing and sintering Fe-Mo-Ni-Graphite	5140 steel pin; 39.2 N; 1 m/s; RT, μ: 0.36, W: 3.13 × 10^{-6} mm³/Nm; 320°C, μ: 0.36, W: 3.29 × 10^{-6} mm³/Nm; 450°C, μ: 0.32, W: 6.44 × 10^{-7} mm³/Nm
	Cold isostatic pressing and sintering PM212, Ni/Co/Cr_3C_2-15 wt% Ag–15 wt% BaF_2/CaF_2	440 C steel ball; 20 N; 0.3 m/s; RT-600°C, μ: 0.3–0.5, W: ~ 10^{-6} mm³/Nm;
Ni matrix composites	Cold isostatic pressing and sintering PM212, Ni/Co/Cr_3C_2-15 wt% Ag–15 wt% BaF_2/CaF_2	R41 alloy disk; 4.9 N; 2.7 m/s; RT, μ: 0.35 ± 0.05, W: 3.2 ± 1.5 × 10^{-5} mm³/Nm (PM212 pin), 7.2 ± 2.0 × 10^{-5} mm³/Nm (R41 disk); 350°C, μ: 0.38 ± 0.02, W: 3.9 ± 1.8 × 10^{-5} mm³/Nm (PM212 pin), 3.5 ± 1.0 × 10^{-6} mm³/Nm (R41 disk); 760°C, μ: 0.35 ± 0.06, W: 3.6 ± 0.9 × 10^{-6} mm³/Nm (PM212 pin), 1.0 ± 0.6 × 10^{-5} mm³/Nm (R41 disk); 850°C, μ: 0.29 ± 0.03, W: 4.1 ± 2.0 × 10^{-6} mm³/Nm (PM212 pin), 5.0 ± 1.0 × 10^{-6} mm³/Nm (R41 disk)
	Hot isostatic pressing PM212, Ni/Co/Cr_3C_2-15 wt% Ag–15 wt% BaF_2/CaF_2	R41 alloy disk; 4.9 N; 2.7 m/s; RT, μ: 0.37 ± 0.04, W: 1.8 ± 0.5 × 10^{-5} mm³/Nm (PM212 pin), 0.45 ± 0.11 × 10^{-5} mm³/Nm (R41 disk); 350°C, μ: 0.32 ± 0.07, W: 2.5 ± 0.3 × 10^{-5} mm³/Nm (PM212 pin), 0.85 ± 0.4 × 10^{-6} mm³/Nm (R41 disk); 760°C, μ: 0.31 ± 0.04, W: 0.7 ± 0.4 × 10^{-6} mm³/Nm (PM212 pin), 2.2 ± 0.8 × 10^{-5} mm³/Nm (R41 disk); 850°C, μ: 0.29 ± 0.04, W: 8.3 ± 1.0 × 10^{-6} mm³/Nm (PM212 pin), 0 mm³/Nm (R41 disk)
	Hot pressing sintering PM300, NiCr-20 wt%Cr_2O_3-10 wt% Ag–10 wt% BaF2/CaF2	low carbon steel 1011; 100–400 N; 0.183–0.732 m/s; 540°C, μ: ~ 0.3
	Medium frequency induction heating Ni alloy-graphite-Ag	TZM alloy disk; 20 MPa; 10 m/s; RT-700°C, μ: 0.18–0.31, W: ~ 2 × 10^{-6} mm³/Nm
	Hot pressing sintering Ni alloy-Ag-CeF_3	Hastelloy C pin; 39.2 N; 0.5 m/s; RT-700°C, μ: 0.11–0.25
	Hot pressing sintering Ni alloy-graphite-CeF_3	Hastelloy C pin; 49 and 98 N; 1.5 m/s; RT-700°C, μ: ~ 0.3, W: ~ 10^{-5} mm³/Nm
	Hot pressing sintering Ni-Cr-W-Fe-C-MoS_2	Si_3N_4 disk; 50 N; 0.8 m/s; RT-600°C, μ: 0.14–0.27, W: 1.0–3.5 × 10^{-6} mm³/Nm
	Hot pressing sintering NiCr-Al_2O_3-$SrSO_4$-Ag	Al_2O_3 ball; 10 N; 0.1 m/s; RT-1000°C, μ: 0.28–0.48, W: 3.19–0.546 × 10^{-5} mm³/Nm
	Hot pressing sintering NiCr-$BaMoO_4$	Si_3N_4 ball; 5 N; 0.13 m/s; 600°C, μ: 0.26–0.30, W: 10–5 ~ 10^{-6} mm³/Nm
	Hot pressing sintering NiCr-$BaCr_2O_4$	Al_2O_3 ball; 5 N; 0.13 m/s; 800°C, μ: 0.27, W: 4.5 × 10^{-6} mm³/Nm

Table 9.1 contd. ...

...Table 9.1 contd.

Materials	Synthetic method and materials composition	Tested conditions and tribological results
	Hot pressing sintering NiCrMoAl-12%Ag-10%CaF$_2$/BaF$_2$	Inconel 718; 10 N; 1.0m/s; RT-800°C, μ: < 0.25; W: 1.9–13.1× 10^{-5} mm^3/Nm
	Hot pressing sintering NiCrMoTiAl-12.5%Ag-(5–15%) CaF$_2$/BaF$_2$	Si$_3$N$_4$ ball; 5 N; 1 m/s; RT-900°C, μ: 0.23–0.31, W: 1.1–43.0 × 10^{-5} mm^3/Nm
	Hot pressing sintering NiCrMoTiAl-12.5%Ag-(5–15%) CaF$_2$/BaF$_2$	Inconel 718; 5 N; 1 m/s; RT-900°C, μ: 0.23–0.34, W: 0.8–39.4 × 10^{-5} mm^3/Nm
	Hot pressing sintering NiCrMoTiAl-Ag-MoS$_2$-CaF$_2$	Si$_3$N$_4$ ball; 5 N; 1 m/s; RT-700°C, μ: 0.16–0.40, W: 1–29.4 × 10^{-5} mm^3/Nm
	Hot pressing sintering NiCrMoTiAl-12.5%Ag-5%CaF$_2$/BaF$_2$	Si$_3$N$_4$ ball; 5 N; 0.8 m/s; Vacuum; RT-800°C, μ: < 0.25, W: ~ 10^{-5} mm^3/Nm
	Hot pressing sintering NiCrMoAl-12.5%Ag-(5–10%) CaF$_2$/BaF$_2$	Si$_3$N$_4$ ball; 5 N; 0.785 m/s; Vacuum; RT-800°C, μ: < 0.2; W: ~ 10^{-6} mm^3/Nm
	Hot pressing sintering NiMoAl-Cr$_2$O$_3$-Ag$_2$Mo$_2$O$_7$	Inconel 718 alloy disk; 2 N; 0.287 m/s; Vacuum; RT-700°C, μ: < 0.85–0.35; W: 90^{-2} × 10^{-5} mm^3/Nm
	Hot pressing sintering NiMoAl-Cr$_2$O$_3$-Ag$_2$MoO$_4$	Inconel 718 alloy disk; 2 N; 0.287 m/s; Vacuum; RT-700°C, μ: < 0.94–0.26; W: 50^{-1} × 10^{-5} mm^3/Nm

RT: room temperature; μ: friction coefficient; W: wear rate

The major lubrication problem with cryogenic components' tribosystems is that the temperatures are far below the operation range of any liquid lubricant. Although some authors reported that most base oils and greases can withstand temperatures dipping below 0°C, and many can tolerate temperatures below –10°C. Practically, lubrication by oils and greases could be applicable at temperatures down to –100°C. Nonetheless, if temperatures drop into the critical zone when the lubricant begins to stiffen or becomes extremely viscous, the application will not be lubricated properly and some equipment may seize up under these conditions (Klüber Lubrication, 2017). The lubricant will start to exhibit the properties similar to a solid and, consequently, catastrophic results for the tribosystem are expected. Furthermore, the lubricants themselves are not able to build up hydrodynamic lubrication at cryogenic temperatures.

In general, lubricating oils of high viscosity index are recommended for low temperature operating conditions. However, merely selecting a lubricant with a high viscosity index is not always the best solution to low-temperature problems because lubricating oil is composed of one or more base oils and additives which can affect low-temperature performance. Table 9.3 represents properties of based oils in low temperatures. Greases, on the other hand, provide excellent low-temperature performance. This is due to the fact that friction at the point of tribological contact (the lubrication point) increases grease temperature sufficiently to maintain lubrication.

At low temperatures and in cryogenic environments, liquid lubricants can solidify or become highly viscous and be not effective. On the other end, solid lubricants have usually been found to be better than liquid lubricants or greases. The most common solid lubricants for cryogenic temperature are Polytetrafluoroethylene (PTFE), Polycarbonate (PC), Tungsten disulphide (WS$_2$), and Molybdenum disulphide (MoS$_2$). In addition, ice could be a possible lubricant for deformation in cryogenic environments. The advantage of this lubrication technique would work for any cryogenic system which has parts exposed to an ambient humid environment. This provides a method of self-lubrication, in the sense that no active mechanism is needed to supply the lubricant.

Table 9.2: Tribological properties of selected intermetallic matrix composites used as solid-lubricating materials. After Zhu et al. (Zhu et al., 2019) with permission from Elsevier publishing.

Materials	Synthetic method and composite composition	Tested conditions and tribological results
Ni$_3$Al matrix solid-lubricating composites	Hot pressing sintering Ni$_3$Al-Cr/Mo/W-Ag-CaF$_2$/BaF$_2$	Si$_3$N$_4$ ball; 10 N; 0.2 m/s; RT-1000°C, μ: < 0.35, W: 10^{-6}–10^{-4} mm^3/Nm
	Hot pressing sintering Ni$_3$Al-Ag-BaMoO$_4$	Si$_3$N$_4$ ball; 20 N; 0.2 m/s; RT-800°C, μ: 0.29–0.38, W: 10^{-5}–10^{-4} mm^3/Nm
	Hot pressing sintering Ni$_3$Al-Ag-BaCrO$_4$	Si$_3$N$_4$ ball; 20 N; 0.2 m/s; RT-800°C, μ: 0.28–0.35, W: 10^{-5}–10^{-4} mm^3/Nm
	Spark plasma sintering Ni$_3$Al-WS$_2$-Ag-hBN	Si$_3$N$_4$ ball; 10 N; 0.234 m/s; RT-800°C, μ: 0.25–0.52, W: ~ 10^{-4} mm^3/Nm
	Spark plasma sintering Ni$_3$Al-Ti$_3$SiC$_2$-Graphene	Si$_3$N$_4$ ball; 12 N; 0.2 m/s; RT-750°C, μ: 0.26–0.57, W: 3.1–6.5 × 10^{-6} mm^3/Nm
	Spark plasma sintering Ni$_3$Al-Ti$_3$SiC$_2$-WS$_2$	Si$_3$N$_4$ ball; 10 N; 0.2 m/s; RT-800°C, μ: 0.18–0.39, W: 1.5–3.7 × 10^{-5} mm^3/Nm
NiAl matrix solid-lubricating composites	Hot pressing sintering NiAl-Cr-Mo-ZnO	Si$_3$N$_4$ ball; 10 N; 0.2 m/s; 800–1000°C, μ: ~ 0.35, W: 10^{-6}–10^{-5} mm^3/Nm
	Hot pressing sintering NiAl-Cr-Mo-CuO	Si$_3$N$_4$ ball; 10 N; 0.2 m/s; 800–1000°C, μ: ~ 0.2, W: 10^{-6}–10^{-5} mm^3/Nm
	Hot pressing sintering NiAl-Cr-Mo-CaF$_2$	Si$_3$N$_4$ ball; 10 N; 0.2 m/s; 800–1000°C, μ: ~ 0.2, W: 10^{-6}–10^{-5} mm^3/Nm
	Hot pressing sintering NiAl-Cr-Mo-CaF$_2$-Ag	Si$_3$N$_4$ ball; 10 N; 0.2 m/s; RT-1000°C, μ: 0.2–0.4, W: 10^{-5}–10^{-4} mm^3/Nm
	Hot pressing sintering NiAl-NbC-AgVO$_3$	RT ~ 900°C, μ: 0.25–0.6, W: 10^{-5} ~ 10^{-4} mm^3/Nm
	Hot pressing sintering NiAl-Mo-AgVO$_3$	718 pin; 2 N; 0.287 m/s; RT-900°C, μ: 0.1–0.3, W: 10^{-5}–10^{-4} mm^3/Nm
	Spark plasma sintering NiAl-Ti$_3$SiC$_2$-MoS$_2$	Si$_3$N$_4$ ball; 10 N; 0.2 m/s; RT-800°C, μ: 0.12–0.29, W: 4.1–6.0 × 10^{-5} mm^3/Nm
	Spark plasma sintering NiAl-Ti$_3$SiC$_2$-WS$_2$	Si$_3$N$_4$ ball; 10 N; 0.2 m/s; RT-800°C, μ: 0.26–0.60, W: 3.5–6.5 × 10^{-5} mm^3/Nm
	Spark plasma sintering NiAl-PbO	Si$_3$N$_4$ ball; 10 N; 0.2 m/s; RT-800°C, μ: 0.10–0.50, W: 4.1–9.2 × 10^{-5} mm^3/Nm
TiAl matrix solid-lubricating composites	Spark plasma sintering TiAl-Ag	Si$_3$N$_4$ ball; 12 N; 0.8 m/s; RT-900°C, μ: 0.25–0.21, W: 2.4–2.8 × 10^{-4} mm^3/Nm
	Spark plasma sintering TiAl-graphene-Ti$_3$SiC$_2$	Si$_3$N$_4$ ball; 10 N; 0.20 m/s; RT-800°C, μ: 0.34–0.56, W: 0.87–2.87 × 10^{-4} mm^3/Nm
	Spark plasma sintering TiAl-Ag-Ti$_3$SiC$_2$	Si$_3$N$_4$ ball; 10 N; 0.234 m/s; RT-600°C, μ: 0.32–0.43, W: 1.23–4.13 × 10^{-4} mm^3/Nm
	Spark plasma sintering TiAl-Ag-TiB$_2$	Si$_3$N$_4$ ball; 12 N; 0.3 m/s; RT-600°C, μ: 0.28–0.36, W: 1.0–2.0 × 10^{-4} mm^3/Nm
	Spark plasma sintering TiAl-Ag-V$_2$O$_5$	Si$_3$N$_4$ ball; 12 N; 0.3 m/s; RT-600°C, μ: 0.19–0.61, W: 1.87–2.0 × 10^{-4} mm^3/Nm
	Spark plasma sintering TiAl-Ag-BaF$_2$/CaF$_2$-Ti$_3$SiC$_2$	Si$_3$N$_4$ ball; 10 N; 0.234 m/s; RT-600°C, μ: 0.33–0.45, W: 3.25–4.07 × 10^{-4} mm^3/Nm
Fe$_3$Al matrix solid-lubricating composite	Hot pressing sintering Fe$_3$Al-Ba$_{0.25}$Sr$_{0.75}$SO$_4$	Si$_3$N$_4$ ball; 10 N; 0.01 m/s; 600–800°C, μ: 0.19–0.29, W: ~ 10^{-5} mm^3/Nm

RT: room temperature; μ: friction coefficient; W: wear rate

Table 9.3: Properties of based oils in low temperatures.

Base oil	Viscosity index	Pour point
Mineral Oil	80–100	Poor low temperature behavior.
Polyalphaolefin (PAO)	150–250	Good low temperature behavior up to –40 to –50°C.
Esters	140–175	Excellent low temperature behavior up to –50°C.
Polyalkylene glycol (PAG)	150–270	Good low temperature behavior up to –40 to –50°C.
Silicone Oils	190–500	Excellent low-temperature behavior up to –50°C.
Perflourinated polyether (PFPE)	50–350	Some types can go up to –70°C.

In addition to classical lubrication mechanisms, non-contact systems, such as gas-lubricated or magnetic bearings, are investigated for cryogenic conditions. The main advantages are that they do not create frictional heat, are maintenance free, and usually have a very long lifetime. Also, magnetic bearings can carry high loads and allow for active control. Gas lubrication has advantages similar to magnetic suspension, but, depending on the application, requires extremely high surface quality, a supply of pressurized gas, or extremely high rotational speed (Gradt, 2013).

Trautmann et al. (Trautmann et al., 2005) investigated the dynamic friction behavior of PC, PTFE and MoS_2 at room temperature, 26°C, and low temperature, –60°C, using a split Hopkinson pressure bar (SHPB) and specimens of varying thicknesses. At ambient temperature, PTFE and MoS_2 were found to be good lubricants, but not as good as petroleum jelly, which reduced friction to the lowest level (almost zero). At cryogenic temperatures, experimental results suggested that MoS_2 is a better lubricant than PTFE. Mechanical properties of PC remain unchanged after sharp changes in temperature. This makes it a promising material for cryogenic applications in liquid nitrogen, hydrogen and helium environments.

9.2.3 Lubrication Under Extreme Pressures

Extreme pressure conditions are another lubrication challenge. In many applications, there exists a lubricating (boundary) condition that is typical for most failures due to adhesive wear. Typical examples of such conditions are gears and metalworking applications (cutting fluids). To overcome these extreme conditions, extreme pressure (EP) additives are polar molecules introduced to the lubricating fluid in order to prevent this adhesive wear and protect the components. These additives are lubricant components that chemically react with the rubbing surface, forming a sacrificial coating that prevents both metal surfaces from welding together under the high temperature and high pressure that occurs during boundary condition. It is worth noting that the nature of the action of the additive is very much surface dependent, the choice of EP additive may well need to be based on the type of the surface being lubricated.

Currently, the most common EP additives contain chlorine, sulphur or phosphorus (Stachowiak, 2005). Organochlorine additives are known to give tribofilms of $FeCl_2$ and $FeCl_3$ and organosulphur FeS/$FeSO_4$/Fe_2O_3 composite films through a complex sequence of thermal and physical changes at the metal/additive interface.

Recently, the addition of different nanoparticles to lubricants and cutting fluids has shown great advantages (Peña-Parás et al., 2019). The main advantages of using nanoparticle additives are (Minh et al., 2017; García et al., 2018): (1) reducing of the friction coefficient, decreasing cutting forces in machining and lowering the energy consumption of the process; (2) extending the life of the manufacturing tools by lowering thermal stress for both anti-wear and extreme pressure applications; and (3) improving the surface finish of the manufactured parts.

Novel toroidal forms of carbon nanotubes (CNTs) are carbon nanotori, consisting of circular CNTs formed by connecting their two ends. Through molecular dynamics simulation, their modulus has been predicted to range from 0.125–1.5 TPa. Due to these outstanding properties and their similarity to CNTs, Peña-Parás et al. (Peña-Parás et al., 2019) proposed carbon nanotori as lubricant additives. Varying concentrations of nanolubricants of carbon nanotori in distilled water (DW) were prepared and tested in order to determine their tribological properties under EP and AW conditions. For the EP analysis, nanolubricants showed a substantial decrease in frictional torque and a delay in scuffing initiation, particularly at concentrations of 1.2–2.0 wt.%. The scuffing load, seizure load and pressure loss-limit of DW were also increased of up to 114%, 143%, and 950%, respectively. These outstanding improvements are attributed to the carbon nanotori's mechanical properties that were able to support the EPs due to a load-bearing effect during the test. Table 9.4 presents the scuffing and seizure loads for DW and carbon nanotori lubricants obtained by the EP test. The data proved that the scuffing load increased from 810 N for DW to a maximum of 1820 N, with 2.0 wt.% carbon nanotori. The seizure load (at which lubricant seizure occurs) was found at 2963 N for DW; with 0.1 wt.%, 0.5 wt.%, and 1.0 wt.% increased to 3043 N, 3600 N, and 4020 N, respectively.

Figure 9.1 shows the proposed tribological mechanisms of carbon nanotori. Similar to the mechanisms for carbon nanotubes at extreme-pressures, carbon nanotori are able to provide a load-bearing effect. This is due to their high elastic modulus that effectively reduced metal-metal contact, evidenced by the marked reduction in surface roughness and increase in load-carrying capacity, particularly at concentrations above 1.2 wt.%. At lower contact pressures, such as for the anti-wear test, carbon nanotori may provide a rolling/sliding mechanism, gradually decreasing friction and wear with increasing nanoparticle content, thereby providing a smoother surface finish.

Table 9.4: Scuffing load and seizure load values for nanolubricant. After Peña-Parás et al. (Peña-Parás et al., 2019), with permission from Elsevier publishing.

Lubricant	Scuffing load	Seizure load
Distilled water (DW)	810	2963
DW + 0.1 wt.% carbon nanotori	825	3043
DW + 0.5 wt.% carbon nanotori	1257	3600
DW + 1.0 wt.% carbon nanotori	1316	4020
DW + 1.2 wt.% carbon nanotori	1327	7200 (no seizure)
DW + 1.5 wt.% carbon nanotori	1481	7200 (no seizure)
DW + 2.0 wt.% carbon nanotori	1820	7200 (no seizure)

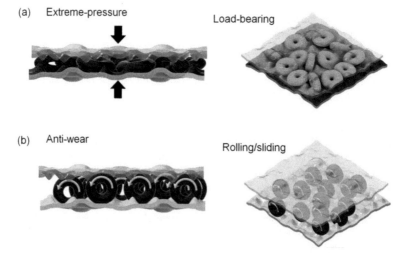

Figure 9.1: Representation of tribological mechanism of carbon nanotori (a) load-bearing under extreme-pressures, and (b) rolling/sliding for anti-wear conditions. After Peña-Parás et al. (2017), with permission from Elsevier publishing.

9.2.4 Lubrication in Vacuum (Space Tribology)

Working in a vacuum environment conditions has become indispensable in modern tribological research. It finds wide applications in space equipment, pumps, microelectronics, particle accelerators, and others. Typically, liquid or grease lubricants are used to combat friction and wear in most tribological applications. However, in a vacuum environment there are some limitations for using classical lubricants, due to the fact that the lubricant may either freeze, evaporate or decompose and, hence, become ineffective. Moreover, there is no oxygen and water in a vacuum (or in space), so no friction reducing oxide layers could grow to provide some form of a lubricating solid film on the surfaces. Therefore, special lubrication mechanisms have been adopted for working in a vacuum environment.

Space tribology is a subset of the lubrication-in-vacuum field, denoting the specialized discipline dealing with friction and wear phenomena related to satellites, spacecraft and space station components (Bhushan, 2000). Such components are designed to operate in space missions, where the environment is extreme. These extreme working conditions include very large temperature gradients, from very low temperatures, below –200°C, up to +250°C when exposed to the sun.

Environmental interactions will have to be considered carefully in the selection and design of the required durable lubrication system. Several environmental factors may be hazardous to performance integrity. Potential threats common to both the Moon and Mars are low ambient temperatures, wide daily temperature swings (thermal cycling), solar flux, cosmic radiation and large quantities of dust (Miyoshi, 2007). Table 9.5 represents minimum, mean and maximum surface temperatures of Earth, the Moon and Mars.

Many liquid lubricants have been used in space: Silicones, mineral oils, perfluoropolyalkylethers (PFPAE), polyalphaolefins, polyolesters and multiply-alkylated cyclopentanes. Also, Greases are used for a variety of space applications: Low- to high-speed, angular-contact ball bearings, journal bearings, and gears. The primary reason for using grease is its ability to act as a reservoir for supplying oil to contacting surfaces. It can also act as a physical barrier to prevent oil loss by creep or by centrifugal forces.

In either vacuum or space environments, solid lubricants may be the best choice for controlling friction and wear.

Solid lubricants are used in space to lubricate various mechanical components, such as rolling-element bearings, journal bearings, gears, bushings, electrical sliding contacts, clamps and latches, bolts, seals, rotating nuts, robotic and telescoping joints, backup bearings for gas and magnetic bearings, fluid transfer joints, various release mechanisms, valves, and harmonic drives (Fusaro, 1994). Solid lubricants have very small evaporation rate in vacuum, so not only can it overcome the above shortcomings of classical lubricants but also it has the stability and better lubricating behavior in vacuum. The following types of solid lubricants are used for these space applications:

(1) Soft metal films: Gold, silver, lead, indium and barium.

(2) Lamellar solids: Molybdenum disulfide, tungsten disulfide, cadmium iodide, lead iodide, molybdenum diselenide, intercalated graphite, fluorinated graphite and pthalocyanines (Ziegelheim et al., 2019).

(3) Polymers: PTFE, polyimides, fluorinated ethylene-propylene, ultra-high-molecular weight polyethylene, polyether ether ketone, polyacetal, and phenolic and epoxy resins.

(4) Other low-shear-strength materials: Fluorides of calcium, lithium, barium and rare earths; sulfides of bismuth and cadmium; oxides of lead, cadmium, cobalt and zinc.

The most common way to utilize a solid lubricant is to apply it to a metal surface as a film or surface coating. Surface coating technology have been developed for vacuum conditions. These coatings consist of a thin layer of soft film, typically molybdenum disulphide, artificially deposited on the surfaces. Coatings of solid lubricant are built up atom by atom yielding a mechanically strong surface layer with a long service life and the minimum quantity of solid lubricant.

Previous works suggested that the lubricity of a solid lubricant is controlled by a number of parameters (Bhushan, 2000). For example, both graphite and MoS_2 have layered crystal structures, but the extent of

Table 9.5: Minimum, mean and surface temperatures of Earth, the Moon and Mars.

	Temperature								
	K			°C			°F		
	Min.	Mean	Max.	Min.	Mean	Max.	Min.	Mean	Max.
Earth	148	288	331	−89	15	58	−128	59	136
Moon	126	250	373	−147	−23	100	−233	−9	212
Mars	161	213	265	−112	−60	−8	−170	−76	17

their lubricity and durability is largely controlled by extrinsic factors, such as the presence or absence of vapors or gas in the test environment.

Graphite, HBN and H_3BO_3 function extremely well in humid air. The lubricity of graphite persists up to 400°C, while H_3BO_3 begins to decompose at about 170°C. These two solids do not provide lubrication in dry or vacuum environments. In dry air, inert atmospheres, or vacuum, graphite's lubricity degrades rapidly. The friction coefficient increases to as high as 0.5, and it wears out quickly.

In contrast, MoS_2 and other transition-metal MX_2 (where M is Mo, W, Nb, Ta, etc., and X is sulfur, selenium, or tellurium) dichalchogenides work best in vacuum or dry running conditions. However, they degrade rather quickly in moist and oxidizing environments. The friction coefficients of self-lubricating metal dichalcogenides are typically in the range of 0.002 to 0.05 in vacuum or dry and inert atmospheres, but increase rapidly to 0.2 in humid air. It is generally agreed that no solid can provide very low friction and wear, regardless of test environment and/or conditions. Among advanced solid lubricants, amorphous carbon (a–C) is considered as a very promising candidate for space applications due to its reasonable thermal stability, chemical inertness and superlubricity under certain conditions (Igartua et al., 2015). However, the challenging drawback of a-C is the lack of coatings that can satisfactorily perform both under vacuum conditions and under atmospheric pressure.

9.2.5 Lubrication in High Dust and Dirty Areas

High dust areas and dirt environments can weigh profoundly on a lubricant due to the high risk of particle contamination. These contaminants readily form a grinding paste, causing failure of tribosystems and subsequently damage of equipment. This type of contamination most frequently takes place when airborne or stagnant particles gain access to the lubrication system through open ports and hatches, especially in systems with negative pressure. Studies have found that half of a bearing's loss of usefulness can be attributed to mechanical wear. This wear, which occurs through surface abrasion, fatigue and adhesion, is often the result of particle contamination (Noria Corporation, 2019). This particle-induced wear makes up roughly 80% of all wear that happens over the equipment's lifespan.

9.2.6 Lubrication in Radiation Environments

In radiation environments, liquid lubricants can decompose. Suitable solid lubricants can extend the operation of systems beyond 10^6 rads while maintaining relatively low coefficients of friction.

9.2.7 Lubrication in Corrosive Environments

Under intermittent loading conditions or in corrosive environments, liquid lubricants become contaminated. Changes in critical service and environmental conditions—such as loading, time, contamination, pressure, temperature and radiation—also affect liquid lubricant efficiency. When equipment is stored or is idling for prolonged periods, solid lubricants provide permanent, satisfactory lubrication.

9.2.8 Lubrication in Limited Weight Applications

In weight-limited spacecraft and rovers, solid lubrication has the advantage of weighing substantially less than liquid lubrication. The elimination (or limited use) of liquid lubricants and their replacement by solid lubricants would reduce spacecraft weight and, therefore, have a dramatic impact on the mission extent and craft maneuverability.

9.2.9 Lubrication in Natural and Artificial Joints

It is widely accepted that the general principles of tribology could be used to understand the lubrication of natural and artificial joints in the body. The natural synovial joint is covered by a soft delicate layer of articular cartilage and is lubricated with synovial fluid. Whilst the cartilage is delicate, the loads experienced by our joints are high and can exceed many times our body weight during normal daily activities (Stewart, 2010; Nečas et al., 2019).

When load is applied, the cartilage does not act like a shock absorber, as it does not absorb impact energy; rather, it deforms under the loads applied to it and acts to distribute the load over a wider area, thus reducing contact stress. This deformation also makes the contact between the articulating surfaces more conforming, thus making it easier to achieve fluid film lubrication that protects the cartilage from direct contact. If the film thickness is much smaller than the combined roughness of the articulating surfaces, then a boundary lubrication regime exists and surface asperity interaction cannot be avoided.

In artificial joints, sliding surfaces generally consist of Co–Cr alloys, ceramics or ultrahigh-molecular-weight polyethylene (UHMWPE), all of which have certain advantages and disadvantages. Wear of UHMWPE has been an important issue in recent years, since PE particles can activate the body's immune system, leading to a response that ultimately causes loosening of the implant in the bone, requiring a further surgery (Stachowiak, 2005).

In case the lubricant film thickness is of a similar magnitude to the combined surface roughness, a mixed lubrication regime will exist with variable amounts of asperity contact.

In terms of joint replacement bearing types, metal on UHMWPE is considered to be boundary lubricated as the relatively soft polyethylene surface has a high roughness, therefore, surface asperity contact and wear cannot be avoided.

Currently, some developments of the tribological property of artificial joints have been attempted by adding lubricants into the artificial joints. Polyethyleneglycol (PEG) was considered for combination with synovial fluid for increasing the wear resistance of UHMWPE in artificial knee joint, but no significant result was achieved (Kobayashi et al., 2014). Some other lubricants were also used for improving tribological behaviors, such as charged polymer brushes-grafted hollow silica nanoparticles, and poly (3-sulfopropyl methacrylate potassium salt) brushes grafted poly (N-isopropylacrylamide) microgels.

Polyvinylpyrrolidone (PVP) is one of the biocompatible water-soluble polymers and have been used in articular cartilages. Besides its good biocompatibility, PVP aqueous solution also has good lubricating properties. Guo et al. (Guo et al., 2016) introduced PVP as a lubricating additive in artificial joints. In this study five different PVP concentrations from 0 to 20 wt% were tested as well as two different solvents. It was hypothesized that PVP could be an effective lubricant for artificial knee joints and the change of the PVP concentration would have a significant influence on the lubricating properties of the PVP solutions. Test results of the PVP solutions with different concentrations are shown in Fig. 9.2. It is postulated that the friction coefficient was reduced significantly (about 50%) by increasing PVP, from 0.042 to 0.090 for the deionized water solution and from 0.060 to 0.12 for the bovine serum solution. It is concluded that, due to the good lubricating property and biocompatibility of the PVP in the artificial joint pairs, PVP could be considered as a potential lubricating additive for the artificial joints in the future.

9.3 Coating Challenges

Many aspects of solid lubricant coating are already described. However, in the present section further significant developments in coating technologies will be revised.

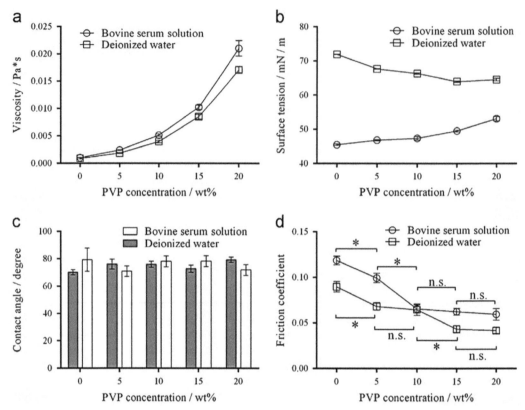

Figure 9.2: Comparisons of the tribological characteristics of different PVP solutions in the friction pair of Ti6Al4V balls and UHMWPE discs. (a) Viscosity; (b) surface tension; (c) static contact angle; (d) friction coefficient. After Guo (Guo, 2017) with permission from Elsevier publishing.

Surface coatings are applied on components using several techniques, such as weld cladding, thermal spraying and electrochemical techniques (Bakshi and Harimkar, 2015). Thermal spray coatings are exclusively applicable for larger parts while thin coating techniques are available for relatively smaller parts for machining and tribological applications.

Surface coating processes include Physical Vapor Deposition (PVD) and Chemical Vapor Deposition (CVD) coatings. These advanced processes facilitate the deposition of ceramic, binary alloy, multilayered, nanocomposite, hybrid and duplex surfaces to fulfil any desired operating requirements. The early uses for hard PVD and CVD ceramic coatings were dominated by metal-cutting and forming tool applications.

Surface coatings are commonly applied in components and devices in order to improve tribological properties, material performance, durability, strength and resistance in harsh operating conditions. For example, turbine blades used in jet engines can be made of materials like nickel-based superalloys. In some cases, the blades may have to be coated with a thermal barrier coating (TBC) in order to minimize exposure of the blade material to high temperatures. Also, the effectiveness and performance of hard to superhard coatings, such as diamond-like carbon, WC/C, TiC/C, VC, diamond films and other solid-film lubricants, have been validated in a variety of severe contact conditions. Table 9.6 introduces some properties of high-temperature coating materials.

The coating processes are divided into the following four categories (Holmberg and Mattews, 1994): (1) gaseous state processes, (2) solution state processes, (3) molten or semi-molten state processes and (4) solid state processes.

Two important characteristic parameters for the coating processes are the thickness of the coatings that can be achieved and the deposition temperature.

Table 9.6: Some properties of coating materials.

Compounds	Melting point (°C)	Hardness (HV)	Service temperature (°C max.)
Al_2O_3	2050	2000	> 1000
B_4C	2500	3000–4000	500–600
Cr-C	1500–1900	1700–2200	800
Cr_2O_3	2400	5000	800–1000
Cr-N	1700	2500	700–800
cBN	2700	3000	> 1000
Diamond	3800	8000	600
DLC		1500–6000	< 400
TiAlN	–	3200	800
TiBN	–	5500	700
TiCN	3150	2000–3000	< 500
Ti-N	3000	3000–4000	< 600
VC	2200	3000	500–600
W-C	2000	2500	500–600
WV+C		1000	< 400

The materials used as a coating layer range from purely ceramic to purely organic in character and have very different properties, but can provide critical benefits to the tribosurface. The advantages of applying surface coating are (Miyoshi, 2007):

- Extreme abrasion and wear resistance
- Ultrahard surface
- High impact strength
- Remarkably low surface energy
- Increased thermal conductivity and thermal transfer
- High nonstick (release) properties
- Lowest friction (energy consumption) attainable
- Permanent dry lubrication to prevent galling
- Erosion protection
- Radiation protection
- Nontoxicity
- Chemical protection
- Precision conformance over complex geometry
- Excellent corrosion resistance
- Wide temperature range

Probably, the most challenging research problems regarding durable surface coatings for working in extreme conditions are:

(1) Dedicating the resources required to develop the coating application process.
(2) The technique used in coatings.
(3) What are their strengths and surface energies?
(4) How do they break down? And how do they self-heal?
(5) How can we extend their lifetimes?

Selection of the most suitable coating material and deposition method requires an understanding of the operational requirements, together with testing and validation of the coated surface. Such tests are required in order to ensure the achievement of the necessary fitness-for-purpose characteristics.

Previously, coatings were achieved by a thin layer of single metallic or ceramic material, but currently there has been a significant interest towards designing, processing and characterizing innovative coating technology, as illustrated in Fig. 9.3. The following are some of the promising technologies in surface coating (Miyoshi, 2007; Stachowiak, 2005):

- Smart coatings (carbon nanotubes CNs, clays, UV screening particles; titanium, cerium and zinc oxides, and antimicrobial particles) (Clingerman, 2014).
- Fluorinated carbon nano-onions.
- Multinanolayered, composite coatings (e.g., $MoS_2/WS_2/C$ and MoS_2/WS_2).
- Functionally graded, multilayered inorganic coatings (e.g., TiCx/C and TiCx).
- Multilayered composite (superlattice and nanolayered) coatings (e.g., WC/C, MoS_2/C, WS_2/C, and TiC/C).
- Multiphase (hybrid and duplex) coatings.
- Nanocrystalline diamond and Nanocomposite (nc) coatings.
- Large-area diamondlike carbon coatings.
- TiO_2 grown on 55Ni-45Ti and titanium-based alloys.
- Hard binary compound coatings, including ceramics and advanced coatings (e.g., BN, B_4C, VC, AlN, CNx, TiO_2, SiO_2, etc.).
- Soft metal coatings and polytetrafluoroethylene (PTFE) deposited by advanced deposition techniques.

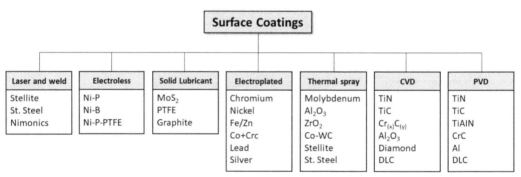

Figure 9.3: Different types of coating technologies.

9.3.1 Space Applications

The challenges of space operation have motivated significant research efforts aimed at the development of multifunctional surface coatings for use in extreme working conditions. Vail et al. (Vail et al., 2009) investigated single-walled carbon nanotubes (SWCNT) coating in order to improve the tribological behavior of PTFE. It is concluded that SWCNT-PTFE nanocomposites are a good multifunctional coatings candidate for extreme environments. It is reported that the use of SWCNT fillers in PTFE increased strength by ~ 50%, increased tensile strain by ~ 8,000%, increased wear resistance by ~ 2,000% and increased friction coefficients by ~ 50%. Figure 9.4 represents the friction coefficient and wear rate of SWCNT-PTFE nanocomposites.

9.3.2 Biomedical Applications

In the field of biomedical applications, there are many challenges for coating technology. For example, to prolong the service lifetime of an artificial joint, it is imperative to improve mechanical and tribological

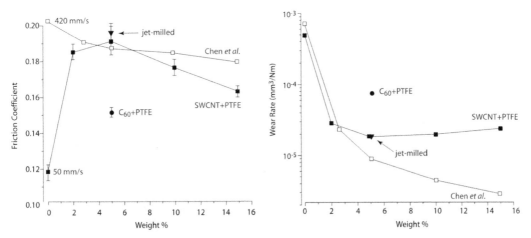

Figure 9.4: Effect of filler loading on friction coefficient (left) and wear rate (right) of SWCNT-PTFE nanocomposites. Confidence intervals on friction coefficient data represent the standard deviation during the test. Confidence intervals on wear rate data represent the standard deviation during the test. Confidence intervals on wear rate data represent the experimental uncertainties in both worn volume and sliding distance and are smaller than data markers in all cases. Normal load was 250 N and sliding speed was 50.8 mm/s. After Vail et al. (2009), with permission from Elsevier publishing.

properties of the prosthetic materials. This improvement could be achieved by adding a surface coating layer. Zhang et al. (Zhang et al., 2018) highlighted the application of TiCuN solid-solution coating with Cu dopant compared to pure TiN and TiN/Cu nano-composite coatings. In this work, a low concentration (5.2%) of Cu atoms is incorporated into TiN to obtain TiCuN solid-solution coating, and structural, mechanical, tribological, and biocompatibility properties are explored in order to further understand the effect of Cu doping. It is reported that the hardness of TiCuN solid-solution coating as 35.6 ± 1.2 GPa is 1.4 times and 4.6 times larger than TiN and TiN/Cu, respectively, as shown in Fig. 9.5a. The wear rate for the TiCuN solid solution as $(4.2 \pm 0.2)\ 10^{-9}$ m^3/Nm is an order of magnitude less than TiN and TiN/Cu nano-composite, as shown in Fig. 9.5c. SEM examination of the worn surface, as shown in Fig. 9.5d, indicated that pure TiN and TiN/Cu nano-composite coatings exhibit poor wear resistance. The evidence for this, is the ploughing grooves that can be observed, exhibiting serious abrasive wear, which also can be confirmed by the distribution of Ti and Si (from worn-out substrate surface). In contrast, the wear track for TiCuN solid solution is narrow and smooth with minimal Si exposure. Such variations in the wear track are consistent with the aforementioned wear rates for the different coatings.

Stainless steel material is extensively used in medical surgery applications due to its properties. Eventually, most of the biomedical materials should have merits, such as anti-corrosion, high mechanical resistance and high durability. Therefore, functional coatings could be applied for achieving such properties.

Graphene is considered as one of the most promising 2D nanomaterials due to its outstanding optical, electronic, mechanical, physicochemical and thermal properties (Cardenas et al., 2014; Reina et al., 2017). Nowadays, graphene sheets are applied as reinforcements and coatings for enhancing medical implants. ElSawy et al. (ElSawy et al., 2019) developed a novel method for synthesis of new graphene sheet-based coating for stainless-steel stent application. In this facile method, 6 μm and 10.6 μm thick graphene sheets have been directly exfoliated in chitosan solution and then coated on stainless-steel. Then, graphene sheets decorated with TiO$_2$ nanoparticles (TiO$_2$NPs) were coated on stainless-steel stent. The experimental tests were procured using three samples coded as SSP-CH, SSP-CH-GRP and SSP-CH-GRP-TiO$_2$NPs. These codes are related to the treatment of stainless-steel stent with coating dispersion of chitosan-graphene (CHGR) and their TiO$_2$NPs decorated composites (CH-GRP-TiO$_2$NPs) coatings. Results suggested that the stent coated by graphene reflects promising positive mechanical and hematological properties compared to the uncoated ones, as shown in Table 9.7.

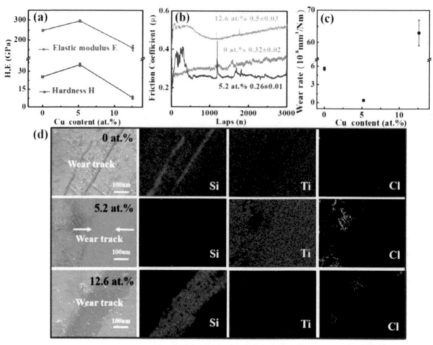

Figure 9.5: (a) Hardness H and elastic modulus E, (b) Friction coefficient curves as a function of sliding cycle, (c) Wear rate and (d) SEM micrographs and SEM-EDS mapping images on the worn surface for the samples with 0 at% Cu, 5.2 at% Cu and 12.6 at% Cu. After Zhang et al. (2018), with permission from Elsevier publishing.

Table 9.7: Mechanical properties of different uncoated and coated samples. After ElSawy et al. (ElSawy et al., 2019), with permission from Elsevier publishing.

Sample code	Tensile strength (MPa)	Stiffness (N/m 10^4)	Young Modulus (MPa 10^2)
SS	320	135	565
SSP-CH	150	35	211
SSP-CH-GRP	197	86	374
SSP-CH-GRP-TiO$_2$NPs	160	54	235

9.3.3 Plasma Sprayed Slag Coatings

Among different surface coating techniques, plasma thermal spraying stands out as one of the most widely-used techniques. This technique is involved in surface modification by augmentation of wear resistance, which may affirm the great versatility and its application to a wide spectrum of materials. In this technique, it is possible to spray all kinds of metallic and nonmetallic materials, like metal oxides, carbides, nitrides and silicides.

The coatings with considerable amounts of hard materials can protect against a variety of wear media, including abrasive, adhesive and corrosive ones. This could be obtained by considering some common conventional materials, like nickel, iron, cobalt and molybdenum-based alloys, carbides of ceramic and tungsten. However, extensive investigations have recently been initiated in order to explore the coating potential of slag for different engineering and structural applications (Mantry et al., 2013; Pati, 2015a; Pati, 2015b; Binici and Aksogan, 2011; Sutar et al., 2015). An attempt has been made by Mantry et al. (Mantry et al., 2013) to use the composite coatings of copper slag and Al powder in a suitable combination on aluminium and mild steel substrates in order to improve the surface properties of these ductile metal-alloy substrates. It was found that the adhesion strength of the coating was significantly affected by the plasma torch input power level. The adhesion strength of the coating was also found to increase with the addition of aluminium to the copper slag. This is attributed to the formation of Al-diffusion layers within the splats, thereby enhancing the cohesive forces between splats. Finally, it is concluded that the new coating has reasonable hardness and may be recommended for tribological applications.

Sutar et al. (Sutar et al., 2015) evaluated the wear behavior of varying percentages of fly ash with pure red mud coating at different operating powers, subject to normal laboratory conditions. The powder mixture concomitants of red mud and different percentage of fly ash were prepared manually. In addition, pure red mud powder was also considered as a coating material for the sake of comparison on the basis of percentage of fly ash addition. Coating of the various combinations of mixed powders was conducted on one side cross section of the mild steel substrate. The spraying process was performed by adopting a conventional atmospheric plasma spraying (APS) set up. The plasma input power was varied from 6 to 15 kW by controlling the gas flow rate, voltage and arc current. The powder feed rate was maintained at a constant 10 gm/min by using a turntable type volumetric powder feeder. The tribological experiment was conducted in the pin on disc tribometer. The rig was used to evaluate the wear behavior of the coatings against hardened ground steel disc (En-32) having hardness of 65 HRC and surface roughness (Ra) 0.5 μm. It is observed that, for the early stage, the wear rate increases slowly and then drastically improves with sliding distance for all coating types and finally becomes stagnant, as represented in Fig. 9.6. The operating power level proved to be a coating property variable which enhances the coating resistance, however, reaching an optimum value indicated some other dominating parameters.

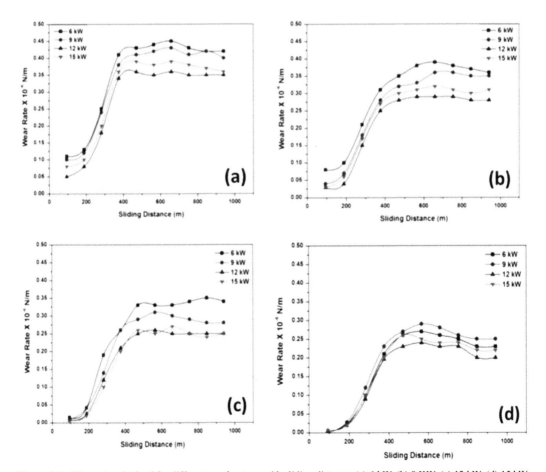

Figure 9.6: Wear rates obtained for different coating type with sliding distance. (a) 6 kW, (b) 9 KW, (c) 12 kW, (d) 15 kW. After Sutar et al. (2015).

9.3.4 Polymer Coatings for Extreme Working Conditions

There are extensive works done in order to investigate advanced materials that can withstand extreme working conditions. These include the development of polymers, polymer composites, and blends that can perform well under extreme operating conditions. A practical example of extreme working conditions are the tribo-components in space applications. These components are subjected to thermal cycling, ultra violet radiation, microgravity, vacuum and cryogenic temperatures, which make the material selection challenging. Another challenge is found in the conditioning and refrigeration industry, where it is common to have oil-less compressors, mainly due to the negative thermodynamic effect from the liquid lubricant/refrigerant mixture in the refrigeration cycle. The properties of the lubricant can be affected under boundary/mixed lubrication conditions, causing scuffing on the sliding components.

Recently, polymers have been introduced as tribological protective coatings, such as erosion/corrosion resistant coatings for automotive applications, transparent protective coatings for touch panel screens and optical applications, and in many other tribo-applications. Thin polymer coatings would have a lower cost because of their easy and inexpensive deposition methods and lower material requirements. Polymer coatings, in pure or blended (composite) form, can be applied to almost all materials by chemical, physical or even mechanical bonding or interlocking. Polymer 'soft' coatings are commonly called antifriction coatings or sometimes bonded coatings (Pradeep, 2013). These coatings are similar to the paints in many aspects, such as application methods and properties. Some of the advantages of these types of coatings are:

-   Reduction of the coefficient of friction with a very small variation.
-   Used in a wide temperature range, from −200 to +650°C.
-   Elimination of oil and grease usage.
-   Improving the running-in wear.
-   Protection against corrosion.

Thermal spray and cold dynamic spray are the main methods employed to coat polymers onto substrates. The main advantages of thermal spraying method are: (a) the feasibility to coat large components of complex geometries, (b) eliminating the use of solvents, by using of as raw material, and (c) maintenance and repair possibility.

Nevertheless, there are other techniques, such as spin coating and dip coating, that can be used for the deposition of uniform thin polymer films (from a few nanometers to a few micrometers). Also, deposition of polymeric coatings for tribological applications has proven to be effective and inexpensive compared to other techniques. Figure 9.7 shows typical cross-sectional images of different polymer coatings obtained by electrostatic deposition. It can be seen that the typical thickness of the coatings is of the order of tenths of microns.

Based on their success in several tribological applications, many studies are focused on the tribological performance of PEEK and PTFE blends under different operating conditions (Gheisari et al., 2019; Lan et al., 2019). Although both PEEK and PTFE coatings show low coefficient of friction (0.2 for PEEK and less than 0.15 for PTFE), the wear mechanisms differ significantly compared to their bulk forms, which are dominated by mild plowing and delamination. Aromatic thermosetting copolyester (ATSP) composite was also introduced as a very promising matrix phase for high performance coating. Figure 9.8 shows surface coated samples cut out from bearing pads and Fig. 9.9 depicts the wear and friction performance of some PTFE-based and PEEK-based coatings.

Polymers are used in a number of applications in space applications. In particular, polymeric surface coatings are excellent candidates for extreme dry sliding conditions, where oil lubricants and greases are ineffective. It is noted that fibre-reinforced aluminium composites, materials protected by preceramic coatings and several types of thermal control paints display excellent resistance to the effects of space environment (Krishnamurthy, 1995). However, cryogenic and elevated temperatures encountered in space applications set a limit for a lot of polymer materials. For example, Polytetrafluoroethylene (PTFE) embrittles and discolors on exposure to ultra-violet (UV) radiation, crack and craze during thermal cycling

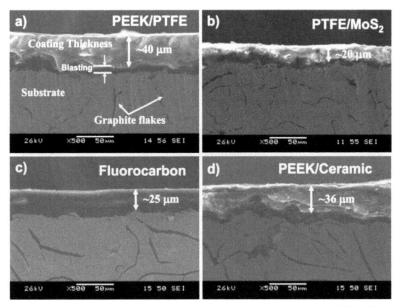

Figure 9.7: Cross sectional SEM images of typical polymer composite coatings: (a) PEEK/PTFE, (b) PTFE/MoS$_2$, (c) Fluorocarbon, (d) PEEK/Ceramic deposited on gray cast iron by electrostatic deposition. After Nunez et al. (2019), with permission from Elsevier Publishing.

Figure 9.8: Samples cut out from bearing pads lined with: (a) PTFE composite, (b) PEEK composite, (c) ATSP composite layer. After Wodtke (2016) with permission from Elsevieer publishing.

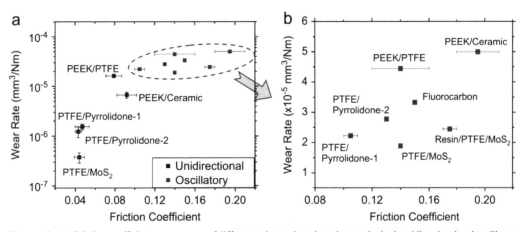

Figure 9.9: (a) Friction coefficient vs. wear rate of different polymer-based coatings under both unidirectional and oscillatory testing and (b) zoom-in of only oscillatory testing results. Reprinted from Yeo (2012), with permission from Elsevier publishing.

Table 9.8: Some damaging effects of space environment upon polymeric materials. After Krishnamurthy (Krishnamurthy, 1995), with permission from Springer publishing.

Environment	Damage
Ultraviolet radiations	Creation of lattice defects in crystalline materials Chain scission of organic materials, free radical formation Crosslinking of organic materials
Charged particle radiation (electron/proton)	Creation of lattice defects in crystalline materials Chain scission of organic materials Crosslinking of organic materials Secondary radiation damages
Vacuum	Volatilization of low vapor pressure fractions and materials Diffusion Vacuum welding
Thermal	Mechanical degradation, softening or embrittlement Chemical degradation Acceleration/deceleration of above environmental effects
Micrometeoroid	Mechanical Fracture/Puncture
Atomic Oxygen	Oxidation Surface erosion Cracking/crazing

can lose its protectiveness for atomic oxygen. The atomic oxygen problem is the critical one for space station materials and is yet to be fully solved. Coatings need to be periodically refurbished in orbit in order to compensate for the space environments.

The combined interaction of the environments causes more severe damages to the polymer coatings than the individual environments as is present in space. The general damaging effects of these environments on the polymers are summarized in Table 9.8. Therefore, the choice of polymeric coating materials for a prescribed mission is based on laboratory simulation studies, previous and present spacecraft data and theoretical considerations of inherent material properties. The duration and position in space must also be taken into account for design.

9.4 Surface Treatment Challenges

As a part of friction surfaces, surface treatment (or surface modification) processes are applied in order to control friction, improve surface wear and corrosion resistance, and change the physical or mechanical properties of the component. Through this technique, metal surfaces can be modified by different treatment processes, such as ion implantation, shot peening, laser surface treatment, thermochemical diffusion and chemical conversion (Bhushan and Gupta, 1991; Cartier et al., 2003). A partial list of these elements is summarized in Table 9.9.

Ion implantation is a surface modification process in which ions are injected into the surface region of a substrate. It is a potential enhancement method for plating processes, such as chrome plating. This process commonly uses products like carbons, nitrides, borides and carbides, etc. (Kennedy et al., 2005). In carburizing, carbon is diffused into the surface of the steel at a certain temperature. A high carbon content is produced at the surface due to rapid diffusion and the high solubility of carbon in austenite. When the steel is then quenched and tempered, the surface becomes a high-carbon tempered martensite, while the ferritic center remains soft and ductile (Bhadeshia and Honeycombe, 2006). Also, Nitrogen provides a hardening effect similar to that of carbon. In carbonitriding, a gas containing carbon monoxide and ammonia is generated, and both carbon and nitrogen diffuse into the steel (Askeland et al., 2011). Finally, only nitrogen diffuses into the surface from a gas in nitriding.

Shot peening is a surface modification process that can improve substrate material properties of hardness, wear resistance, corrosion resistance and fatigue, thus improving the product's service life. By using shot peening, the service life of springs can increase by 400% to 1200%, depending on the extent of peening already imparted on the spring. Shot peening processes can also increase the fatigue life of gears by over 500%. Figure 9.10 illustrates the general process of shot peening, where a stream of round hardened steel shot or similar is propelled onto the surface during the process. The presence of this surface compressive stress serves to retard the initiation and growth of fatigue cracks.

Laser surface treatment is a form of thermal hardening, in which a high-power laser beam is scanned over a surface in order to cause melting to a limited depth. This process is followed by a rapid cooling of the surface resulting in a hard quenched microstructure with a fine grain size formed on re-solidification (Stachowiak and Batchelor, 1993).

Hardening is a heat-treating process involving the heating of an electrically conductive workpiece by inducing a current from a magnetic field generated by an external coil. This process is used for gear and sprocket teeth, axles, crankshafts, piston rods, etc. It usually develops a maximum case depth of around 0.25 in, and results in surface hardnesses of 50–60 HRC (Schmid and Jeswiet, 2018).

Chemical conversions are operations where the component is heated in an atmosphere containing elements (such as carbon, nitrogen or boron) that alter the composition, microstructure and properties of surfaces. For steels with sufficiently high carbon content, surface hardening takes place without using any of these additional elements. The chemical approaches to hardening are summarized in Table 9.10.

Duplex (or hybrid) surface treatment is a sequential combination of surface treatment and coating deposition used to produce a tribological surface with optimized tribological properties.

Surface hardening usually involves a diffusion treatment (e.g., Nitriding, carburizing, chromizing and oxidizing) (Hakami et al., 2017). This treatment is followed by a coating process using plasma-assisted PVD or plasma-enhanced CVD. Usually, the surface hardening layer is much deeper than the thickness of the coating layer. Surface hardening depths depend on diffusion of the species being used. In some cases, however, more than two surface technologies may be established.

Table 9.9: Methods used in surface treatment processes.

Diffusion		Ion implantation	Laser surface treatment	Chemical conversion
Nitriding	Boriding	Argon	Hardening	Anodizing
Tuftride	Tiduran	Nitrogen	Glazing	Phosphating
Sulfinuz	Tifran	Carbon	Surface alloying	Chromating
Sursulf	Tiodize	Chromium		Oxidizing
Sulf-BT	Delsum			Sulfurizing
Noskuff	Zinal			
Forez	Metalliding			
Carburizing	Siliconizing			

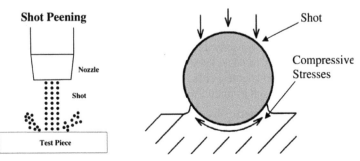

Figure 9.10: Illustration of shot peening process.

Table 9.10: Summary of heat-treating processes involving chemical diffusion into surfaces. After Schmid et al. (2018) with permission from John Wiley and Sons publishing.

Process	Metals hardened	Element added to surface	Procedure	General characteristics	Typical applications
Carburizing	Low-carbon steel (0.2% C), alloy steels (0.08-0.2% C)	C	Heat steel at 870°-950°C in an atmosphere of carbonaceous gases (gas carburizing) or carbon-containing solids (pack carburizing). Then quench.	A hard, high-carbon surface is produced. Hardness 55 to 65 HRC. Case depth up to 1.5 mm. Some distortion of part during heat treatment.	Gears, cams, shafts, bearings, piston pins, sprockets, clutch plates
Carbonitriding	Low-carbon steel	C and N	Heat steel at 700°-800°C in an atmosphere of carbonaceous gas and ammonia. Then quench in oil.	Surface hardness 55 to 62 HRC. Case depth 0.07 to 0.5 mm. Less distortion than in carburizing.	Bolts, nuts, gears
Cyaniding	Low-carbon steel (0.2% C), alloy steels (0.08-0.2% C)	C and N	Heat steel at 760°-845°C in a molten bath of solutions of cyanide (e.g., 30% sodium cyanide) and other salts.	Surface hardness up to 65 HRC. Case depth 0.025 to 0.25 mm. Some distortion.	Bolts, nuts, screws, small gears
Nitriding	Steels (1% Al, 1.5% Cr, 0.3% Mo), alloy steels (Cr, Mo), stainless steels, high-speed tool steels	N	Heat steel at 500°-600°C in an atmosphere of ammonia gas or mixtures of molten cyanide salts. No further treatment.	Surface hardness up to 1100 HV. Case depth 0.1 to 0.6 mm and 0.02 to 0.07 mm for high speed steel.	Gears, shafts, sprockets, valves, cutters, boring bars, fuel-injection pump parts
Boronizing	Steels	B	Part is heated using boron-containing gas or solid in contact with part.	Extremely hard and wear resistant surface. Case depth 0.025 to 0.075 mm (0.001 to 0.003 in.).	Tool and die steels

An example for duplex treatment are the low friction coatings, such as DLC, nitrides and oxides can be used as the soft top coating layer, which can not only increase wear resistance but also reduce interfacial shear stress, and decrease the tendency for debonding of top coatings (Fu, 2010). A deep, case-hardened layer can be used as the bottom layer, which can significantly improve the coating/film adhesion. This concept is conceived to enhance the substrate load-bearing capacity, providing a gradual change of mechanical properties between the relatively "soft" substrate and the hard ceramic coating (Tschiptschin, 2013).

9.4.1 Applications in Automotive Industry

In the automotive industry, where the components are designed to work in extreme operating conditions, surface treatment is one of the optimal solutions in reduction of friction, wear and corrosion. This includes the enhancement of the tribological behavior of rubbing surfaces of powertrain parts and engine interiors.

Surface enhancement engineering basically involves changes to the surface of a material by additive processes, such as thermal spray, PVD, plasma enhanced CVD, nitriding or nitrocarburizing (IONITOX) (Pawlowski, 2008). Also, hard carbon overlays become a solution for increased demands in transmission parts (e.g., for synchronization rings). Table 9.11 shows selected properties of the different treatments and typical applications. The term "Spraying" in the table does not stand only for plasma spraying, but stands for the whole group of thermal spraying.

Cylinder bores: The most frequently used tribological solution for engine blocks in aluminum cast alloys is the insert of cast iron sleeves. Currently, the internal plasma spray coatings are used in the production of a variety of gasoline and diesel engines. These coatings have the potential to reduce the friction of the piston groups by about 30%, as well as up to 3% reduction in fuel consumption.

Ball pivot: High corrosion resistance surfaces are generated using IONITOX process, in which a combination of gas nitriding and plasma activation processes are developed. Before the oxidation, the plasma activation step is carried out. Corrosion stability is reached due to a dense oxide layer (Fe_3O_4) of a thickness of about 2 μm.

Table 9.11: Selected properties of the different treatments and typical applications. After Vetter et al. (Vetter et al., 2005) with permission from Elsevier publishing.

	PVD	PVD/ PECVD	Thermochemical heat treatment		Spraying		Bonding
	MeNCO	DLC	PN/PNC	PNC+OX	Metal	Metal+ ceramics	Carbon onlay
Thickness (µm)	2–50	1–5	DZ/CZ 50–500	DZ/CZ+ oxide 1–3	100–400	200–500	400–700
Hardness (micro)	1600–3500	1000–3000	600–1200	500–800	300–600	400–1000	Not measured
Friction against steel	Dry 0.5–0.8	Dry 0.1–0.3	Dry 0.7–0.9	Dry 0.6–0.8	Lub. 0.05–0.15	Lub. 0.05–0.1	Lub. 0.11–0.13
Deposition temperature (°C)	50–500	60–250	350–800	400–570	50–150	50–150	160–280
Typical application	Piston ring Pumps Decoration	Injection systems Tappest Gears	Piston rings Synchronize Clutches	Ball pivots Gears Pumps	Piston rings Synchronize Shift forks	Piston rings Bores Break discs	Shift forks Synchronize Clutches

Piston rings: In addition to the different substrate materials, different surface treatments are used. Spraying, galvanizing, nitriding, PVD coatings and a combination of nitriding plus PVD coating are used, depending on the application.

9.4.2 Surface Treatment of Polymers

In general, polymers are widely applied in many tribological applications. In particular, thermoplastics polyethylene terephthalate (PET) and polyether-etherketone (PEEK) are of interest because of high strength, stiffness and good dimensional stability. However, they often experience difficulties in gluing and instability or overload during sliding. Depending on the selection of adequate parameters, different types of plasma treatment are used for enhancing and decreasing adhesion or surface hardness of these polymers. In addition, the surface wettability is an important parameter to control adhesion, lubrication and/or interactions with molecules. Therefore, the formation of polar groups at the surface after plasma treatment can increase the surface energy, thereby improving the polymer surface wettability.

Recently, Al-Maliki et al. (Al-Maliki et al., 2018) studied different types of plasma treatment in order to improve the physical and chemical surface characteristics of PEEK and PET polymers. In their study, the PEEK and PET polymer surfaces were modified by cold atmospheric plasma treatment using a Dielectric Barrier Discharge (DBD) equipment operating under controlled air atmosphere.

Coefficients of friction (COF), wear and bulk temperature measurements during dry sliding tests on pristine and plasma-treated PET and PEEK samples under different normal loads are shown in Fig. 9.11. For pristine polymers, COF show significant running-in phenomena with a peak value during the first couple of meters. This can be explained by the presence of a contaminated hydrocarbon layer on the untreated polymers. In general, the friction for pristine PEEK is slightly higher compared to pristine PET for the same normal loads, which can be attributed to the higher mechanical strength and stiffness of PEEK, providing higher sliding resistance. The plasma-treated PET and PEEK present lower friction than pristine polymers, except for the PEEK at highest normal load. It can be observed, however, that the differences in coefficients of friction between untreated and plasma-treated polymers become smaller at high loads.

The wear and deformation Δh of the pristine samples gradually increase at higher loads and are lower for PEEK than for PET at all load levels. This is due to the higher mechanical strength and higher stiffness in combination with smaller contributions of deformation for PEEK. After the plasma treatment, the Δh values for PET increase, in opposition to the lower friction after plasma treatment, while they remain almost similar for PEEK, in opposition to the higher friction after plasma treatment.

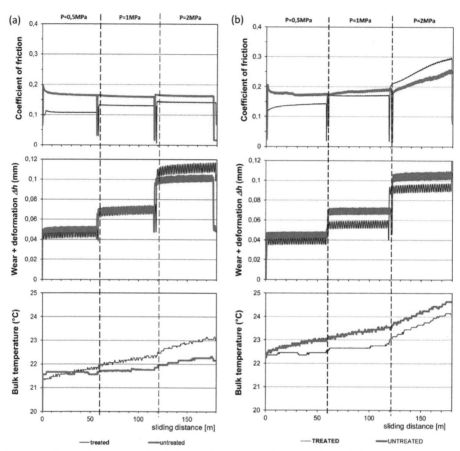

Figure 9.11: Tribological testing under dry sliding conditions at 0.5, 1, and 2 MPa, including on-line measurements for coefficients of friction, wear and displacement Dh, temperature for (a) PET, (b) PEEK. After Al-Maliki et al. (2018), with permission from John Wiley and Sons publishing.

References

Al-Maliki, H., Zsidai, L., Samyn, P., Szakál, Z., Keresztes, R. and Kalácska, G. 2018. Effects of atmospheric plasma treatment on adhesion and tribology of aromatic thermoplastic polymers. Polymer Engineering & Science 58(S1): E93–E103. Doi: 10.1002/pen.24689.

Askeland, D.R., Fulay, P.P. and Wright, W.J. 2011. The Science and Engineering of Materials, SI Edition: Cengage Learning.

Bakshi, S.R. and Harimkar, S.P. 2015. Surface engineering for extreme conditions. JOM 67(7): 1526–1527. Doi: 10.1007/s11837-015-1481-x.

Bhadeshia, H.K.D.H. and Honeycombe, Sir R. 2006. 8—The heat treatment of steels: Hardenability. pp. 167–181. *In:* Bhadeshia, H.K.D.H. and Honeycombe, Sir R. (eds.). Steels (Third Edition), Oxford: Butterworth-Heinemann.

Bhushan, B. and Gupta, B.K. 1991. Handbook of Tribology: Materials, Coatings, and Surface Treatments. United States: McGraw-Hill, New York, NY (United States).

Bhushan, B. 2000. Modern Tribology Handbook. New York: CRC Press.

Binici, H. and Aksogan, O. 2011. The use of ground blast furnace slag, chrome slag and corn stem ash mixture as a coating against corrosion. Construction and Building Materials 25(11): 4197–4201. Doi: https://doi.org/10.1016/j.conbuildmat.2011.04.057.

Cardenas, L., MacLeod, J., Lipton-Duffin, J., Seifu, D.G., Popescu, F., Siaj, M., Mantovani, D. and Rosei, F. 2014. Reduced graphene oxide growth on 316L stainless steel for medical applications. Nanoscale 6(15): 8664–8670. Doi: 10.1039/C4NR02512A.

Cartier, M., Curtis, J.M., Polak, T.A. and Wilcox, G.D. 2003. Handbook of Surface Treatments and Coatings: Professional Engineering.

Clingerman, M. 2014. Smart Coatings, Definitions and Opportunities.

ElSawy, A.M., Attia, N.F., Mohamed, H.I., Mohsen, M. and Talaat, M.H. 2019. Innovative coating based on graphene and their decorated nanoparticles for medical stent applications. Materials Science and Engineering C 96: 708–715. Doi: https://doi.org/10.1016/j.msec.2018.11.084.

Fu, R.Y.Q. 2010. 15—Duplex surface treatments of light alloys. pp. 501–545. *In*: Dong, H. (ed.). Surface Engineering of Light Alloys, Woodhead Publishing.

Fusaro, R.L. 1994. Lubrication of Space Systems. Prepared for the Society of Tribologists and Lubrication Engineers Annual Meeting, Pittsburgh, Pennsylvania.

García, G.E., Trigos, F. and Maldonado-Cortés, D. 2018. Optimization of surface roughness on slitting knives by titanium dioxide nano particles as an additive in grinding lubricant. Int. J. Adv. Manuf. Technol. 96 (9-12): 4111–4121. Doi: doi.org/10.1007/s00170-018-1834-z.

Gheisari, R., Lan, P. and Polycarpou, A.A. 2019. Tribological interactions of advanced polymeric coatings. In: Reference Module in Materials Science and Materials Engineering. Elsevier.

Gradt, T. 2013. Cryogenic solid lubrication. pp. 641–647. *In*: Wang, Q.J. and Chung, Y.-W. (eds.). Encyclopedia of Tribology, Boston, MA: Springer US.

Guo, Y., Hao, Z. and Wan, C. 2016. Tribological characteristics of polyvinylpyrrolidone (PVP) as a lubrication additive for artificial knee joint. Tribology International 93: 214–219. Doi: https://doi.org/10.1016/j.triboint.2015.08.043.

Hakami, F., Pramanik, A. and Basak, A.K. 2017. Duplex surface treatment of steels by nitriding and chromizing. Australian Journal of Mechanical Engineering 15(1): 55–72. Doi: 10.1080/14484846.2015.1093256.

Holmberg, K. and Matthews, A. (ed.). 1994. Chapter 2 Surface Coating Methods. Tribology Series. Elsevier 28: 7–32. Doi: https://doi.org/10.1016/S0167-8922(08)70752-1.

Igartua, A., Berriozabal, E., Zabala, E., Pagano, F., Minami, I., Doerr, N., Gabler, C., Nevshupa, R., Roman, E., Nielsen, L.P., Louring, S. and Muntada, L. 2015. Lubricity and tribochemical reactivity of advanced materials under high vacuum. 16th European Space Mechanisms and Tribology Symposium ESMATS 2015, Bilbao, Spain.

Kennedy, D., Xue, Y. and Mihaylova, M. 2005. Current and future applications of surface engineering. The Engineers Journal (Technical) 59: 287–292.

Klüber Lubrication. 2017. Lubricant Challenges in Extreme Cold Environments.

Kobayashi, M., Koide, T. and Hyon, S.-H. 2014. Tribological characteristics of polyethylene glycol (PEG) as a lubricant for wear resistance of ultra-high-molecular-weight polyethylene (UHMWPE) in artificial knee joint. Journal of the Mechanical Behavior of Biomedical Materials 38: 33–38. Doi: https://doi.org/10.1016/j.jmbbm.2014.06.003.

Krishnamurthy, V.N. 1995. Polymers in space environments. pp. 221–223. *In*: Prasad, P.N., Mark, J.E. and Fai, T.J. (eds.). Polymers and Other Advanced Materials: Emerging Technologies and Business Opportunities, Boston, MA: Springer US.

Lan, P., Nunez, E.E. and Polycarpou, A.A. 2019. Advanced polymeric coatings and their applications: Green tribology. In: Reference Module in Materials Science and Materials Engineering. Elsevier.

Mantry, S., Jha, B.B. and Satapathy, A. 2013. Evaluation and characterization of plasma sprayed Cu Slag-Al composite coatings on metal substrates. Journal of Coatings 2013: 7. Doi: 10.1155/2013/842865.

Menezes, L.P., Ingole, S.P., Nosonovsky, M., Kailas, S.V. and Lovell, M.R. 2013. Tribology for Scientists and Engineers. New York: Springer-Verlag.

Minh, D.T., The, L.T. and Bao, N.T. 2017. Performance of Al_2O_3 nanofluids in minimum quantity lubrication in hard milling of 60Si2Mn steel using cemented carbide tools. Advances in Mechanical Engineering 9(7): 1687814017710618. Doi: 10.1177/1687814017710618.

Miyoshi, K. 2007. Solid lubricants and coatings for extreme environments: State-of-the-Art survey. Hanover, USA: NASA Center for Aerospace Information.

Nečas, D., Vrbka, M., Gallo, J., Křupka, I. and Hartl, M. 2019. On the observation of lubrication mechanisms within hip joint replacements. Part II: Hard-on-hard bearing pairs. Journal of the Mechanical Behavior of Biomedical Materials 89: 249–259. Doi: https://doi.org/10.1016/j.jmbbm.2018.09.026.

Noria Corporation. 2019. How Extreme Operating Conditions Affect Lubrication. https://www.machinerylubrication.com/Read/31232/extreme-conditions-lubrication.

Pati, P.R. 2015a. A Study on Utilization of LD Slag in Erosion Resistant Coatings and Polymer Composites. Ph.D., Department of Mechanical Engineering, National Institution of Technology.

Pati, P.R. and Satapathy, A. 2015b. Development of wear resistant coatings using LD slag premixed with Al_2O_3. J. Mater. Cycles. Waste. Manag. 17(1): 135–143. Doi: 10.1007/s10163-014-0234-1.

Pawlowski, L. 2008. Applications of Coatings. In The Science and Engineering of Thermal Spray Coatings. Doi: 10.1002/9780470754085.ch9.

Peña-Parás, L., Maldonado-Cortés, D., Kharissova, O.V., Saldívar, K.I., Contreras, L., Arquieta, P. and Castaños, B. 2019. Novel carbon nanotori additives for lubricants with superior anti-wear and extreme pressure properties. Tribology International 131: 488–495. Doi: https://doi.org/10.1016/j.triboint.2018.10.039.

Reina, G., González-Domínguez, J.M., Criado, A., Vázquez, E., Bianco, A. and Prato, M. 2017. Promises, facts and challenges for graphene in biomedical applications. Chemical Society Reviews 46(15): 4400–4416. Doi: 10.1039/C7CS00363C.

Schmid, S.R. and Jeswiet, J. 2018. Surface treatment and tribological considerations. Energy Efficient Manufacturing. Doi: doi:10.1002/9781119519904.ch7 10.1002/9781119519904.ch7.

Stachowiak, G.W. and Batchelor, A.W. (ed.). 1993. 9—Solid lubrication and surface treatments. Tribology Series. Elsevier 24: 485–526. Doi: https://doi.org/10.1016/S0167-8922(08)70583-2.

Stachowiak, G.W. 2005. Wear—Materials, Mechanisms and Practice. U.K: WILEY.

Stewart, T.D. 2010. Tribology of artificial joints. Orthopaedics and Trauma 24(6): 435–440. Doi: https://doi.org/10.1016/j.mporth.2010.08.002.

Sutar, H., Mishra, S.C., Sahoo, S.K., Chakraverty, A.P., Maharana, H. and Satapathy, A. 2015. Friction and wear behaviour of plasma sprayed fly ash added red mud coatings. Physical Science International Journal 5(1): 61–73.

Trautmann, A., Siviour, C.R., Walley, S.M. and Field, J.E. 2005. Lubrication of polycarbonate at cryogenic temperatures in the split Hopkinson pressure bar. International Journal of Impact Engineering 31(5): 523–544. Doi: https://doi.org/10.1016/j.ijimpeng.2004.02.007.

Tschiptschin, A.P. 2013. Duplex coatings. In: Wang, Q.J. and Chung, Y.-W. (eds.). Encyclopedia of Tribology, Boston, MA: Springer US.

Vail, J.R., Burris, D.L. and Sawyer, W.G. 2009. Multifunctionality of single-walled carbon nanotube–polytetrafluoroethylene nanocomposites. Wear 267(1): 619–624. Doi: https://doi.org/10.1016/j.wear.2008.12.117.

Vetter, J., Barbezat, G., Crummenauer, J. and Avissar, J. 2005. Surface treatment selections for automotive applications. Surface and Coatings Technology 200(5): 1962–1968. Doi: https://doi.org/10.1016/j.surfcoat.2005.08.011.

Zhang, Y.J., Qin, Y.G., Qing, Y.A., Deng, R.P., Jin, H., Li, R.Y., Rehman, J., Wen, M. and Zhang, K. 2018. TiCuN solid solution coating: Excellent wear-resistant biocompatible material to protect artificial joint. Materials Letters 227: 145–148. Doi: https://doi.org/10.1016/j.matlet.2018.05.061.

Zhu, S., Cheng, J., Qiao, Z. and Yang, J. 2019. High temperature solid-lubricating materials: A review. Tribology International 133: 206–223. Doi: https://doi.org/10.1016/j.triboint.2018.12.037.

Ziegelheim, J., Lombardi, L., Cesanek, Z., Houdkova, S., Schubert, J., Jech, D., Celko, L. and Pala, Z. 2019. Abradable coatings for small turboprop engines: A case study of nickel-graphite coating. Journal of Thermal Spray Technology. Doi: 10.1007/s11666-019-00838-4.

Chapter 10

Simulation and Modeling of Tribo-Systems

10.1 Introduction

The extent to which we understand the tribological processes enables appropriate design and selection of materials. Over the decades, several tribological models and simulations have been proposed for different applications. The fundamental objective of such modeling and simulation is to provide tribologists with the characterisation of systems' behaviour, evaluate influences and tendencies, detect critical conditions, and focus on promising solutions. Most of these models are applicable for the particular material pair, contact geometry, operating conditions, and specific environmental and lubricating conditions. Also, a model is usually derived from the wear and friction data of a specific system and, therefore, only applies to that data set (Al-Bender and De Moerlooze, 2011). Owing to these complexities, no comprehensive and practicable model that shows all of the experimentally observed tribological aspects in one formulation is available.

In fact, it is necessary to improve our understanding of friction and wear in tribo-systems by using theoretical investigations. Analytical models are also necessary for recognizing the behavior at tribological interfaces, partly to avoid using numerical simulations as "black boxes" where the distinctions of the phenomena involved are lost, and partly because full computational models often require prohibitively high computational costs (Vakis et al., 2018; Barber, 2020).

Yet, still there a need for accompanying experimental investigations in tribometrology and analytical tools, like modern investigation procedures for lubrication engineering, surface characterization and wear analyses (Franek et al., 2007). Friction and wear response depends markedly on a variety of tribo-system parameters and any change in any of them will affect the system outcome. In particular, different mechanisms of wear occur simultaneously and, therefore, it is difficult to assign the unique contributor from each mechanism. Finally, tribological behaviour is not a material property, it should be recognized as a system function.

This chapter presents diverse methods which are currently applied in order to figure out and predict tribological behavior. This will include mathematical models, finite element modeling and artificial neural networks approach. It is worth mentioning that wear mapping has been previously introduced in Section 4.3.

10.2 Empirical Equations and Mathematical Models

10.2.1 Empirical Equations

Basically, most current friction and wear mathematical models are empirical ones. These equations are usually developed from experimental data by, for example, fitting curves to the plotted data or by estimating equations by least squares. Empirical models based on a number of variables are often applicable for the specific scope of conditions. To generalize the model, other variables should be included. A common generalization method is to conduct tribological tests with all variables fixed except the one of interest. The resulting slopes of the fitted curves become, then, exponents in an equation of the type $\Psi = A^a B^b C^c \ldots Z^z$. These exponents are usually not precise because the data used to obtain them are derived from tests using very small ranges of the other variables. Also, in the general case, each of the variables and exponents is independent of all others.

Yet, many empirical equations exist in the field of tribology. For example Park et al. (Park et al., 2008) introduced empirical equations in order to estimate failure-time for fretting-corrosion of tin-plated contacts. In this study, failure-time estimation models have been developed as empirical equations. The authors investigated the effect of temperature as a corrosion acceleration factor and fretting frequency as a fretting acceleration factor on fretting-corrosion of electric contacts.

The temperature effect is generally explained using the Arrhenius reaction rate Equation (10.1):

$$r = A \exp (-Ea/kT) \; or \; \ln r = -Ea/kT + \ln C \tag{10.1}$$

where r, C, Ea, k and T are speed of reaction, non-thermal constant, the activation energy (eV), the Boltzmann constant (8.623×10^{-5} eV/K) and the temperature in Kelvin, respectively. Activation energy (Ea) is a factor that determines the slope of reaction rate (r) with temperature (T) and it describes the acceleration effect that temperature has on the rate of a reaction. The Arrhenius reaction rate equation is assumed to describe the failure-time in a fretting-corrosion reaction with temperature by

$$\ln t = \frac{A}{T} + B \tag{10.2}$$

where t, B, A and T are failure-time in electrical contact, non-thermal constant (intercept), a factor or slope that describes failure-time sensitivity for temperature and the temperature in Kelvin, respectively. The failure-times with temperature are plotted and their linear fitting is conducted. The empirical equation is developed as $\ln t = 2094/T + 0.75$ by linear fitting.

On the other hand, the type of empirical equation in a fretting-corrosion reaction with fretting frequency is given as

$$\ln t = A \, Hz + B \tag{10.3}$$

where t and Hz are failure-time in electrical contact and fretting frequency, respectively. A is the slope that describes life-time sensitivity for fretting frequency change. B is an unknown constant which is not related to fretting frequency but is related to temperature.

The three-dimensional failure-time estimation models for planes A and B through extrapolation are shown in Fig. 10.1. The second-order equation is derived from the model for failure-time prediction and given as

$$\ln t = A\left(\frac{1}{T}\right)^2 + B\left(Hz\right)^2 + C\left(\frac{1}{T}\right) + D\left(Hz\right) + E\left(\frac{1}{T}\right)\left(Hz\right) + F \tag{10.4}$$

where t, T and Hz are failure-time, temperature and fretting frequency, respectively. For the proposed failure-time estimation models, the values of coefficients A, B, C, D, E and F are shown in Table 10.1.

Marko et al. (Marko et al., 2017) also made an effort to investigate and characterize the evolution of transient tribological wear in the presence of sliding contact, which is often characterized experimentally via the standard ASTM D4172 four-ball test. Based on the results of a Monte Carlo study, the authors developed new empirical equations for the wear rate as a function of asperity height and lubricant thickness. The dimensionless normalized wear volume V_N is calculated from

$$V_N = 0.2763 \cdot \exp[-1.6754 \cdot \lambda_w] \tag{10.5}$$

where the dimensionless parameter λ_w is given by

$$\lambda_w = \frac{h + W_P}{\sigma} \tag{10.6}$$

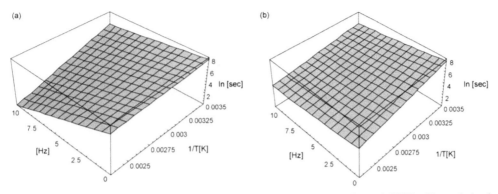

Figure 10.1: The failure-time estimation models (a) for plane A and (b) for plane B. After Park (2008) with permission from Elsevier publishing.

Table 10.1: Failure-time equations as the function of temperature and fretting frequency. After Park (2008), with permission from Elsevier publishing.

$\ln t = A\left(\dfrac{1}{T}\right)^2 + B(\text{Hz})^2 + C\left(\dfrac{1}{T}\right) + D(\text{Hz}) + E\left(\dfrac{1}{T}\right)(\text{Hz}) + F$		
	Model A (for plane A)	**Model B (for plane B)**
A	-2.49256×10^{-7}	-1.76983×10^{-7}
B	9.92903×10^{-4}	5.65363×10^{-3}
C	927.169	2460.34
D	-1.44215	4.48596×10^{-2}
E	355.834	-50.6892
F	4.50746	-8938.72×10^{-1}

where h (m) is the lubricating oil thickness, σ (m) is the RMS surface roughness, and W_p (m) is the yield/plasticity length given by

$$W_P = R' \cdot \left(\frac{G_{yield}}{E'}\right)^2 \tag{10.7}$$

where R' (m) is the reduced radius of the ball bearing, G_{yield} (Pa) is the ultimate yield stress, E' (Pa) is the reduced Young's modulus. The obtained empirical equations closely represented the experimental data and successfully modelled the sliding contact, as shown in Fig. 10.2.

Another example of empirical equations has been derived earlier by Hamrock et al. (1976c; 1976a, b) to describe the influence of the dimensionless numbers on the dimensionless minimum and central film thickness in elastohydrodynamic lubrication EHL as

$$h_{min} = 3.63U^{0.68}G^{0.49}W^{-0.073}[1 - \exp(-0.68k)] \tag{10.8}$$

$$h_C = 2.69U^{0.67}G^{0.53}W^{-0.067}[1 - 0.61\exp(-0.73k)] \tag{10.9}$$

where h_{min} is the minimum film thickness, h_C is the central film thickness, U is a dimensionless speed parameter, G is a dimensionless material parameter, W is a dimensionless load parameter, and k is an ellipticity parameter that can be calculated from

$$k = 1.03\left(\frac{R_y}{R_x}\right)^{0.64} \tag{10.10}$$

where R_y/R_x is the radius of curvature ratio.

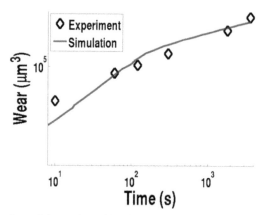

Figure 10.2: Wear (μm^3) experimental data and matching simulation results, for neat mineral oil at a bulk lubricant oil temperature of T = 51°C. After Marko et al. (2017).

Although Equations (10.8) and (10.9) are approximate, they show the effect of the different physical parameters. As expected, the entrainment velocity has a large effect on film thickness, the exponent is 0.67–0.68. The higher the velocity, the more lubricant will be entrained and the thicker the film will become. The viscosity has the same exponent and will, thus, have as great an influence on the film thickness as the velocity. However, it is important to be aware that these empirical expressions are only valid within a limited parameter range.

10.2.2 *Mathematical Models*

Mathematical models could be defined as the equation(s) that describe or simulate the response of some entity of unknown internal composition to some input variables. Simply put, a mathematical model could be considered as a "black box". Mathematical models may help to explain a system and to study the effects of its different components, and to make predictions about the behaviour of a system in some circumstances. The excellence of the modelling process depends on how well the developed model correlates to the theoretical bases and agrees with results of repeatable experiments.

In order to properly model the friction and wear of materials, it is essential to understand specific mechanical and tribological features. Unfortunately, most models are correlational in nature and, therefore, system-specific, i.e., the model only works for the particular material pair, contact geometry, operating condition, and specific environment and lubricating conditions (Lansdown, 1996). Obviously, the model can never be a complete description of a system, but methods have been developed to improve its utility.

Today's computer-based modelling enabled us to describe the behavior of single-part components fairly accurately. Nevertheless, technical systems in which the components are connected by mechanical interfaces still cannot be directly modelled with reliable results. Another way to view this problem is to note that component properties, such as structural strength, can be predicted fairly well, but system responses at the interfaces, such as friction and wear, cannot. Consequently, one way to obtain more accurate and reliable modelling of complex technical systems is by developing models that describe the characteristics of mechanical interfaces under different conditions. These models should address a number of different aspects, such as contact stiffness, damping, friction, wear and lubrication (Andersson et al., 2008; Dmytrychenko et al. 2020).

Ma et al. (Ma et al., 2006) developed a one-dimensional mixed lubrication, wear, and friction model for the piston rings and cylinder liner. Wear is computed from the surface asperity contact pressure. The variation in the elastohydrodynamic lubrication and the oil viscosity with the change in temperature and pressure are included in their model.

Wear model: The major wear mechanism of the cylinder liner wear is abrasion; therefore, the Archard wear model is selected for this research. Although the Archard model is for unlubricated surfaces, it is well suited in mixed lubrication conditions because only asperity contact pressure p_c is used to compute wear. The wear coefficient K_w is not constant during the break-in period and is given by

$$K_w = (K_{w0} - K_{w\infty}) \exp(-\beta t) + K_{w\infty} \qquad (10.11)$$

where

K_{wo} wear coefficient at time 0
$K_{w\infty}$ wear coefficient at steady state
β time dependence of the break-in
t time, sec.

The cumulative wear depth of the piston ring w_{pr} at the point y is obtained by

$$w_{pr}(y) = \Sigma_{\text{cycle}}\, \delta w_{pr}(y) \qquad (10.12)$$

and

$$\delta w_{pr} = \int_0^{4\pi} K_w \frac{p_c V_p}{H_{pr}\omega}\, d\theta \qquad (10.13)$$

where

V_p velocity of piston, m/sec
H_{pr} hardness of piston ring, N/m²
θ the crank angle degree
ω engine crankshaft angular velocity, rad/sec

Friction model (for Liner/Ring): The friction force is computed by the model (Equation 10.13) developed by Patir et al. (1979) where the friction between the contacting asperities causes the friction force between the ring face and the cylinder wall in the mixed lubrication regime.

$$F_f = -\pi \cdot D \int_{yt}^{yb} \left\{ \frac{\mu V_p}{h} [(\emptyset_f - \emptyset_{fs}) + 2V_{r1}\, \emptyset_{fs}] + \mu_f P_c \right\} dy \qquad (10.14)$$

The shear stress factors ϕ_f and ϕ_{fs} are correction factors that represent the influence of the surface roughness and the surface pattern on the viscous shearing friction force.

The lubrication, wear, and friction of the cylinder bore and the piston rings are simulated using the above models. Then, the predicted cylinder wear and the ring pack friction are compared with the experimental results in order to validate the model (Ma et al., 2006; Zabala et al., 2017).

Figure 10.3 presents the comparison between the measured cylinder wear and the predicted wear. It is clear that the predicted cumulative wear matches well with the experimental results over the engine break-in period. Also, Fig. 10.4 shows the ring pack friction behavior, where the cycle averaged ring pack friction is represented as a function of cylinder wall temperature.

10.3 Finite Element Analysis (FEA)

In recent years, Finite Element (FE) analysis had provided us with an effective method for predicting the tribological behavior of rubbing surfaces. FE is necessary to be employed in some cases, such as in nonlinear material behavior or when the dimensions of a body are not large compared with the regions of contact.

The method uses a complex system of points called nodes which make a grid called a mesh. This mesh is programmed to contain the material and structural properties, which define how the structure will react to certain operating conditions. The mesh acts like a spider web, where a mesh element extends from each node to the adjacent nodes. This web of vectors is what carries the material properties to the object creating many elements. The simple equations that model these finite elements are then assembled into a larger system of equations that models the entire problem.

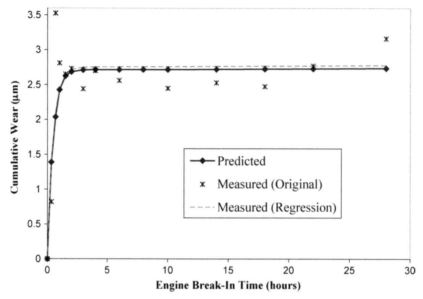

Figure 10.3: Measured and predicted bore wear at TDC during engine break-in. After Ma et al. (2006).

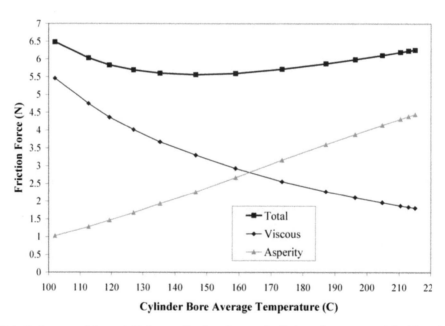

Figure 10.4: Cycle-averaged ring pack friction as a function of averaged cylinder wall temperature. After Ma et al. (2006).

Generally, there are two types of analysis that are used: Linear modeling, and non-linear modelling. Linear modelling conserves simplicity and allows the analysis to be run on a relatively normal computer. However, it tends to yield less accurate results. On the other hand, non-linear modelling produces more accurate results, while sacrificing the ability to run on all but the fastest computers effectively. Today, there are numerous available FE codes which are appropriate for the solution of tribological problems (e.g., ABAQUS, DYNA3D, MARC, NASTRAN, RADIOSS, ANSYS). Also, there are modern programs that have graphical interfaces to assist in the generation of meshes.

10.3.1 FEA of Wear in Pin-on-disk Configuration

Assume that the wear rate of a pin, in sliding contact with rotating disk, obeys Archard's law. The simulation of the wear profile can be performed using FE, applying the wear equation to update the geometry. For a certain sliding speed, the 2-D profile of the contacting surface can be modelled as a function of two factors influencing wear: Contact pressure and sliding distance. These factors were taken into account by the proportionality constant of the wear equation, which can be determined experimentally. The simulation is composed of two components: (1) FE code used to determine the contact pressure at each node and (2) wear algorithm that interprets the nodal position changes. Supplementary subroutines can be implemented to the FE program in order to permit its application to dynamic contact wear (Abdelbary, 2014).

There are several methods that can be applied in order to create the FE mesh. For convenience, the "direct generation" method is used in meshing pin-on-disk tribosystems. Accordingly, the coordinate location of every node and the size, shape and connectivity of every element are determined manually. Although this technique is time consuming, it allows for complete control over nodal and element numbering. The mesh of the pin and disk is created using different types of elements, as shown in Fig. 10.5. Wear elements, are made of uniform-sized quadrilateral elements. It should be emphasized that, in order to obtain the best numerical results, the size of the elements should remain uniform and their shape to be of the original parent element. Contact elements line the base of the pin face. Coarse mesh is implemented in the rest of the pin, tied to the finer mesh through a transition zone made of triangular elements. Rectangular elements are created across the width of the pin and characterized by stretched length in order to ensure that the scale of the model is exactly matching that of the real pin. The rotating disk is simulated as rigid surface and meshed using a single type of two-node target elements.

In an ideal case, the boundary conditions to be applied are: (1) nodes holding the pin, (2) the whole system can move freely along the vertical axis, (3) the contact pressure is applied directly to the top of the pin, and (4) the contact pressure is derived from the normal load. The load is applied as a uniformly distributed pressure on the uppermost nodes of the pin. The magnitude of the applied load affects the amount of lift between the two surfaces, but not to the extent of separation. A single horizontal translation is used to simulate the relative motion of the rotating disk in contact with the stationary pin.

The simulation procedure can be summarized in the flowchart shown in Fig. 10.6. After defining the load-step of a particular cycle, the solution can start. Then, the wear algorithm accesses the solution database and finds the status of each contact element, deciding whether it is fully contacted or closed-gap. The contact pressure, at the nodes of these elements, is then used in the wear equation in order to calculate the incremental wear height of each node for that cycle. This height is then added to the cumulative wear height and stored in a file for the next cycle. Finally, the FE model can predict pressure distribution at the contact zone that would explain the experimental wear profile. Although the simulation procedure is quite complex, analysis and recommendations, with respect to methods of increasing the accuracy of the results while keeping the computing time to a reasonable limit, were presented.

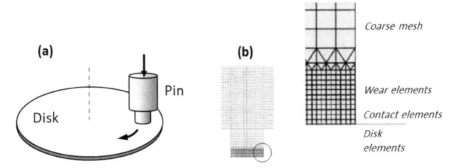

Figure 10.5: (a) Schematic of the pin-on-disk configuration, and (b) FE model of the pin. After Benabdallah et al. (2006), with permission from Elsevier publishing.

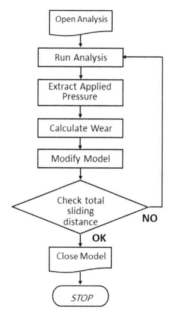

Figure 10.6: Flow chart representing the basic steps of typical script used in calculating wear. After Benabdallah et al. (2006), with permission from Elsevier publishing.

10.3.2 FEA of Frictional Heating and Contact Temperature (Thermal FEA)

Many available FE codes allow for solution of heat conduction problems (Kennedy and Hussaini, 1987; Roda-Casanova and Sanchez-Marin, 2019). During sliding contact, the principle of the finite element method for surface temperature determination is as follows: Finite element meshes are defined for the two contacting bodies. Those meshes share nodes within the real area of contact, and those contact nodes enable temperature matching for the contacting surfaces throughout the contact region. The finite element meshes should be very fine in the contact region where temperature gradients are highest (Bhushan, 2000).

An example of thermal FEA is that used to determine the temperature field of polymer spur gears during their operation (Roda-Casanova and Sanchez-Marin, 2019). The temperature field of the gear studied in operating conditions is determined through the thermal analysis of a two-dimensional finite element model of the gear. The thermal FEA can be performed in either transient or steady-state conditions. In steady-state, the gear is assumed to reach its thermal equilibrium, and consequently the temperature field across the gear geometry remains constant with time. Henceforth, the temperature field across the gear is obtained by solving the steady-state heat conduction equation.

In transient conditions, the gear is assumed not to reach its thermal equilibrium, and consequently, its temperature field across the gear geometry varies with time. For the first time increment, an initial temperature must be defined for each node of the finite element model, enabling the transient analysis to be performed. For the construction of the finite element model, the geometry of the gear is generated and then discretized into linear quadrilateral finite elements, as shown in Fig. 10.7.

The heat supplied to the gear as a consequence of frictional dissipation is simulated through a moving heat source that acts over its contact profile, and is represented by a heat flux function. This heat flux function is converted into a set of nodal heat fluxes Q_j, which are applied over each node j of the contact profile of the gear. In the transient analysis, the nodal fluxes are time-dependent and can be calculated using the following equation:

$$Q_j(t) = b \cdot \int_{\xi_{j-1}}^{\xi_j} \frac{\xi - \xi_{j-1}}{\xi_j - \xi_{j-1}} \cdot q(\xi, t) \cdot d\xi + b \cdot \int_{\xi_j}^{\xi_{j+1}} \frac{\xi - \xi_{j+1}}{\xi_j - \xi_{j+1}} \cdot q(\xi, t) \cdot d\xi \qquad (10.15)$$

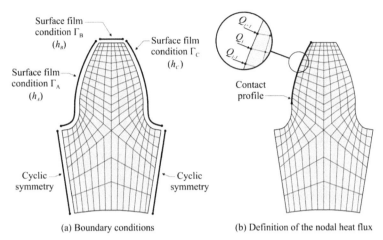

(a) Boundary conditions   (b) Definition of the nodal heat flux

Figure 10.7: Definition of finite element model for thermal analysis. After Roda-Casanova et al. (2019), with permission from Elsevier publishing.

where

ξj is the intrinsic coordinate of the node under consideration,

$\xi j+1$ is the intrinsic coordinate of the following node,

$\xi j-1$ is the intrinsic coordinate of the previous node,

b represents the face width of the gears,

$q\,(\xi)$ is the average heat flux

In the steady-state analysis, all the variables that define the finite element model must be time-independent, including the nodal fluxes. For this reason, the heat flux function that is used in Equation (10.15) to determine the nodal fluxes is replaced by an averaged heat flux function $q\,(\xi)$ given by

$$\overline{q}(\xi) = \frac{E(\xi)}{\Delta t_{cycle}} \tag{10.16}$$

The temperature field across the geometry of the gear tooth is obtained after the analysis of this FE model. It is postulated that the obtained approach is convenient for determining the temperature field of the polymeric gears after reaching their thermal equilibrium. Also, this approach allows one to determine the temperature field of the gears in unsteady thermal conditions.

It can help the tribologists to study how the material selected for the gears affects the thermal behavior of the transmission. The approach can also help in proposing topological modifications of the gears in order to minimize the generation of frictional heat and power loss, or in estimating the temperature of the gears under certain lubrication conditions.

10.4 Dimensional Analysis

Dimensional analysis (or Buckingham's dimensional analysis) is based on the hypothesis that the dimension of the body or system is the property that can be measured and can be used to describe its physical state. This method can be used to characterize a phenomenon in terms of the relationships among dimensionless variables. Dimensionless variable (DV) is a unit-less value resulting from multiplying or dividing combinations of physical variables, parameters and constants of any particular system. The number of independent DVs in a physical system is the difference between the number of variables and the number of fundamental dimensions (Buckingham, 1914). Since there was no generally-accepted equation for the prediction of wear, dimensional analysis of the factors influencing the wear process may be expected to provide a better characteristic expression.

Dimensional analysis of wear has been of interest for decades. The objective is to consider those variables which dominate the wear process, and to express them in the form of an equation to determine wear rate. Fundamental variables that are considered in developing the wear equation are wear volume, sliding variables, such as load P, speed v, time T, counterface roughness α, material properties (modulus of elasticity E), physical property (surface energy γ) and thermal properties (thermal conductivity K and specific heat C_p). Using dimensional analysis, Kar and Bahadur (Kar and Bahadur, 1974) developed wear Equation (10.17) for poly-oxymethylene (POM). The equation was based specifically on wear tests from the considered materials, but the model could be extended to other materials. However, it did not include any effects of counterface roughness.

$$V = kP^x v^{y-z} T^y \gamma^{3-y+z} E^{-3-x+y} \left(\frac{C_p}{K} \right)^z \qquad (10.17)$$

where x, y and z are experimentally determined exponents.

Equation (10.16) was modified by Viswanath and Bellow (Viswanath and Bellow, 1995) in order to include the effect of counterface surface roughness, and to be applicable for a wide variety of polymers. The four primary dimensions used are mass (M), length (L), time (T) and temperature (θ). The dependent variable wear volume V is expressed as

$$\Psi(V, W, T, \alpha, C_p, \gamma, E, v, K) = 0 \qquad (10.18)$$

where Ψ is some arbitrary function.

Using Buckingham *Pi* theorem, the number of variables was reduced from nine to five dimensionless groups:

$\left(\dfrac{VE^3}{\gamma^3} \right)$ represents the significance of interface contact and deformation.

$\left(\dfrac{WE}{\gamma^2} \right)$ represents the influence of normal load on interface contact.

$\left(\dfrac{vTE}{\gamma} \right)$ represents the distance over which formation and deformation of junctions occur.

$\left(\dfrac{\alpha E}{\gamma} \right)$ determines the apparent contact area that supports the load and can influence the type of wear process that occurs.

$\left(\dfrac{C_p \gamma}{\gamma K} \right)$ a factor controlling thermal contribution during interface formation and is also influenced by speed.

All these groups were combined to obtain a relationship between the main dependent variable, the volume V of the worn polymer material, and the variables influencing the wear of polymers.

$$\left(\frac{VE^3}{\gamma^3} \right) = \Psi \left(\frac{WE}{\gamma^2} \cdot \frac{vK}{\gamma C_p} \cdot \frac{TEC_p}{K} \cdot \frac{\alpha E}{\gamma} \right) \qquad (10.19)$$

The function Ψ can be determined by conducting a number of experiments where one operating variable is varied at a time, while maintaining the other variables constant.

Equation (10.19) was simplified and rewritten with exponents *p, q, r* and *s* for the dimensionless groups and was expressed with *k,* as a proportionality dimensionless wear coefficient. The following equations can be used to evaluate the wear coefficients for different polymers, and determine whether linear or non-linear relation is the most appropriate relationship between the wear volume and the other factors.

For linear relationship

$$V = \frac{k_w W v T \alpha}{\gamma} \qquad (10.20)$$

For non-linear relationship

$$V = k_w W^p v^q T^r \alpha^s E^{-3+p+r+s} (C_p/K)^{r-q} \tag{10.21}$$

(The exponents p, q, r and s can be determined experimentally)

A comparison of linear and non-linear relationships showed that the wear volume was better represented by a non-linear proportionality with the operating variables.

Similarly, dimensional analysis was applied to develop a relationship for the abrasive wear behaviour of polymers (Rajesh and Bijwe, 2005). The worn volume V of polymers was expressed in terms of the following variables:

$$V = f\{F_N \cdot \phi \cdot V_C \cdot t_F \cdot E \cdot G_C \cdot \sigma^* \cdot C^* \ and \ \psi\} \tag{10.22}$$

where, F_N is the normal load, ϕ is the mass of the abrasive particles required for significant wear, V_C is the crack growth velocity, t_F is the time to failure under tensile stress, E is the Young's modulus, G_C is the fracture energy, σ^* is the fracture stress, C^* is the critical crack length and ψ is the ductility factor. The dimensionless groups are:

$$\left(\frac{V}{\varphi^3}\right) \cdot \left(\frac{V_C \phi^{1/2}}{\varphi^{3/2} \sigma^{1/2}}\right) \cdot \left(\frac{F_N}{\sigma_* \varphi^2}\right) \cdot \left(\frac{E}{\sigma_*}\right) \cdot \left(\frac{G_C}{\varphi \sigma_*}\right) \cdot \left(\frac{t_F \varphi^{1/2} \sigma_*^{1/2}}{\phi^{1/2}}\right) \cdot \left(\frac{C_*}{\varphi}\right)$$

The final form of wear equation is:

$$V = k \, F_N^a \, \phi^b \, E^{a-1} \, \sigma^{-(a+b)} V_C^b \, t_F^{-b} \, \varphi^{-3b/4} \tag{10.23}$$

where k is the wear coefficient, a and b are unknown exponents and can be calculated, experimentally, from the logarithmic plots of the corresponding dimensionless variables.

The dimensional analysis is applied to dry sliding contact to determine a relationship between wear and entropy flow (Amiri et al., 2012).

$$K = \mu \frac{\dot{w} H}{T \dot{S}} \tag{10.24}$$

where

K wear coefficient,

T temperature,

S total entropy,

H hardness,

w wear volume

The term $T S$ represents the heat transfer to the wearing material. Thus, the physical meaning of the nondimensional wear coefficient can be expressed as the ratio of the frictional power required to degrade a given material of hardness H to the energy transferred to the material as heat. The obtained relation is verified by considering the sliding contact in a disk-on-disk configuration for two sets of contacting materials: Bronze SAE 40 on steel 4140 and cartridge brass on steel 4140. A dimensional analysis is used to correlate seven tribological variables, which are wear rate, temperature, coefficient of friction, rate of entropy flow, hardness, normal load, and sliding velocity. Having determined three dimensionless groups, test results are utilized to assess the correlation between these groups. It is shown that the entropy flow can effectively characterize the behavior of the sliding system during steady-state wear condition. Wear coefficient for two different materials sliding against steel is assessed via measurement of entropy flow.

The average value of the wear coefficient obtained from dimensional analysis for the brass on steel tribo-pair is $K_{Brass} = 6.3 \times 10^{-4}$ and for bronze on steel tribo-pair is $K_{Bronze} = 3.43 \times 10^{-4}$. The published values (Amiri et al., 2012) of the wear coefficient for brass on steel are $K_{Brass} = 4.3\text{–}6 \times 10^{-4}$, and for bronze on steel $K_{Bronze} = 2.02 \times 10^{-4}$. Comparison of the Archard's wear coefficient calculated using the present model with the published values reveals a good quantitative agreement.

According to the former discussion, the dimensional analysis method could be considered as an acceptable method for representing the wear of polymers in terms of operating variables, material properties and the wear mechanism.

10.5 Artificial Neural Networks (ANNs)

Artificial Neural Network (ANN) is inspired by the biological nerve system and is being used to solve a wide variety of complex scientific and engineering problems. Recently, it has emerged as a good candidate for mathematical models, due to its capabilities of nonlinear behavior, learning from experimental data and generalization. This computational technique is very useful for simulating correlations that is difficult to describe with a physical model. ANN has the ability to model complex nonlinear, multi-dimensional relationships, without any prior assumptions about their nature, and the network is built directly from experimental data by its self-organizing capabilities.

ANN is based on the structure and functioning of the biological nervous system, where neurons are the basic unit or building blocks of the brain. A neuron receives many input signals but it produces only one output signal at a time. Inspired by these biological neurons, ANN is composed of simple elements operating in parallel. ANN is the simple clustering of the primitive artificial neurons. This clustering occurs by creating layers, which are then connected one to another. An ANN can be trained to perform a particular function by adjusting the values of the connections (weights) between the elements. The input/output data allows the neural network to be trained in a way that minimizes the error between the real output and the estimated (neural net) output. After the network has learned to solve the example problems, it can be used for different purposes among which are prediction and identification.

In recent years, modelling of ANNs has been introduced into the field of polymers and composites tribology. In such models, known measured details, i.e., material compositions, mechanical properties and testing conditions can be used as inputs and the expected wear characteristics, such as the coefficient of friction and the specific wear rate of a virtual case, are calculated as output results. The pioneering investigations of neural network techniques to predict tribological parameters have been presented by Rutherford et al. (Rutherford et al., 1996) and Jones et al. (Jones et al., 1997). Subsequently, a lot of researches investigated the potential of neural networks to predict and analyze the tribological behavior under various parameters.

10.5.1 Fundamentals of Neural Network Approach

An ANN is conventionally constructed with three layers, namely, the input layer, the hidden layer(s) and the output layer. The input layer accepts data and the output layer generates outputs. There may be one or more hidden layers consisting of a number of neurons. This number is fixed, in the input and output layers, to be equal to that of input and output variables, respectively, whereas the hidden layer may contain more than one layer, and in each layer the number of neurons is flexible. Each neuron sums its input signals, modified by the interconnection weights. For the convenience of description, the structure of the ANN is expressed as:

$$N_{in} - [N_1 - N_2 - \cdots - N_h]_h - N_{out} \tag{10.25}$$

where, N_{in} and N_{out} refer to the number of input and output variables, respectively, h denotes the number of hidden layers, and N_1, N_2 and N_h are the numbers of neurons in each hidden layer. For example, 6-[5-3]$_2$-2 describes a network containing one input layer with 6 input parameters, 2 hidden layers with 5 and 3 neurons, respectively, and one output layer with 2 output parameters.

Training is the act of continuously adjusting the connection weights until they reach distinctive values that allow the network to produce outputs that are close enough to the actual desired outputs.

The learning algorithm is based on a gradient search to optimize the performance function of the network. The algorithm is generally evaluated by means of squared errors (E) between the predicted and desired values.

$$E = \frac{1}{2L} \sum_{i=1}^{L} [d(t) + p(t)]^2 \tag{10.26}$$

where, L equals the number of training samples, $d(t)$ is the desired output value, and $p(t)$ refers to the target output value predicted by the ANN for the tth sample.

To obtain an optimized neural network construction, the dataset is divided into a training dataset and a test dataset. The first is used to adjust the weights of all the connecting nodes until the desired error level is reached. Then, by using the test dataset, ANN performance could be evaluated on the basis of the coefficient of determination (B). Higher coefficients indicate an ANN with better output predictive capabilities.

$$B = 1 - \frac{\sum_{i=1}^{M}(O(P)^{(i)} - O^{(i)})^2}{\sum_{i=1}^{M}(O^{(i)} - O)^2} \tag{10.27}$$

where, $O(P)^{(i)}$ is the ith predicted property characteristic, $(O)^{(i)}$ is the measured value, O is the mean value of $(O)^{(i)}$, and M is the number of test data points.

10.5.2 *Prediction of Wear and Friction of Aluminum Composites*

The ANN prediction method has been used in several tribological applications, such as friction and wear (Nasir et al., 2010; Shebani and Iwnicki, 2018; Yao et al., 2018; Zhang et al., 2002; Veeresh Kumar et al., 2018). It is widely accepted that ANN is able to predict the output parameters to different levels of accuracy. However, it was suggested that the ANN performance is controlled by a few elements: Training function, input data and the number of hidden layers.

Kumar et al. (Kumar et al., 2012) developed an ANNs model to relate the effects of dry sliding wear and frictional behaviour of aluminium (A380)–fly ash composites reinforced with fly ash particles of different size ranges. The input/output datasets collected during the machining trials consist of four input parameters, namely, applied load L, sliding speed S, size of fly ash particle reinforcement R and percentage reinforcement P. The output dataset consists of two parameters, namely, wear rate and coefficient of friction. In this study, a standard back propagation multilayer feed forward network of structure [4-6-2] was designed using MATLAB.

The network was trained automatically with the MATLAB function 'train' with the 'weights' and 'biases' initialized to random values. Prior to the training phase, the data set was divided randomly into training and test data sets. 98 Data sets were used as the training set, while 10 data sets were used to test the network. The Levenberg–Marquard algorithm was used in order to determine the optimum network generalization. During training, the 'weights' and 'biases' of the network are adjusted so as to minimize the mean square error between the expected and the predicted values.

Figure 10.8 shows the performance of the network for predicting the wear rate and coefficient of friction using the entire data set. It is postulated that the correlation coefficient between the predicted and experimental values using the entire data set for wear rate and coefficient of friction are 0.986 and 0.984, respectively, which is a good sign for the model to be accurate.

10.5.3 *Prediction of Wear of Polyamide*

ANN approach is applied to study the wear of Polyamide 66 under different conditions. A systematic parametric study was carried out in order to predict the wear rates of the polymer under different loading conditions, surface defects and different sliding media. Figure 10.9 gives a schematic example of an ANN model configuration. The network was trained, using the Levenberg-Marquardt (LM) algorithm, based on an experimental dataset of independent wear tests (Abdelbary et al., 2012). The choice of LM algorithm was due to its faster training, especially with moderate size networks.

The dataset used included the testing conditions (load F or F_{mean}, maximum load F_{max}, cyclic frequency f, number of surface crack n_c, and fluid film thickness h_o) while the output parameter was the wear rate WR. The selection of input and output parameters used in the ANN can be expressed as follows:

a) Applied Load (F): The choice of this parameter was due to its effect on wear of polymeric materials, it represents both constant and fluctuating applied load (F). It is important to note that the value of (F) in fluctuating load tests is the mean value of the load cycle (F_{mean}).

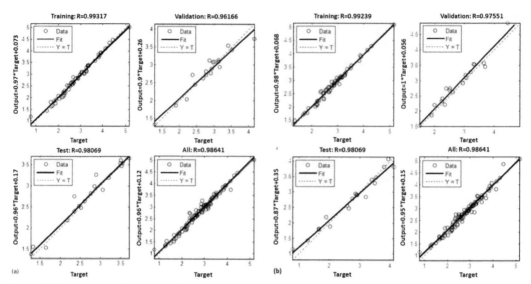

Figure 10.8: Correlations between predicted and experimental values: (a) wear rate, and (b) coefficient of friction. After Kumar et al. (Kumar et al., 2012), with permission from T&F publishing.

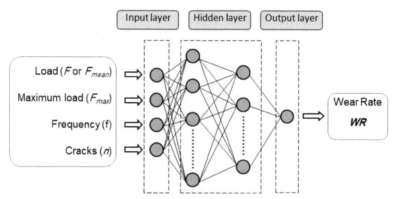

Figure 10.9: Schematic description of an ANN configuration.

b) Load Ratio (R): The effect of load ratio (R) was obviously detected in the wear tests performed under cyclic loading conditions. The governing equation of this parameter is given by:

$$R = F_{min}/F_{max} \qquad (10.28)$$

In constant load tests, this parameter is equal to unity, i.e., $R = 1$.

c) Cyclic Loading Frequency (f): This parameter is directly related to the time dependent nature of loading condition. It is clear that, under constant loading conditions, the value of (f) is equal to zero.

d) Number of Surface Cracks (n_c): Such a parameter is very easy to be evaluated on the wear test. Although there are very few tests considering the number of surface cracks, it was found that it has a significant effect on wear rate under dry sliding conditions. Also, it is clear that for uncracked polymer, the value of (n_c) will be equal to zero.

e) Fluid Film Thickness (h_o): This parameter is introduced to the dataset of lubricated wear tests. It refers to the type of lubricant condition (boundary or hydrodynamic). The governing equation was introduced previously in Chapter Five.

f) Linear Wear Rate (WR): This parameter can be calculated from the wear curve using Equation (10.29), for both the running-in and steady state portions on the wear curve.

$$WR = \frac{V}{X} \quad mm^3/m \tag{10.29}$$

where, V is the volume loss and X represents the sliding distance.

Normally, eighty percent of the data is used for training; the remaining 20% of the dataset is utilized for testing. The maximum number of iterations (epochs) in the training process is 1000. The designed ANN model is tested with a group of data not included in the training dataset to evaluate the network strength.

Effect of input variables

The choice of ANN input variables is fundamental in identifying the optimal functional form of statistical models. The task of selecting input variables is common to the development of all statistical models, and is largely dependent on the discovery of relationships within the available data, to identify suitable predictors of the model output. In the field of tribology, the effect of input data is considered by introducing two simple ANN configurations, $\{4\text{-}[25]_1\text{-}1\}$ and $\{5\text{-}[25]_1\text{-}1\}$, in order to predict the wear rate of surface cracked polymer, in dry and wet sliding conditions, respectively (Abdelbary, 2011). The first ANN is chosen for predicting the wear rate in dry sliding based on the input parameters (F, R, f, and n_c). The second ANN is employed to predict the wear rate in wet sliding based on the input parameters (F, R, f, n_c, and h_o). Figures 10.10 to 10.13 present the predicted wear rates for training and test datasets.

In fact, the above NN configurations sometimes gave good output results with high predictive quality, i.e., about 30% of the B-values were in the range of 0.95–0.99. However, the results are not stable, which means that when repeating the random selection of test datasets, the quality could be even very poor. Moreover, the predicted wear rates of the test dataset are relatively different from the measured values, as shown in Figs. 10.10(b) and 10.11(b). The reason for the unstable quality of the simple ANN is not clear, but one should always consider that ANN is inspired by our biological neural system. In human beings, someone can learn very fast and catch the right solution even only if a little information is available. The increased number of input parameters and training epochs may result in improving the ANN performance. Consequently, in order to enhance the performance of the ANN of dry sliding wear, the running-in wear rate ($WR_{running\text{-}in}$) dataset is added to the input parameters. Figure 10.12, shows the comparison between

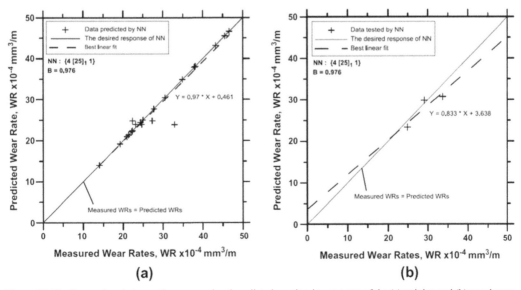

Figure 10.10: Comparison between the measured and predicted running-in wear rate of the (a) training and (b) test dataset, Dry sliding. Input: F, R, f, nc and Output: WRrunning-in. After Abdelbary (Abdelbary, 2011).

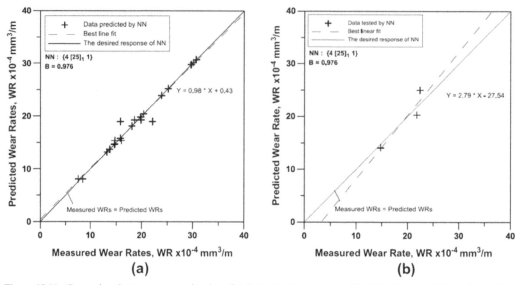

Figure 10.11: Comparison between measured and predicted steady state wear rate of the (a) training and (b) test dataset, Dry sliding. Input: F, R, f, nc and Output: WRsteady state. After Abdelbary (Abdelbary, 2011).

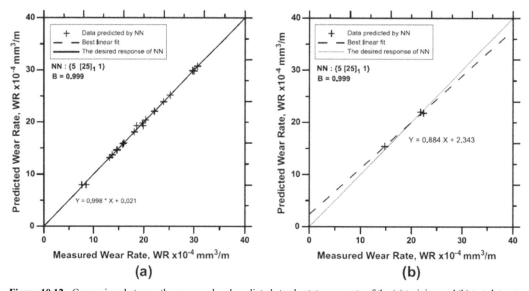

Figure 10.12: Comparison between the measured and predicted steady state wear rate of the (a) training and (b) test dataset, Dry sliding. Input: F, R, f, nc, WRrunning-in and Output: WRsteady state. After Abdelbary (Abdelbary, 2011).

the measured and predicted wear rates for the training and test datasets. The present ANN indicated a higher coefficient of determination ($B = 0.999$) as well as better *WR* prediction, as shown in Fig. 10.12(b), which agrees with the previous suggestion about the effect of the input size. To find a general NN that can predict the wear of surface cracked polymer under different types of loading and sliding conditions, all datasets (dry and wet tests) are collected as input parameters in {5-[40]$_1$-1} configuration, as shown in Fig. 10.13(a). Note that the lubrication parameter (h_o) in dry tests was equal to zero. Predicted *WRs* of this ANN showed relatively lower performance ($B = 0.961$), as shown in Fig. 10.13(b). It is believed that the ANN configuration still needs further adaptation in order to achieve better prediction quality.

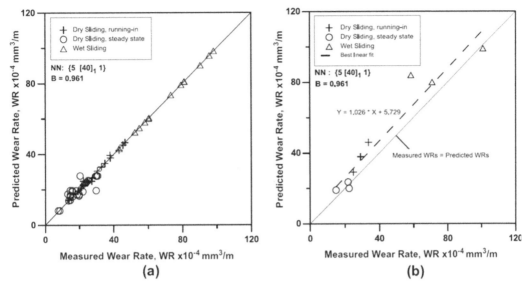

Figure 10.13: Comparison between the measured and predicted wear rate of the (a) training and (b) test dataset, Dry and wet sliding. Input: F, R, f, nc, ho and Output: WRrunning-in or WRsteady state. After Abdelbary (Abdelbary, 2011).

Effect of ANN configuration

Not only the training algorithm and input variables, but also the configuration has a strong influence on the ANN predictive quality. In order to find a powerful ANN configuration, five different ANNs introduced in previous literatures are selected and modified in order to establish the optimized construction, as shown in Table 10.2.

After multiple trial computations using MATLAB 7.8.0 Software, the best topology {5-[20−10-10]$_3$-1} NN is employed. *LM* training method is applied until a normalized mean squared error between the measured data and that obtained from the ANN was minimal. Figures 10.14 and 10.15 show a comparison of the predicted quality for various ANN configurations, and a determination of the measured and predicted wear rates using suggested ANN.

A similar study is performed in order to investigate the effect of NN configuration on the network performance, where two ANN models, {5-[20 5]$_2$-1} and {5-[20 10 10]$_3$-1}, with two and three hidden layers, respectively, are employed. Predicted data showed that the response of the ANN is improved by increasing the number of hidden layers, as shown in Fig. 10.16.

Complex ANN configuration (three layers NN) is more convenient in predicting the wear behavior of polymer that has a very complex relationship with the input parameters. Or, put simply, one can say that, in designing ANNs, the complexity of the network configuration should be selected in the light of the size of the input parameters. Finally, the {7-[20 10 10]$_3$-1} ANN is introduced to predict the wear rates while the maximum load (F_{max}) and running-in wear ($WR_{running-in}$) rate are used as an input parameters. Excellent performance ($B = 1$) is obtained, especially in predicting the test data, as shown in Fig. 10.17.

It is concluded that the predictive quality of the ANN can be further improved by enlarging the input training datasets and by optimizing the network construction. A well-trained ANN is expected to be very helpful for optimum design of polymeric materials, in particular tribological applications. When introducing load parameters (F or F_{mean}, f, R), surface cracks (n_c), as input parameters to a suitable ANN, acceptable wear rates are predicted. This successful prediction could be beneficial towards reducing the number of tribo-experiments and could be used to enlarge the dataset.

The ANN is employed in order to predict the number of *WR* values that are not included in the experimental work. Virtual load frequencies *f* and the number of surface cracks *n* are used as input parameters and the predicted *WRs* are the output. Figure 10.18 shows the predicted 3D profile of the wear rate *WR* of nylon 66 as a function of *f* and *n*. The predicted profile demonstrates the dependence of wear properties

Table 10.2: Configuration of ANNs adopted from literatures.

Ref.	ANN	Neurons Type			Training algorithm
		I/P	Hidden	O/P	
(Zhang et al., 2002) (Jiang et al., 2007)	{9–[15–10–5]$_3$ –1}	tan-sigmoid	tan-sigmoid	pure-linear	LM
(Gyurova and Friedrich, 2011)	{7–[9–3]$_2$ –1}	tan-sigmoid	tan-sigmoid	pure-linear	CGB
(Abdelbary et al., 2012)	{5–[20–10–10]$_3$ –1}	tan-sigmoid	tan-sigmoid	pure-linear	LM
(LiuJie et al., 2007)	{2–[8]$_1$ –1}	–	tan-sigmoid	pure-linear	LM
(Jiang et al., 2008)	{9–[12–6–3]$_3$ –1}	–	–	–	CGB
(Helmy, 2004)	{35–[8–5]$_2$ –1}	tan-sigmoid	pure-linear	pure-linear	LM

LM   Levenberg-Marquardt
CGB   Powell–Beale Conjugate Gradient

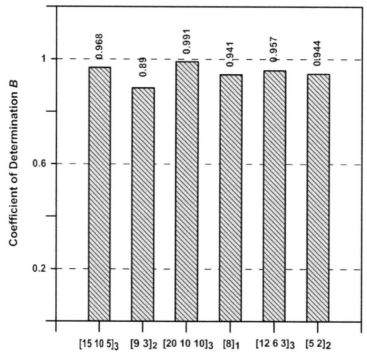

Figure 10.14: Comparison of the prediction quality for various ANNs configurations. After Abdelbary (Abdelbary, 2011) (The given number above each column is B factor).

on the fluctuating frequency of applied load, as well as the density of surface defects. Predicted wear data show a good agreement with the measured points.

Moreover, the proposed ANN shows an acceptable accuracy (3.16% average error) in predicting wear rates (*WR*) of the same polymer sliding against a stainless steel counterface, and subjected to constant and cyclic loads in dry conditions (Abdelbary, 2003), as shown in Table 10.3. The application of the ANN results in improving the curve fitting of the relation between the wear rate and frequency (*f*), as presented in Fig. 10.19. About 5% increase in R-squared value is obtained by introducing seven additional predicted (*WR*)

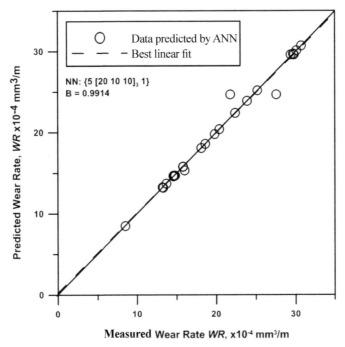

Figure 10.15: Determination of the measured and predicted WR of dataset for ANN configuration. After Abdelbary (Abdelbary, 2011).

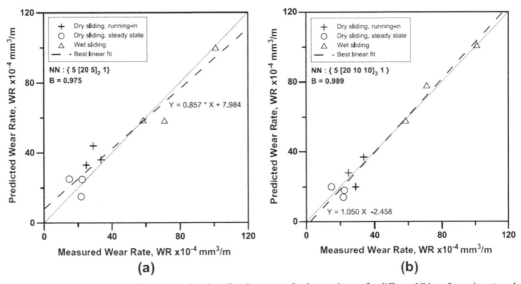

Figure 10.16: Determination of the measured and predicted wear rate for the test dataset for different NN configurations (a and b), Dry and wet sliding. Input: F, R, f, nc, ho and Output: WRrunning-in or WRsteady state. After Abdelbary (Abdelbary, 2011).

values using seven virtual frequencies within and beyond the experimental domain. This issue illustrates the advantage of ANNs in saving time and cost, and strengthens the application of ANN modelling.

In summary, the successful introduction of ANNs in tribological applications could be beneficial toward predicting the wear rate of polymers and their composites. Once a well-trained ANN model has

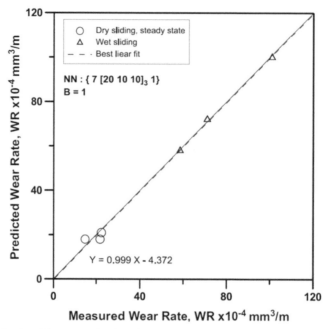

Figure 10.17: Determination of the measured and predicted wear rate of the test dataset for the final general ANN, Dry and wet sliding. Input: F, Fmax, R, f, nc, ho, WRrunning-in and Output: WRsteady state. After Abdelbary (Abdelbary, 2011).

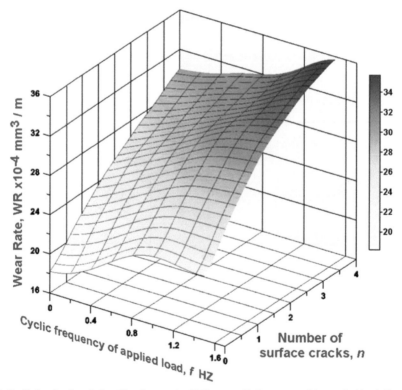

Figure 10.18: Prediction for the relationship of wear rate (WR) vs. cyclic frequency of the applied load (f) and number of surface cracks (n). After Abdelbary (Abdelbary, 2011).

Table 10.3: Data adopted from Abdelbary (Abdelbary, 2012) and used to evaluate the predictive accuracy of the present ANN.

Test	Input parameters				Output WR (x 10^{-4} mm³/m)		
	F (N)	F_{max} (N)	R	f (Hz)	Measured	Predicted	Error (%)
1	35	35	1	0	3.50	3.50	0
2	70	70	1	0	3.70	3.68	0.54
3	90	90	1	0	5.70	5.92	−3.86
4*	105	105	1	0	6.10	5.52	9.51
5	40	75	0.06	0.38	6.00	5.99	0.17
6	70	105	0.33	0.38	5.39	5.09	5.57
7	90	125	0.44	0.38	9.00	9.03	−0.33
8	105	140	0.50	0.38	6.30	6.29	0.17
9	150	185	0.62	0.38	4.05	4.32	−6.67
10	70	105	0.33	0.70	5.25	5.19	1.14
11*	90	125	0.44	0.70	13.5	13.47	0.22
12	70	105	0.33	1.38	7.42	7.45	0.40
13	90	125	0.44	1.38	15.30	15.31	0.07
14	90	105	0.71	0.70	5.40	5.93	−9.81
15	90	112	0.61	0.70	5.67	5.82	−2.65
16	90	150	0.20	0.70	3.60	3.64	−1.11
17	90	175	0.03	0.70	5.40	5.47	−1.30
18	90	112	0.61	1.38	5.85	5.69	2.74
19	90	150	0.40	1.38	5.94	5.59	5.89
20*	90	175	0.03	1.38	5.49	6.10	−11.11
Average error (arithmetic mean)							**3.16%**

* Data used to test the NN performance

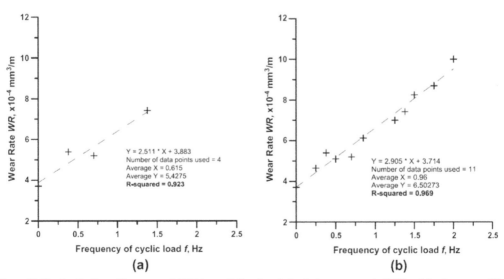

Figure 10.19: Application of the proposed ANN in predicting the relation between wear rate (WR) and the frequency of the applied cyclic load (f). (a) Relation between frequency of the cyclic load and the measured wear rates as presented in Abdelbary (Abdelbary, 2011), and (b) Relation between frequency of the cyclic load and the corresponding wear rates (measured and predicted data using the proposed NN).

been obtained, new data can be predicted without performing too many time-consuming experiments. The ANN model is able to construct a proper fitting function that correctly reproduces not only the characteristic tribological properties for these examples, but is also able to interpolate correctly for unseen cases from the same knowledge domain. It is believed that, using a suitable dataset and selected neural architectures, a good, trained ANN can be generalized to model the wear phenomena under virtual conditions.

References

Abdelbary, A. 2011. Effect of vertical cracks at the surface of polyamide 66 on the wear characteristics during sliding under variable loading conditions. Ph.D., Mechanical Engineering Department, Alexandria University.

Abdelbary, A., Abouelwafa, M.N., El Fahham, I.M. and Hamdy, A.H. 2012. Modeling the wear of polyamide 66 using artificial neural network. Materials & Design 41: 460–469. Doi: https://doi.org/10.1016/j.matdes.2012.05.013.

Abdelbary, A. 2014. 8—Prediction of wear in polymers and their composites. pp. 185–217. *In*: Abdelbary, A. (ed.). Wear of Polymers and Composites, Oxford: Woodhead Publishing.

Al-Bender, F. and De Moerlooze, K. 2011. Characterization and modeling of friction and wear: an overview. Sustainable Construction and Design 2(1): 19–28.

Amiri, M., Khonsari, M.M. and Brahmeshwarkar, S. 2012. An application of dimensional analysis to entropy-wear relationship. Journal of Tribology 134(1): 011604-011604-5. Doi: 10.1115/1.4003765.

Andersson, S., Söderberg, A. and Olofsson, U. 2008. A random wear model for the interaction between a rough and a smooth surface. Wear 264(9): 763–769. Doi: https://doi.org/10.1016/j.wear.2006.12.075.

Barber, J.R. 2020. Contact problems involving friction. *In*: Paggi, M. and Hills, D. (eds.). Modeling and Simulation of Tribological Problems in Technology. CISM International Centre for Mechanical Sciences. Vol. 593. Springer. Cham. Doi: https://doi.org/10.1007/978-3-030-20377-1_2.

Bhushan, B. 2000. Modern Tribology Handbook. New York: CRC Press.

Buckingham, E. 1914. On physically similar systems; illustrations of the use of dimensional equations. Physical Review 4(4): 345–376. Doi: 10.1103/PhysRev.4.345.

Dmytrychenko, N., Khrutba, V., Savchuk, A. and Hlukhonets, A. 2020. Using mathematical, experimental and statistical modeling to predict the lubricant layer thickness in tribosystems. *In*: Palagin, A., Anisimov, A., Morozov, A. and Shkarlet, S. (eds.). Mathematical Modeling and Simulation of Systems. MODS 2019. Advances in Intelligent Systems and Computing. Vol. 1019. Springer. Cham. Doi: https://doi.org/10.1007/978-3-030-25741-5_5.

Franek, F., Vorlaufer, G., Edelbauer, W. and Bukovnik, S. 2007. Simulation of tribosystems and tribometrology. Tribology in Industry 29(1&2).

Gyurova, L.A. and Friedrich, K. 2011. Artificial neural networks for predicting sliding friction and wear properties of polyphenylene sulfide composites. Tribology International 44(5): 603–609. Doi: https://doi.org/10.1016/j.triboint.2010.12.011.

Hamrock, B.J. and Dowson, D. 1976a. Isothermal elastohydrodynamic lubrication of point contacts: Part I—Theoretical formulation. Journal of Lubrication Technology 98(2): 223–228. Doi: 10.1115/1.3452801.

Hamrock, B.J. and Dowson, D. 1976b. Isothermal elastohydrodynamic lubrication of point contacts: Part II—Ellipticity parameter results. Journal of Lubrication Technology 98(3): 375–381. Doi: 10.1115/1.3452861.

Hamrock, B.J. and Dowson, D. 1976c. Isothermal Elastohydrodynamic Lubrication of Point Contacts, III Fully Flooded Results. In NASA Technical Note.

Helmy, A. 2004. Neural network wear prediction models for the polymethylmethacrylate PMMA. Alexandria Engineering Journal 43: 401–407.

Jiang, Z., Zhang, Z. and Friedrich, K. 2007. Prediction on wear properties of polymer composites with artificial neural networks. Composites Science and Technology 67(2): 168–176. Doi: https://doi.org/10.1016/j.compscitech.2006.07.026.

Jiang, Z., Gyurova, L.A., Zhang, Z., Friedrich, K. and Schlarb, A.K. 2008. Neural network based prediction on mechanical and wear properties of short fibers reinforced polyamide composites. Materials & Design 29(3): 628–637. Doi: https://doi.org/10.1016/j.matdes.2007.02.008.

Jones, S.P., Jansen, R. and Fusaro, R.L. 1997. Preliminary investigation of neural network techniques to predict tribological properties. Tribology Transactions 40(2): 312–320. Doi: 10.1080/10402009708983660.

Kar, M.K. and Bahadur, S. 1974. The wear equation for unfilled and filled polyoxymethylene. Wear 30(3): 337–348. Doi: https://doi.org/10.1016/0043-1648(74)90148-3.

Kennedy, F.E. and Hussaini, S.Z. 1987. Thermo-mechanical analysis of dry sliding systems. Computers & Structures 26(1): 345–355. Doi: https://doi.org/10.1016/0045-7949(87)90264-1.

Krishnan, R.K., Mohanasundaram, K.M., Arumaikkannu, G. and Ramanathan, S. 2012. Artificial neural networks based prediction of wear and frictional behaviour of aluminium (A380)–fly ash composites. Tribology—Materials, Surfaces & Interfaces 6(1): 15–19. Doi: 10.1179/1751584X11Y.0000000025.

Lansdown, A.R. 1996. Lubrication and lubricant selection—A practical guide. Neale, M.J., Polak, T.A. and Priest, M. (eds.). Tribology in Practice Series. London and Bury St Edmunds, UK: Mechanical Engineering Publications.

LiuJie, X., Davim, J.P. and Cardoso, R. 2007. Prediction on tribological behaviour of composite PEEK-CF30 using artificial neural networks. Journal of Materials Processing Technology 189(1): 374–378. Doi: https://doi.org/10.1016/j.jmatprotec.2007.02.019.

Ma, Z., Henein, N.A. and Bryzik, W. 2006. A model for wear and friction in cylinder liners and piston rings. Tribology Transactions 49(3): 315–327. Doi: 10.1080/05698190600678630.

Marko, M.D., Kyle, J.P., Wang, Y.S. and Terrell, E.J. 2017. Tribological investigations of the load, temperature, and time dependence of wear in sliding contact. PLoS ONE 12(4). Doi: 10.1371/journal.pone.0175198.

Nasir, T., Yousif, B.F., McWilliam, S., Salih, N.D. and Hui, L.T. 2010. An artificial neural network for prediction of the friction coefficient of multi-layer polymeric composites in three different orientations. Proceedings of the Institution of Mechanical Engineers, Part C: Journal of Mechanical Engineering Science 224(2): 419–429. Doi: 10.1243/09544062JMES1677.

Park, Y.W. and Lee, K.Y. 2008. Development of empirical equations for fretting-corrosion failure-time of tin-plated contacts. Wear 265(5): 756–762. Doi: https://doi.org/10.1016/j.wear.2008.01.010.

Patir, N. and Cheng, H.S. 1979. Application of average flow model to lubrication between rough sliding surfaces. Journal of Lubrication Technology 101(2): 220–229. Doi: 10.1115/1.3453329.

Rajesh, J.J. and Bijwe, J. 2005. Dimensional analysis for abrasive wear behaviour of various polyamides. Tribology Letters 18(3): 331–340. Doi: 10.1007/s11249-004-2759-2.

Roda-Casanova, V. and Sanchez-Marin, F. 2019. A 2D finite element based approach to predict the temperature field in polymer spur gear transmissions. Mechanism and Machine Theory 133: 195–210. Doi: https://doi.org/10.1016/j.mechmachtheory.2018.11.019.

Rutherford, K.L., Hatto, P.W., Davies, C. and Hutchings, I.M. 1996. Abrasive wear resistance of TiN/NbN multi-layers: Measurement and neural network modelling. Surface and Coatings Technology 86-87: 472–479. Doi: https://doi.org/10.1016/S0257-8972(96)02956-8.

Shebani, A. and Iwnicki, S. 2018. Prediction of wheel and rail wear under different contact conditions using artificial neural networks. Wear 406-407: 173–184. Doi: https://doi.org/10.1016/j.wear.2018.01.007.

Vakis, A.I., Yastrebov, V.A., Scheibert, J., Nicola, L., Dini, D., Minfray, C., Almqvist, A., Paggi, M., Lee, S., Limbert, G., Molinari, J.F., Anciaux, G., Aghababaei, R., Restrepo, S.E., Papangelo, A., Cammarata, A., Nicolini, P., Putignano, C., Carbone, G., Stupkiewicz, S., Lengiewicz, J., Costagliola, G., Bosia, F., Guarino, R., Pugno, N.M., Müser, M.H. and Ciavarella, M. 2018. Modeling and simulation in tribology across scales: An overview. Tribology International 125: 169–199. Doi: https://doi.org/10.1016/j.triboint.2018.02.005.

Veeresh Kumar, G.B., Pramod, R., Rao, C.S.P. and Shivakumar Gouda, P.S. 2018. Artificial neural network prediction on wear of Al6061 alloy metal matrix composites reinforced with $-Al_2O_3$. Materials Today: Proceedings 5(5, Part 2): 11268–11276. Doi: https://doi.org/10.1016/j.matpr.2018.02.093.

Viswanath, N. and Bellow, D.G. 1995. Development of an equation for the wear of polymers. Wear 181-183: 42–49. Doi: https://doi.org/10.1016/0043-1648(95)90006-3.

Yao, W., Bao, J., Yan, Y., Liu, T. and Wang, N. 2018. Life predictions of brake friction pair based on physical models and statistical analysis. Recent Patents on Mechanical Engineering 11(1): 58–66. Doi: http://dx.doi.org/10.2174/2212797611666180216160021.

Zabala, B., Igartua, A., Fernandez, X., Priestner, C., Ofner, H., Knaus, O., Abramczuk, M., Tribotte, P., Girot, F., Roman, E. and Nevshupa, R. 2017. Friction and wear of a piston ring/cylinder liner at the top dead centre: Experimental study and modelling. Tribology International 106: 23–33. Doi: https://doi.org/10.1016/j.triboint.2016.10.005.

Zhang, Z., Friedrich, K. and Velten, K. 2002. Prediction on tribological properties of short fibre composites using artificial neural networks. Wear 252(7): 668–675. Doi: https://doi.org/10.1016/S0043-1648(02)00023-6.

Chapter 11

Failure of Tribo-Systems

11.1 Introduction

In general, there is no unique definition of what constitutes a failure or, in particular, the difference between damage and failure. However, failure can be defined as any identifiable deviation from the original condition which is unsatisfactory to a particular user.

In tribo-systems, failure of tribological components has been earlier mentioned in the Jost Report (Jost, 1966) as a troublesome economic milestone in machinery and manufacturing systems. Since then, tribologists worked on two major roles; identifying the causes of failure, and consequently suggesting how to prevent failure. Also, maintenance engineers must work closely together with tribologists in monitoring the health of machinery and the performance of tribological components.

The main concern when discussing the failure of tribo-systems, is the failure at the contact between mating surfaces. Failure at the contact surfaces can be described as a progressive loss of quality of the surface as a result of shearing and tearing away of particles. Failure of tribo-surfaces takes several forms (or patterns). It could be, for example, a flat spot, as when a locked wheel slides on a rail. More generally, the deterioration in surface quality is distributed over an entire active surface because of a combination of sliding and rolling actions, as on gear teeth. It may occur in dry conditions or in the presence of a lubricant, where a lubricating film is not sufficiently developed, for complete separation of the contacting surfaces (Stolarski, 1990). Insufficient lubrication may also result in rapid deterioration of surface quality, as in failure of cam shafts.

Misalignments and unanticipated deflections usually result in generation of large amounts of wear particles. This has been detected on the teeth of gears mounted on insufficiently rigid shafts, particularly when the gear is overhung. In addition to these patterns of failure, pitting, surface cracks, spalling and other forms will be introduced in this chapter.

11.2 Bearing Failure

It is widely accepted that failure in many tribological applications is associated with bearing failures. However, there is no single definition that describes the failure of bearings. The International Organization for Standardization defines the failure of rolling bearing as the damage that prevents the bearing meeting the intended design performance, ISO15243:2017 (The International Organization for Standardization, 2017). Also, ISO 281:2007 (The International Organization for Standardization, 2007) suggested that the bearing life would expire when the first evidence of fatigue (or spall) in the material is evident on any of the bearing sub-components.

Broadly speaking, failure modes (or forms) in rolling elements bearing are classified according to the failure mechanism. The major groups include fatigue or fracture, overheating, corrosion, wear, fretting, rust, and plastic flow. Selected modes of bearing failure are discussed below.

11.2.1 Rolling Contact Fatigue (RCF)

Moving elements in rolling bearings are commonly exposed to external loading. In several situations they are also subjected to cyclic stresses which cause fatigue damage. It is widely accepted that Rolling Contact Fatigue (RCF) is the most predominant mode of failure in rolling bearing. This mode is a progressive and a localized material damaging process. The common evidences of this mode include surface pitting and peeling, spalling or flaking and, in rare cases, section cracking. Most probably, all these mechanisms of damage take place simultaneously and compete with each other. Figure 11.1 shows examples of typical failure mechanisms of rolling element bearings.

This mode of failure, RCF, is usually attributed to severe operating conditions when maximum stresses are less than the yield limit of the bearing material. Typical examples include:

1) excessive weight or pressure from the abnormal meshing of gears or machine overloads.
2) rollers fail to fully contact raceways.
3) roller-raceway adjustments that are too tight.
4) misalignment.
5) cocking of the bearing.
6) out-of-round or bowed shaft.
7) sudden impact, such as a wheel striking deep chuckholes.
8) exceeding the life limit of the bearing.

To a large extent, calculation of the fatigue life of rolling bearings is based on the weak point theory introduced by Lundberg and Palmgren (Lundberg and Palmgren, 1974). The preliminary assumption was that fatigue cracking starts at a weak point below the surface of the material. Additionally, the probability of failure is proportional to the stressed volume, the condition of the stressed material, and the change in the condition of the stressed material, as shown in Fig. 11.2.

11.2.2 Overheating

Overheating is frequently attributed to high friction at contacts which, in turn, causes the lubricant to decompose and finally eliminates its ability to lubricate. The most common symptoms of bearing overheating are discoloration of the balls, rings and cages from gold to blue, as shown in Fig. 11.3. In practical working conditions, temperatures in excess of 200°C can anneal the ball and ring materials. The resulting loss in hardness reduces the bearing capacity, causing early failure. In extreme cases, balls and rings will deform. The temperature rise can also degrade or destroy the lubricant. Common culprits are heavy electrical heat loads, inadequate heat paths. and insufficient cooling or lubrication when loads and speeds are excessive. Thermal or overload controls, adequate heat paths, and supplemental cooling are effective solutions.

11.2.3 Wear

As already mentioned in Chapter 4, wear is defined as a gradual material removal from a surface by relative motion. Regarding bearings, the bearing is considered to be failed when wear has increased to a degree that it affects the bearing's normal operation. Generally, in a rolling element bearing, wear is far less severe than that in a sliding bearing. This is particularly true at raceways for cylindrical and tapered roller bearings, where true rolling motion dominates. Three fundamental wear mechanisms are often seen in rolling element bearings: Adhesive wear, abrasive wear and fretting wear (Ai, 2013). These forms of wear can be minimized by the formation of lubrication films through hydrodynamic lubrication or elasto-hydrodynamic lubrication.

Abrasive wear

Abrasive wear involves hard foreign particles acting as rubbing compounds between mating surfaces. Generated wear debris often work and act as micro-cutting tools to bearing surfaces, causing plowings that aggravate the wear process. This form of wear is recognized by observable irregular grooves and furrows

Figure 11.1: Examples of typical failure mechanisms of rolling element bearings: (a) Pitting preceded by spalling; (b) Peeling of ball bearing inner race ring; (c) Flaking damage on inner raceway of a double row tapered roller bearing. After Ai (Ai, 2013) with permission from Springer publishing.

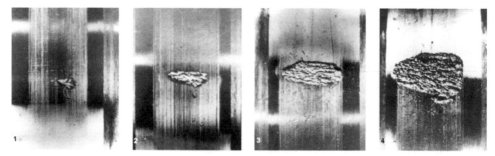

Figure 11.2: Spall started just behind a dent in the raceway and, over a period of time, it became more severe. After Morales-Espejel and Gabelli (Morales-Espejel and Gabelli, 2015), with permission from Evolution SKF.com.

Figure 11.3: Discoloration of a bearing ring due to overheating.

on the surface or by dulling of surfaces by microscopic furrows (Ai, 2013). The result of abrasive wear is a loss of bearing dimension which, in turn, affects bearing performance. Figure 11.4 shows an example of abrasive wear in bearings.

Adhesive wear

Adhesive wear results from material transfer from one surface to the mating surface through welded junctions and tearing. It is usually observed at the interfaces between shafts and bearing inner rings and

Figure 11.4: Grooves caused by debris in abrasive wear.

between housings and bearing outer rings when the lubricant cannot sufficiently prevent metal-to-metal contact. This results in dramatic heating and the damage propagates in the direction of sliding. Adhesive wear appears in various forms, depending on its severity.

i. Mild wear: in many cases the damage will not be visible to the naked eye, however close-up inspection will reveal small and dark furrows which represent weld-spots. The appearance of the bearing will be largely unaffected.

ii. Moderate wear (scoring): occurs when lubricant film thickness is inadequate to separate the contact surfaces. On the micro-scale, discrete asperities weld and tear under high normal applied stresses. Damage tends to form in a line of material transfer from surface to surface.

iii. Severe wear (scuffing and smearing): is a progressive scoring resulting from the welding and tearing of multiple but discrete asperities. Scuffing is identified by severe surface roughening, and is accompanied by perceptibly high friction and temperature. on the other hand, smearing (or galling) involves severe plastic deformation and massive material transfer from surface to surface. Figure 11.5 shows scuffing and smearing in bearing elements.

iv. Seizing: is the final stage of adhesive wear, in which welding between contacting asperities becomes so severe that it seizes the bearing. This stage of adhesive wear results in catastrophic failure. Seizing is rarely seen in rolling element bearings; it is often associated with loss of lubrication.

11.2.4 Corrosion

Chemical wear is often referred to as corrosive wear. The most relevant mechanisms are etching and hydrolysis. Chemical wear can be prevented by applying appropriate coatings or by adding additives to the lubricant and atmospheric corrosion can be prevented by using good oil seals, lubrication and good lubricating bearings. In some cases, special seals may be necessary to avoid grease splattering. Bearing should be filled with the lubricant at every relatively long machine stop. Figure 11.6 shows corrosion of bearings.

11.3 Gear Failure

In almost all mechanical engineering systems, gears are used to transmit force and motion from one mechanical unit to another. Various types of gears have been developed to perform different functions, the most common of these being spur gears, helical gears, straight and spiral bevel gears and hypoid gears. During service, gears can fail for a variety of reasons, usually leading to a total system failure. Also, gear failure can occur in various modes. Many of these modes have been identified, including surface contact fatigue, tooth bending fatigue, wear, pitting, impact, and plastic deformation. The gear failure is explained by means of a hierarchy diagram in Fig. 11.7.

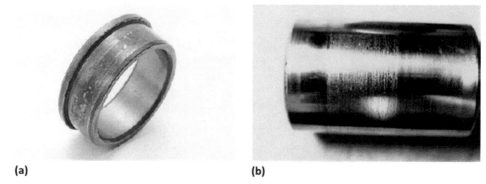

Figure 11.5: (a) Scuffing refers to roughness on the race collar and the ends of roller, and (b) Smearing in which the surface becomes rough and small deposits form.

Figure 11.6: Corrosion of bearing.

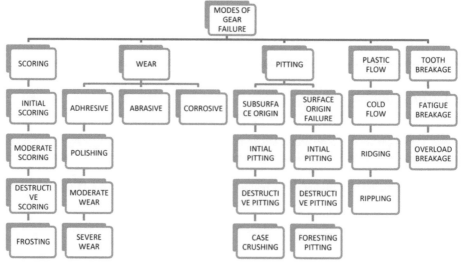

Figure 11.7: Different modes of gear failure. After Yadav (Yadav, 2012).

Shen et al. (Shen et al., 2018) explored the failure mechanism of a planetary gear train by analyzing the actual failing components. Investigation and analysis of failed components demonstrated that the failure development process of planetary gears can be divided into two stages, including fretting wear and fatigue source generation, as shown in Fig. 11.8.

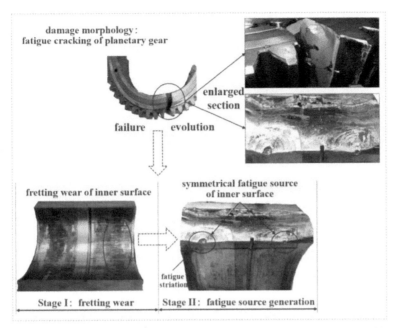

damage morphology:
fatigue cracking of planetary gear

enlarged
section

failure   evolution

fretting wear of inner surface

symmetrical fatigue source
of inner surface

fatigue
striation

Stage I: fretting wear   Stage II: fatigue source generation

Figure 11.8: Failure development process of a planetary gear. After Shen et al. (Shen et al., 2018), with permission from Elsevier publishing.

It is postulated that the fretting wear is the initial stage of gear failure. After the fretting wear, the fatigue crack initiation and the following propagation tend to be associated with the stress raiser. The concentrated stress weakens the material and leads to the fatigue source under the long-term loading.

Pitting is one of the major causes of gear failure and accounts for nearly 60% of gear failures. It is mainly caused by repetitive Hertzian contact stresses of the mating gear teeth surfaces. This may cause fatigue cracks initiating from the surface or nearby subsurface leading to the detachment of a crater (macro-pitting). Craters usually appear at the dedentum tooth flank, where both stress and sliding velocity are high.

Direct measurement of wear and pitting of gears without disturbing the running contact state and lubrication condition is a challenging issue. The most commonly-used methods for continuously observing the wear of gears are particle and vibration monitoring, which can be used as an indicator of gear flank geometry changes due to wear.

Kattelus et al. (Kattelus et al., 2018) detect the progression of gear macro-pitting failure with on-line and in-line particle monitoring from lubricating oil. In these methods, the detached pits or other solid contaminants in lubricating oil are continuously detected with specific sensors at the gearbox outlet providing real time information. The results were compared to the corresponding results obtained with vibration monitoring test and with visual inspection documented by photographs.

Visual inspections and photographs were taken after 2.50, 5, 7.50, 10, 10.76 and 10.84 million cycles, as shown in Fig. 11.9. The results of continuous on-line particle monitoring from oil outlet in the range of 0–10.84 million load cycles are grouped into three different size classes. The comparison of the results obtained from visual inspection and particle on-line monitoring are shown in Fig. 11.10. The overall trend shows that the results remarkably agree.

Failure of polymer gears is different from steel ones because the material properties of polymers are totally different. Most of the failures occurred due to the limitation of the material, such as the load handling capability and thermal properties. An example of polymer gear failure is melting of material which does not occur for steel gears. Types of gear failure also include wear, microstructure surface condition monitoring, weight loss and thermal damage (Md Ghazali et al., 2017).

There are many types of failure that can be categorized under wear, such as cracking or breaking, tooth thickness reduction and debris formation. Cracks often occur at the root of the tooth and propagate, causing

Figure 11.9: Progression of macropitting. Pinion teeth after 2.5 (stage 0) (up left), 10 (stage 1) (up right), 10.76 (stage 2) (lower left) and 10.84 (stage 3) (lower right) million load cycles. After Kattelus et al. (Kattelus et al., 2018), with permission from Elsevier publishing.

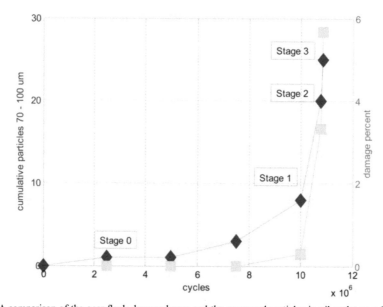

Figure 11.10: A comparison of the gear flank damaged area and the measured particles in oil outlet as a function of load cycles. Visual inspection points are dotted. Stages 0 to 3 represent vibration analysis points. After Kattelus et al. (Kattelus et al., 2018), with permission from Elsevier publishing.

tooth breakage. Detection of wear, microstructure surface condition monitoring, weight loss and thermal damage and temperature detection are the most prominent methods used to predict failure in polymer gears.

11.4 Cam/Follower Failure

Cam/follower systems are used in most internal combustion engines to provide a prescribed motion to a valve train system. The cam moves on its follower in extreme and complex contact conditions of high

Figure 11.11: Typical pitted cam follower surface.

temperatures, no lubrication and high contact stresses. Consequently, the performance of a cam/follower system is highly affected by the contact conditions. It is believed that failure from contact stresses generally leads to localized deformation or fracture by the progressive spreading of a crack (Tounsi et al., 2011). Pitting is a fatigue process that can also cause initiation and propagation of cracks. Rolling contact fatigue (RCF) cracks of the cam/follower system can be categorized into two groups, based on where they started. First, cracks that initiated at the surface and propagated down into the bulk of the cam at a shallow angle to the surface, or cracks initiated below the surface. Second, cracks which initiated in a region of maximum cyclic shear stress. An experimental and analytical research on contact fatigue wear of cam and roller follower performed by Cheng et al. (Cheng et al., 1994) indicated that the end clearance between the roller and rocker had a significant influence on friction and heat generation in the roller follower surface.

Under rolling-sliding contact, the cam/follower system is exposed to surface fatigue failure due to the repetition of contact load. This form of fatigue wear is usually characterized by pitting and spalling of the surface. The surface fatigue wear, however, occurs suddenly and the generated wear particles are removed, resulting in relatively large pits on the surface, as shown in Fig. 11.11.

In general, the typical modes of the follower failure are pitting, polish wear and scuffing. On the other hand, surface fatigue failure of cam follower initiates as superficial micro cracks of irregular shape on the surface as a result of maximum value of contact stress (Nayak et al., 2006). The interacting asperities will wear mildly without affecting the contact conditions. These cracks, under cyclic loading, propagate quickly due to induced compressive stress. The compressive stress increases slowly with the development of plasticly deformed surfaces near the two contacting elements, i.e., at the cam and the follower surfaces. In the later stage of operation, it leads to disintegration of the surface. The destructive nature of surface fatigue cracks increase due to high contact stresses and sliding action, constituting a stress raiser and hastening the fatigue failure of the follower surface. The surface flakes leave the parent surface and damage the soft follower surface, as shown in Fig. 11.12.

11.5 Wheel/Rail Failure

The wheel/rail contact is the critical point of the railway system, especially at high speeds and in the case of heavy freight trains with high axle load. This is due to the high contact forces and the dynamics. Railway track dynamic models typically use the Hertzian Contact Theory to represent the wheel-rail contact. When a train is running on the track, the springs and masses of the train and the track contribute to the dynamic behavior of the whole system. Dynamic effects in the system become more significant when train speed increases and when the axle load is greater. Accordingly, there are several sources of failure in wheel/rail systems; for example: Rail corrugation, wheel out-of-roundness, track stiffness irregularities, and impact loads (Tzanakakis, 2013).

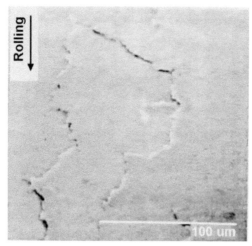

Figure 11.12: Surface flakes of follower surface under cyclic loading. After Nayak et al. (Nayak et al., 2006) with permission from Elsevier publishing.

In general, the failure mechanism in wheel/rail contacts is very complicated. It is believed that RCF, caused by cyclic rolling contact between the wheel and rail head, is the most vital problem in railways. In high-speed rails, the main damage form is governed by fatigue crack growth. Therefore, the prediction of crack initiation and crack propagation of rails has been an important research direction in the field of wheel/rail contact fatigue (Jiang et al., 2017). Recently, many studies have been carried out in order to explore the wear behaviors and wheel/rail RCF life by means of various experimental and numerical methods (Zhao et al., 2018).

Although RCF is considered as the main cause of surface damage in wheel/rail contact, surface damage can be caused by the hard body coming from the external environment. Once a hard body enters the wheel/rail interface and crashes, the damage occurs both on the rail and wheel surfaces in the form of "dents". Based on several experimental investigations, it is suggested that dents have an obvious influence on the fatigue life of a wheel/rail system. Cantini and Cervello (Cantini and Cervello, 2016) detected the vertical RCF cracks around the dents on the wheel tread. It is pointed out that the RCF cracks would initiate and propagate when the diameter of dents reaches a certain size (Seo et al., 2011). For larger dents, cracks initiate and propagate, causing large pieces of the specimen to break off, as shown in Figs. 11.13 and 11.14. Also, a different plastic deformation extent of rail material around dents was observed by Zhao et al. (Zhao et al., 2018), as shown in Fig. 11.15.

Subsequently, in the service phase, surface damage will change the contact condition and affect the service life. Figure 11.16 shows different forms of wheel/rail failure: (a) wheel flats, which can occur due to low adhesion, (b) squats, which are a form of RCF that appear as a discrete defect, and (c) head checks,

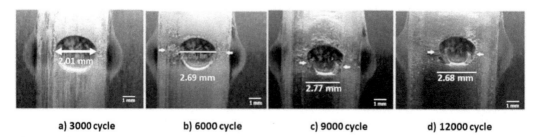

Figure 11.13: Progression of damages on dent specimen surface. After Seo et al. (Seo et al., 2011), with permission from Elsevier publishing.

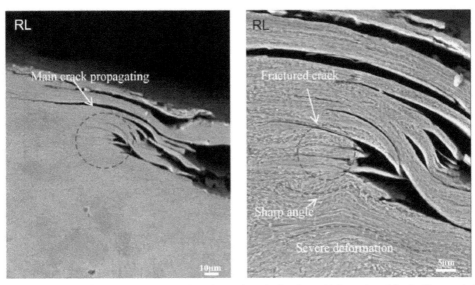

Figure 11.14: SEM micrographs of fatigue crack on the section of rail rollers with large dent. After by Zhao et al. (Zhao et al., 2018), with permission from Elsevier publishing.

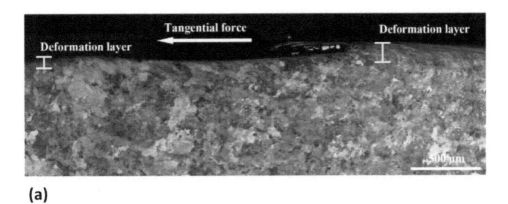

(a)

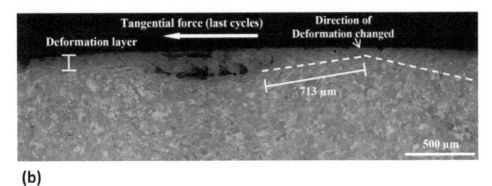

(b)

Figure 11.15: Micrographs of plastic deformation of (a) rail rollers, and (b) wheel rollers with large dent. After by Zhao et al. (Zhao et al., 2018), with permission from Elsevier publishing.

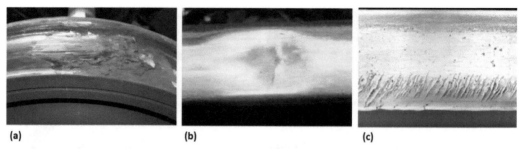

Figure 11.16: (a) Wheel flats, (b) Squats, and (c) Head checks. After Tzanakakis (Tzanakakis, 2013), with permission from Springer publishing.

which are a form of RCF that appear on the rail as a series of cracks along the running band inclined to the gauge corner.

11.6 Failure of Artificial Joints

Total joint replacement is a very essential surgical treatment for severe joint diseases. This arthroplasty involves the replacement of native articulating joints, such as the hip, knee, shoulder and ankle, with artificial components. Various material combinations are utilized in bearing surfaces of joint prostheses. The most commonly-used bearing surface is ultra-high molecular weight polyethylene (UHMWPE), usually articulating against a metallic or a ceramic counterface.

Failure of artificial joints occurs due to adverse reactions in the surrounding tissue to the minute debris that results from corrosion and/or implant debris wearing off a non-corrosive joint. Failure occurs, also, either because the cemented interfaces are not sufficiently strong to bear the loads, or because the interface was originally weak, or because the applied loads have been excessive (Yousif, 2009; Shaik and Sarkar, 2020). From the tribological point of view, the wear of the bearing surfaces of hip joint prostheses is a key problem causing their primary failure. Analysis of retrieved artificial joints demonstrated evidence of surface cracking, abrasion, burnishing and scratching on the articulating surfaces (Bradford et al., 2004). Also, wear is the main reason for an adverse tissue inflammation caused by wear debris of the implant. Most of existing tests and clinical medical studies have demonstrated that the wear debris generated from the wear of the UHMWPE artificial joints will induce a series of harmful biological reactions around the artificial joints and eventually result in aseptic loosening and premature failure (Zhang et al., 2018).

Friction and wear of artificial joints is unpredictable due to complex tribological and biological behaviors and long term wear. Research efforts are currently addressing the evaluation of the determinants affecting the overall wear rate of the artificial joint articulating surfaces, with the aim of reducing wear rate.

Many researches have focused on the reduction of the volume of wear debris, since it is the main factor of joint failure. They collected the worn particles and made the calculations according to the particles volume. Experimental studies of the wear of polymers in wet sliding showed that the resulting wear debris are in the form of submicron-sized particles (Abdelbary, 2011). Polymeric wear debris, detected in the lubricating water fluid, were in the form of suspended particles. This scale of wear particles was detected in total hip joints; most UHMWPE wear debris have been shown to be of the micron or submicron size (Flannery et al., 2008; Zhang et al., 2018). Similar size wear debris were also produced in hip simulator testing using bovine serum as the lubricant.

Cooper et al. (Cooper et al., 1993; Cooper et al., 1994) studied the wear of UHMWPE sliding on metallic and ceramic counterfaces under a wide range of tribological conditions in order to investigate the influence of contact stresses on the macroscopic and microscopic wear mechanisms. The studies were conducted in pin-on-disc and pin-on-plate tribometers, hip joint simulators, and retrieved artificial joints taken from patients. In the body, under cyclic loading, the macroscopic polymer asperity is cyclically deformed at the frequency of the loading cycle and this can produce crack propagation and surface fatigue within 10 μm of the surface under the polymer asperity. It is likely that cyclic loading is one of the main factors causing the failure of artificial joints. Under dynamic loading conditions found in artificial joints, subsurface cracking was found in the highly strained region. Subsurface crack propagation may well have accelerated the failure and removal of material from the highly strained polymer peaks, hence, greatly increasing the macroscopic polymer asperity wear processes. These processes were not only important with respect to the increased volume wear debris produced, but may also produce large wear particles which can cause adverse tissue reactions in the body.

Similar observations were remarked by Bradford et al. (Bradford et al., 2004), in their study of *in vivo* wear mechanisms of highly cross-linked UHMWPE correlated with predictions based on *in vitro* wear testing. All of the retrieved polyethylene acetabular liners that had been examined in their study displayed a combination of surface cracking, abrasion, pitting and scratching, as shown in Figs. 11.17 to 11.19. These findings were clearly different from those observed in hip-simulators which demonstrated only occasional scratches on the surface of the acetabular liners. They concluded that the discrepancy between *in vitro* and *in vivo* findings may be due to the variability of *in vivo* lubrication and cyclic loading or may represent early wear mechanisms that are not well represented by long-term *in vitro* studies.

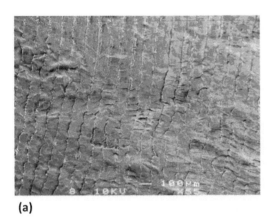

(a)   (b)

Figure 11.17: (a) Low-magnification image showing the evolution of altered machining marks and surface cracking parallel to the marks and (b) High-magnification image showing cracking perpendicular to the machining marks. After Bradford et al. (Bradford et al., 2004), with permission from Wolters Kluwer Health, Inc.

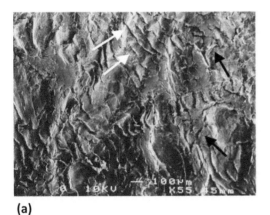
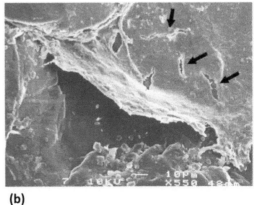

(a)   (b)

Figure 11.18: (a) Low-magnification image of a highly cross-linked liner that was retrieved after seven months *in situ*, demonstrating surface cracking along the articulating surface. The white arrows indicate cracks parallel to the machining marks, and the black arrows indicate cracks perpendicular to the machining marks, and (b) High-magnification image illustrating pitting and surface tearing. After Bradford et al. (Bradford et al., 2004), with permission from Wolters Kluwer Health, Inc.

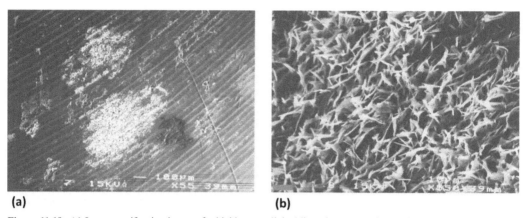

(a)   **(b)**

Figure 11.19: (a) Low-magnification image of a highly cross-linked liner that was retrieved after sixteen months *in situ*, demonstrating an abraded region with obliteration of the machining marks, and (b) Higher-magnification image of the abraded region, consistent with a high-density crystalline surface. After Bradford et al. (Bradford et al., 2004), with permission from Wolters Kluwer Health, Inc.

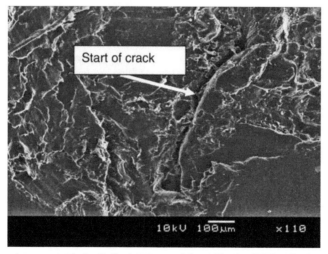

Figure 11.20: Electron microscope analysis of adhesion wear area (magnification 110X). After Burger et al. (Burger et al., 2007), with permission from Elsevier publishing.

Burger et al. (Burger et al., 2007) investigated more than 100 retrieved polyethylene acetabular cups to propose an explicit set of failure criteria in order to undertake a comprehensive analysis of the root cause of implant failure. According to their study, the most common defects identified were:

i.   Mechanical Damage: Caused by an acetabular cup not properly aligned *in vivo*.

ii.   Cracks: Caused by localized stress areas or on the rim of the cup, as shown in Fig. 11.20.

iii. Plastic Flow: Occurring just outside the region of high contact stress, as shown in Fig. 11.21.

iv.   Scratches: Caused by third-body wear, independent of what caused the wear particles or their type.

v.   Adhesion Wear: Probably caused by overheating or lack of lubrication, as shown in Fig. 11.20.

vi.   Flaking and Delamination: Arising in areas where pieces of polymer separate from the base material. It is shown as areas of delamination. This type of defects is associated with a defect within the material and occurs in the high stress or contact stress areas, as shown in Fig. 11.22.

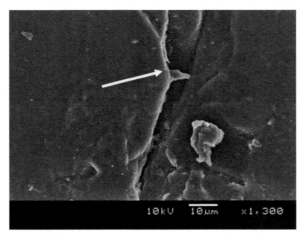

Figure 11.21: Plastic flow in acetabular cup (magnification 1300X). After Burger et al. (Burger et al., 2007), with permission from Elsevier publishing.

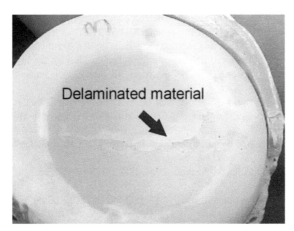

Figure 11.22: Acetabular cup with delaminated material visible on inside of cup on bearing surface. After Burger et al. (Burger et al., 2007), with permission from Elsevier publishing.

References

Abdelbary, A. 2011. Effect of vertical cracks at the surface of polyamide 66 on the wear characteristics during sliding under variable loading conditions. Ph.D., Mechanical Engineering Department, Alexandria University.

Ai, X. 2013. Failure mechanisms of rolling element bearings. pp. 1006–1014. *In*: Wang, Q.J. and Chung, Y.-W. (eds.). Encyclopedia of Tribology, Boston, MA: Springer US.

Bradford, L., Baker, D.A., Graham, J., Chawan, A., Ries, M.D. and Pruitt, L.A. 2004. Wear and surface cracking in early retrieved highly cross-linked polyethylene acetabular liners. JBJS 86(6): 1271–1282.

Burger, N.D.L., de Vaal, P.L. and Meyer, J.P. 2007. Failure analysis on retrieved ultra high molecular weight polyethylene (UHMWPE) acetabular cups. Engineering Failure Analysis 14(7): 1329–1345. Doi: https://doi.org/10.1016/j.engfailanal.2006.11.005.

Cantini, S. and Cervello, S. 2016. The competitive role of wear and RCF: Full scale experimental assessment of artificial and natural defects in railway wheel treads. Wear 366-367: 325–337. Doi: https://doi.org/10.1016/j.wear.2016.06.020.

Cheng, W., Cheng, H.S. and Yasuda, Y. 1994. Wear and Life Prediction of CAM Roller Follower.

Cooper, J.R., Dowson, D. and Fisher, J. 1993. Macroscopic and microscopic wear mechanisms in ultra-high molecular weight polyethylene. Wear 162-164: 378–384. Doi: https://doi.org/10.1016/0043-1648(93)90521-M.

Cooper, J.R., Dowson, D., Fisher, J., Isaac, G.H. and Wroblewski, B.M. 1994. Observations of residual sub-surface shear strain in the ultrahigh molecular weight polyethylene acetabular cups of hip prostheses. Journal of Materials Science: Materials in Medicine 5(1): 52–57. Doi: 10.1007/BF00121154.

Flannery, M., McGloughlin, T., Jones, E. and Birkinshaw, C. 2008. Analysis of wear and friction of total knee replacements: Part I. Wear assessment on a three station wear simulator. Wear 265(7): 999–1008. Doi: https://doi.org/10.1016/j.wear.2008.02.024.

Jiang, X., Li, X., Li, X. and Cao, S. 2017. Rail fatigue crack propagation in high-speed wheel/rail rolling contact. Journal of Modern Transportation 25(3): 178–184. Doi: 10.1007/s40534-017-0138-6.

Jost, P. 1966. Lubrication (Tribology) Education and Research, Technical report. H.M.S.O.

Kattelus, J., Miettinen, J. and Lehtovaara, A. 2018. Detection of gear pitting failure progression with on-line particle monitoring. Tribology International 118: 458–464. Doi: https://doi.org/10.1016/j.triboint.2017.02.045.

Lundberg, G. and Palmgren, A. 1974. Dynamic capacity of rolling bearings. Acta Polytechnica Scandinavica, Mechanical Engineering Series 1(3).

Md Ghazali, W., Idris, D.M.N.D., Sofian, A.H., Siregar, J.P. and Aziz, I.A.A. 2017. A review on failure characteristics of polymer gear. MATEC Web Conf. 90.

Morales-Espejel, G.E. and Gabelli, A. 2015. The progression of surface rolling contact fatigue damage of rolling bearings. SKF Evolution, Issue 2. http://evolution.skf.com/the-progression-of-surface-rolling-contact-fatigue-damage-of-rolling-bearings/.

Nayak, N., Lakshminarayanan, P.A., Gajendra Babu, M.K. and Dani, A.D. 2006. Predictions of cam follower wear in diesel engines. Wear 260(1): 181–192. Doi: https://doi.org/10.1016/j.wear.2005.02.022.

Seo, J., Kwon, S. and Lee, D. 2011. Effects of surface defects on rolling contact fatigue of rail. Procedia Engineering 10: 1274–1278. Doi: https://doi.org/10.1016/j.proeng.2011.04.212.

Shaik, A.B. and Sarkar, D. 2020. Dynamic analysis and life estimation of the artificial hip joint prosthesis. *In*: Venkata Rao, R. and Taler, J. (eds.). Advanced Engineering Optimization Through Intelligent Techniques. Advances in Intelligent Systems and Computing. Vol. 949. Springer. Singapore. Doi: https://doi.org/10.1007/978-981-13-8196-6_25.

Shen, G., Xiang, D., Zhu, K., Jiang, L., Shen, Y. and Li, Y. 2018. Fatigue failure mechanism of planetary gear train for wind turbine gearbox. Engineering Failure Analysis 87: 96–110. Doi: https://doi.org/10.1016/j.engfailanal.2018.01.007.

Stolarski, T. A. 1990. 3—Elements of contact mechanics. In Tribology in Machine Design, edited by T.A. Stolarski, 64–96. Newnes.

The International Organization for Standardization. 2007. ISO 281:2007 Rolling bearings—Dynamic load ratings and rating life.

The International Organization for Standardization. 2017. ISO 15243:2017 Rolling bearings—damage and failures-terms, characteristics and causes.

Tounsi, M., Chaari, F., Abbes, M.S., Fakhfakh, T. and Haddar, M. 2011. Failure analysis of a cam–follower system affected by a crack. Journal of Failure Analysis and Prevention 11(1): 41–50. Doi: 10.1007/s11668-010-9413-0.

Tzanakakis, K. 2013. The wheel-rail interface. pp. 89–112. *In*: Tzanakakis, K. (ed.). The Railway Track and its Long Term Behaviour: A Handbook for a Railway Track of High Quality, Berlin, Heidelberg: Springer Berlin Heidelberg.

Yadav, A. 2012. Different types failure in gears—A review. International Journal of Science, Engineering and Technology Research (IJSETR) 1(5): 86–92.

Yousif, A.E. 2009. Modes of Failure in Natural and Artificial Human Hip Joints. 25th Southern Biomedical Engineering Conference 2009, 15–17 May 2009, Miami, Florida, USA, Berlin, Heidelberg.

Zhang, D., Liu, H., Wang, J., Sheng, C. and Li, Z. 2018. Wear mechanism of artificial joint failure using wear debris analysis. Journal of Nanoscience and Nanotechnology 18(10): 6805–6814. Doi: 10.1166/jnn.2018.15513.

Zhao, X.J., Guo, J., Liu, Q.Y., Butini, E., Marini, L., Meli, E., Rindi, A. and Wang, W.J. 2018. Effect of spherical dents on microstructure evolution and rolling contact fatigue of wheel/rail materials. Tribology International 127: 520–532. Doi: https://doi.org/10.1016/j.triboint.2018.07.001.

Index

T - #0890 - 101024 - C332 - 254/178/15 - PB - 9781032175997 - Gloss Lamination